HANDBOOK OF TERMINOLOGY MANAGEMENT

HANDBOOK OF TERMINOLOGY MANAGEMENT

VOLUME 2

Application-Oriented Terminology Management

Compiled by

SUE ELLEN WRIGHT
Kent State University

GERHARD BUDIN
University of Vienna

JOHN BENJAMINS PUBLISHING COMPANY
AMSTERDAM / PHILADELPHIA

The paper used in this publication meets the minimum requirements of American National Standard for Information Sciences — Permanence of Paper for Printed Library Materials, ANSI Z39.48-1984.

Library of Congress Cataloging-in-Publication Data

Handbook of terminology management / compiled by Sue Ellen Wright, Gerhard Budin.
p. cm.
Contents: v. 2. Application-Oriented terminology management.
1. Technology--Terminology--Handbooks, manuals, etc. I. Wright, Sue Ellen, 1943-. II. Budin, Gerhard.
T11.H27 1997
401'.4--dc21 01-7100
ISBN 90 272 2155 3 (Eur.) / 1 55619 509 5 (US) (Vol. 2; alk. paper)
ISBN 90 272 2149 9 (Eur.) / 1 55619 502 8 (US) (Set; alk. paper) CIP

John Benjamins Publishing Company • P.O.Box 75577 • 1070 AN Amsterdam • The Netherlands
John Benjamins North America • P.O.Box 27519 • Philadelphia PA 19118-0519 • USA

CONTENTS

VOLUME II: APPLICATIONS-ORIENTED TERMINOLOGY MANAGEMENT

INFORMATION BOXES

APPENDICES

List of Illustrations

Chapter 8.4.4, Collocations in Sub-Language Texts

Chapter 8.4.5, The Role of Terminology Management in Artificial Intelligence and Knowledge Acquisition

Infobox 26, The Economics of Terminology Management

For the convenience of readers who may not have Volume I of the Handbook of Terminology Management readily at hand, the Table of Contents from that volume is reproduced below.

VOLUME I: BASIC ASPECTS OF TERMINOLOGY MANAGEMENT

5 TERMINOLOGY TRAINING

INFORMATION BOXES

Note: The authors of the Information Boxes are Sue Ellen Wright and Gerhard Budin, with the exception of Infobox 18, which is by Clove Lynch.

APPENDICES

The Handbook of Terminology Management

Introduction to Volume II

THE NATURE OF A HANDBOOK

In the Introduction to Volume I of this *Handbook of Terminology Management*, the compilers outlined the affinity of this work to handbooks treating, for instance, topics in engineering and the social sciences. Whereas Volume I covered basic, to a great extent timeless, topics in the field, Volume II addresses a range of issues involving pragmatic technical applications in terminology management.

The focus of the project on leading edge technologies has been the source of countless difficulties in completing the volume as planned. First of all, we initially (and perhaps naively) cast our net wide to encompass a broad scope of topics and applications involving the role of terminology in human language technologies and language engineering at a time when these fields were and are still undergoing unprecedented growth and innovation (see Infobox 31:887, *Language Engineering and Language Technology*). Not only did we find that the amount of space originally allotted to certain key topics was inadequate, we also discovered that it was exceptionally difficult to achieve closure in a number of areas, some of which were the direct locus of our own core research. Indeed, our involvement in the development of new technologies, for instance in terminology data interchange, has stood in the way of our timely concentration on this volume. This was notably the case with the chapters in Section 8.1, *COMPUTER APPLICATIONS FOR TERMINOLOGY*. The final approval and publication of ISO 12200:1999, *Computer Applications in Terminology — Machine Readable Interchange Format (MARTIF) — Negotiated Interchange,* has finally provided a sense of completion that underlies the spirit of Chapter 8.1.5, *Terminology Interchange*, but even as this volume goes to press, late-breaking developments in the evolution of terminology interchange for use in integrated computer environments have inspired last-minute changes to that chapter.

To further complicate matters, we found it almost impossible to synchronize the many different topics in this broad and constantly shifting field while at the same time trying to bring each article to a state-of-the-art status simultaneous with that of other articles. In the end, each article represents a certain cut-off point, a kind of snapshot of the topic in question, reflecting the conditions at a particular moment in time. We can only hope that the quality of the contributions will compensate for the impossibility of total temporal coordination.

As a result of the passage of time and evolving intentional aims, some of the annexes listed in Volume I have been revised or eliminated, and there have been a few changes from the authors and titles announced in the earlier volume. We commend the reader to the *Table of Contents* in this volume for current information and apologize for any inconvenience or disappointments that these changes may have caused. At the end of our discussion in Volume I of the nature of a handbook, we noted that this text is intended to be a permanent companion of the working terminologist, a resource that is continuously *at hand*, designed to provide an ever-ready reservoir of useful basic information on the prevalent topics of concern to a broad range of users. We reiterate this goal with respect to Volume II, with the additional caveat that we hope to be able to maintain the currency of the more topical information included here by maintaining an updatable website at: http://appling.kent.edu/htmweb/htmweb.htm

Terminology Management

In Volume I we defined *terminology management* as:

any deliberate manipulation of terminological information

We refer the reader to that volume for further treatment of the scope of this activity. Within the framework of language technology and language engineering, we want to emphasize that terminology management extends beyond *terminology database management* to include all aspects of terminology used as a component of information and quality management, as well as the role that terminology plays in document production and corpus management.

The Scope of Terminology Management

The Organization of this Handbook

As noted in Volume I, the extensive topic range treated in this collection has dictated that this "book" be published in two volumes. The following overview provides a brief summary of the topics treated in Volume II.

Volume II: Applications-oriented Terminology Management

Information Management

As information takes its position as a fundamental production variable (along with the classic variables, capital and labor), information organization and retrieval comprises a core concern. In chapter 6.1, *Computer-Assisted Thesaurus Management*, Roulin outlines the use of electronic tools for the creation of documentary languages used for indexing and querying documents. In 6.2, *Terminology and Indexing* Strehlow provides a brief overview of standard indexing practice and explores the critical principles of precision and recall in 6.3, *Terminology and Information Retrieval*.

Commercial and Industrial Applications

In one of the shifts in writing assignments that has occurred between the two volumes, Wettengel and de Weyer have combined forces in 7.1, *Terminology in Technical Writing*, to analyze ways in which technical writers can use terminological resources to model documents to fit the needs of their target audiences. In contrast to the traditional terminological focus on nominal forms, they stress the role that actions (verbs) play in good technical writing.

The growing role of terminology in industrial management is reflected by a series of articles on industrial operations and inventory control. In 7.2.1, *Terminology as an Organizational Principle in CIM Environments*, Wright describes the central function of terminology as the motivating force behind the assignment of data element names and definitions in computer-integrated manufacturing. Düsterbeck and Hesser (7.2.2) explore *Terminology-Based Knowledge Engineering in Enterprises*, recounting the cost savings that can be achieved through terminologically informed inventory control. Wright's *Terminology and Total Quality Management* (7.2.3) documents the role that parallel-text-oriented terminological research plays in ensuring accurate multilingual terminology. In 7.2.4, *Project-Integrated Terminology Management for Technical Writing and Translation*, Champe presents a case-study reflecting best practices for coordinating terminology research with commercial translation clients. In 7.2.5, *Software Terminology and Localization*, Corbolante and Irmler outline procedures for guaranteeing terminological appropriateness when adapting software to the individual market demands of different global locales.

Computer Applications for Terminology

The articles under the heading of *Database Management Methods* begin with advice on selecting *Terminology Database Management Programs* (TMSs) by Schmitz (8.1.1), who analyzes the essential features of terminology applications and advises the reader on selecting the right one for a given environment. In 8.1.2, *Data Categories for Terminology Management*, Wright discusses the distinguishing characteristics of the data categories used in terminological databases (termbases) and outlines the structure of the ISO 12620 standard on data categories for terminology management, explaining how the individual categories fit into the data model in different working environments. Wright goes on in 8.1.3, *Terminology Management Entry Structures*, to chronicle the evolution of terminological record and its modern manifestation as a terminological entry in a well-structured database environment, particularly in light of developments with respect to meta-models for representing cross-platform structural relationships. The article on *User-Specific Terminological Data Retrieval* by Nkwenti-Azeh (8.1.4) presents the prospect of multi-purpose, user-customizable terminology retrieval from large term banks. In the final article of this section (8.1.5), Melby, Schmitz and Wright provide an introduction to *Terminology Interchange* with the intention of orienting the reader toward efforts to implement a universal terminology exchange format designed to facilitate the importation and exportation of terminological data using Standard Generalized Markup Language (SGML) and eXtensible Markup Language (XML).

Section 8.2, *Database Applications and Products* features three case studies that explore actual working environments within which terminology products are created and maintained. The three contributions in this section also document case studies, providing a snapshot view of a particular working environment viewed at a specific time in its development. Although it is impossible to maintain current updates on this kind of information, the compilers hope that the examples shown here will provide models for others in structuring their own tools and procedures. These case studies include *Preparing Multi-Volume Illustrated Terminological Dictionaries* (8.2.1, Le Néal) and *Practical Considerations for a Term Bank: Termium* (8.2.2, Hutcheson; referenced in the marginal notes as *Termium*). The case studies also include a chapter on *Globalization and Terminology Management* (8.2.3), which describes a highly complex terminological support system within a major corporate environment (Warburton). These topics lead naturally into the consideration of the role of *Terminology and Machine Translation* and the special requirements needed to fully integrate terminological information into machine translation systems (Vasconcellos, 8.3).

Corpus-related Applications addresses the question of automatic terminology generation in articles on the *Corpus Linguistics and Terminology Extraction* (Ahmad and Rogers, 8.4.1) and *Terminology Compilation: Consequences and Aspects of Automation* (Sager, 8.4.2). *Terminological Text Production* (Shreve, 8.4.3), and *Collocations in Sub-language Texts* (Heid, 8.4.4). These discussions underscore the relationship of terminological issues to issues involving discourse and texuality. The *Role of Specialist Terminology in Artificial Intelligence and Knowledge Acquisition* (Ahmad, 8.4.5) provides a structural overview of the importance of terminology in creating artificial intelligence and expert systems.

In 8.5, *Terminology Management Resources on the Internet*, Budin and Wright document the growing wealth of information accessible at sites on the World Wide Web (WWW) or available for downloading from the Internet using File Transfer Protocol (FTP).

Backmatter

The Information Boxes included in Volume II serve a similar function to those in Volume I and pertain specifically to the issues addressed in the second volume. The topics addressed here are generally ones that relate to several of the articles in the book and are more effective as shared resources than if they were treated separately (or severally) in the other articles. Several of the annexes promised for this volume have actually been incorporated into the Infoboxes. In addition to the *List of Abbreviations Used in This Volume,* Volume II contains a listing of current standards and projects that are part of the work program of Technical Committee 37 of the International Organization for Standardization.

Volume I: Basic Aspects of Terminology Management

For the convenience of those who may not have Volume I of this handbook at hand, the introductory collection addresses the classic issues of terminology management, along with overriding practical and external factors, such as terminology training issues and copyright.

Fundamental Principles

The initial segment of Volume I addresses basic concerns involved in the documentation of terminology. *Term Selection* (Wright, 1.1) and *Term Formation* (Sager, 1.2) together provide insight into the most salient of all terminological activities: the identification and creation of terms used or to be used in a discipline and to be documented in terminology management projects. The chapter on *Graphic Representation* (Galinski and Picht, 1.3) supports the thesis that non-verbal means of representation occur in addition to the verbal forms that comprise the primary substance of most terminology collections, and that these forms of representation are proliferating in modern information environments.

Under the heading of *Concept Description*, de Bessé's article on *Definitions* (1.3.1) stresses the importance of complete terminological definitions, as opposed to some of the forms of so-called definitions used in other resources. In *Frames and Display of Definitions* (1.3.2), Strehlow uses frame theory to provide a new view of the components making up a formal definition. Dubuc and Lauriston (1.3.3) underscore the value of using *Contexts* to document the relationship of terms to concepts, the meaning of terms, and the way that they function in texts. Beginning with a basic explanation of the *Representation of Concept Systems* (Wright, 1.4.1), the discussion moves on to Eck, Meyer, and Skuce (1.4.2), who offer a *Systematic Concept Analysis within a Knowledge-Based Approach to Terminology*. Chapters treating the *Multifaceted* and *Multidimensional* aspects of concept classification (Kageura, 1.4.3, and Bowker, 1.4.4) explore efforts to overcome concentration on overly simplistic, two-dimensional systems for representing concept relationships.

Types of Terminology Management

The second major section of the book addresses the many different applications involved in terminology management.

The treatment of *Descriptive Terminology* management explores primarily non-normative terminological activities involving *Technical Translation* (Wright and Wright, 2.1.1), *Medical Science* (Lynch, 2.1.2), *Patent Language* (Lawson, 2.1.3), and the *Social Sciences* (Riggs, Mälkiä, and Budin, 2.1.4). These activities primarily involve documentation on the part of both experts and quasi-specialists (well-versed, non-expert terminologists) who prepare terminology products and online resources for use by both experts and lay practitioners, such as translators, technical writers, and students.

The articles pertaining to *Prescriptive Terminology* management, on the other hand, examine practices and methodologies that have evolved in the arena of normative terminology (Wright, 2.2. 1; Strehlow, 2.2.2; Schrade, 2.2.3; and Bowmann, Michaud, and Suonuuti, 2.2.4), primarily within the framework of national and international standardization, such as the International Organization for Standardization (ISO), and professional societies, such as the International Union for Pure and Applied Chemistry (IUPAC). Merritt's contribution (2.2.5) outlines resources for those working in the field of chemistry and provides a model for the development of highly structured nomenclatures.

Language Policy

Multilingualism in Terminology Management (Budin and Wright, 3.1) examines language policy as applied to countries and regions that must contend with multiple official and unofficial languages and even with the use of multiple scripts, which generates the need to cope with problems involving transliteration and transcription. It also addresses situations involving languages that are facing the challenge of developing technical terminologies in an orderly fashion in order to combat a confusing proliferation of synonyms and potentially ambiguous forms. *Language and Terminology Planning* (Humbley, 3.2) in the francophone environment documents both the history and the current development of French efforts in language planning.

Intellectual Property Rights

Copyright and Terminology (Galinski and Wright, 4.1) outlines the extent to which copyright issues affect terminology collections. In lieu of clearcut protection within the framework of national and international copyright conventions, the authors support the use of a *Guide to Terminology Agreements* as a useful tool for establishing safe and equitable contractual relationships among individuals and entities sharing and exchanging terminological data. Such agreements are especially valuable in flexible electronic environments where the merging and reuse of data gives rise to questions with respect to authorship and exploitation that as yet remain unclear within the framework of emerging trends in copyright protection.

Terminology Training

The chapter on *Terminology Training* (Picht and Acuña, 5.1) outlines the types of terminology training that are being practiced in the world today and proposes modular components that make up the subject matter of courses presented in different venues and for different intentional aims.

ORGANIZATIONAL PHILOSOPHY AND INFORMATION RETRIEVAL MECHANISMS

The topics treated in this book have been chosen with an eye to providing comprehensive coverage of each aspect of terminology management. They do not merely represent yet another collection of potentially unrelated or uncoordinated articles selected at random by their authors for presentation within a particular symposium or series collection. These subjects were carefully assigned to the individual authors, based on the assumption that they have distinguished themselves because of their work in specific subfields of the discipline. The articles have been fine-tuned, and a rich network of cross-references have been integrated into the text to facilitate hypertext-like movement within the book as a whole. These mechanisms are more clearly explained in the following section.

HOW TO USE THIS BOOK

Although the serious or zealous reader is cordially invited to read the book from cover to cover, it is not intended to present a one-time reading experience, but rather to be a permanent reference work that users will return to again and again, both to reconfirm prior knowledge and to familiarize themselves with unfamiliar areas of the field. To this end, we have provided the following devices designed for rapid navigation through the text in search of specific information:

- *Quick-search keywords*
  Marginal references that highlight topic concepts in the individual articles

- SEE

  Term Selection
  1.1.1:13

  SEE boxes cross-reference to other articles in this volume. Articles are indicated by chapter short name and number, followed by a colon and a page reference. For instance, in the above example, *Term Selection* is the title of a chapter, *1.1.1* the number of that chapter, and *13* the page on which the chapter begins. In cases where the entire article is related to a given point in the text, the page reference is for the first page of the article. Where a specific passage in a chapter is referenced, the page number may reflect the page where the related topic is discussed.

- INFO

  Classification
  INFB 9:338

  INFO boxes act as pointers that guide the reader to *Information Boxes* treating basic concepts that are relevant to more than one of the chapters in the book; Infoboxes are indicated by a short name and number, followed by a colon and a page reference. In the example, *Classification* is the short name for Infobox 9: *Classification and Documentary Languages*, which begins on page *338*.

- A detailed, structured subject index designed to allow easy access to topics discussed in the articles, together with an authors' index for easy reference to authors cited in the book

In addition to our many authors, we wish to acknowledge the contributions of our assistants Isabelle Cannas, Anke Förster, Yuthika Prabhu, Elisabeth Skorianz, and Anne Beinchet, as well as the work of our indexer, Michelle Foss. Thanks also go to our careful first readers, Bertie Kaal, Leland D. Wright, Jr., and Juan C. Sager, as well as Kara Warburton. Needless to say, we also want to thank Aban Budin and Leland D. Wright for feeding us well and putting up with the chaos generated by such a massive project.

Sue Ellen Wright
Kent, Ohio, U.S.A.
July, 2000

Gerhard Budin
Vienna, Austria

Section 6

Information Management

Section 6.1

Computer-Assisted Thesaurus Management

CORENTIN ROULIN

1 INTRODUCTION

1.1 Aim and Contents

This chapter provides:

- an overview of how computers can assist thesaurus management (creation, production, and maintenance) and
- assistance in evaluating, choosing, customizing, and implementing a software tool for this purpose

thesaurus use

thesaurus management

After distinguishing between thesaurus management and thesaurus use, the chapter examines data and data structures related to handling thesauri in the form of computer files. Thesaurus management systems must obviously provide functions that allow users to consult the thesaurus, but they must also provide management features for building the thesaurus, updating it, and producing printed and electronic editions. Finally, system designers must provide for an effective interface to the "ideal" thesaurus management system described throughout the course of the chapter.

1.2 Thesauri

documentary language

A thesaurus is a *documentary language,* i.e., the "common language" used in a documentary system. Thesaurus managers use it for representing (*indexing*) the content of the recorded documents (or any type of item that is to be described using thesaurus descriptors in an information system), and users employ it to query the subject they are interested in and thus for *retrieving* relevant recorded documents. In this text, the term *document* is used to designate such an item.

syntax

As a documentary language, the *vocabulary* of a thesaurus consists of terms selected from natural language, as opposed to the artificial codes used in a classification scheme. The *syntax* of such a language is quite basic in most documentary systems. In indexing, the juxtaposition of indexing terms (known as *post-coordination*) is generally the only way to represent both the specific aspects of the subject treated in a given document and the extent of its content. In order to retrieve documents, users enter indexing terms or groups of terms together with Boolean operators (AND, OR, NOT). Functions can also be provided to take advantage of the *a priori* relations that give structure to the vocabulary of a thesaurus.

conceptual structures

A thesaurus is more than a mere list of terms. It is a semantically structured set of concepts. The vocabulary is organized according to relations so that it will be easier to understand, survey, and use in a way that emphasizes the meaning of the terms that comprise it. To ensure its efficiency, a given thesaurus covers a limited number of fields of knowledge in detail and others more superficially. This specialization is marked in the choice of the specific terms introduced (by "pre-coordination", i.e., the introduction of compound terms) and by establishing the relations between these terms.

multilingual thesauri

A *multilingual thesaurus* allows access to common documentary data sources for users who speak different languages. Terms are selected from more than one natural language for each concept present and the terms are treated as linguistically equivalent. Consequently, the semantic relations are, as a rule, strictly identical from one language version to another since they actually link the concepts behind the words.

1.3 Thesaurus Management

thesaurus use

If using a tool includes being able to maintain, improve and customize it according to specific needs, thesaurus management can be viewed as one component of thesaurus use. However, using a thesaurus primarily consists of indexing and retrieving documents, i.e., assigning descriptors to documents and expressing queries using the same vocabulary. User-friendly performance of these two tasks in a given computerized information system is essential, but this function is outside the scope of this chapter.

user and management needs

Both thesaurus users and managers need to consult the thesaurus, which implies searching for terms, displaying concepts, and navigating among these concepts based on the relations programmed into the system. Manager, however, have different needs from users, notably because managers need to view the thesaurus as a dynamic system—one could say as a living object—whereas users want to rely on a stabilized version of the documentary language. This is especially true when a thesaurus is used by more than one group of users, i.e., in more than one information system, which introduces numerous specific requirements for thesaurus management.

1.4 Computer Assistance

paper vs. computer-based systems

Despite today's prevalence of computer-based systems, it is still possible to envisage thesaurus management using pencil and paper only. Experience shows that this means using several pencils, an eraser, a large amount of paper—or, preferably, index cards—and a card file. Thesauri are paper-hungry. This chapter, however, concentrates on software tools designed for professional thesaurus management, both those that are available on the market and those that could be built in the future. It excludes discussing non-specialized tools, such as word processors or spreadsheets, which are sometimes used for this purpose.

1.4.1 *Specialized vs. integrated tools*

integrated tools

Thesaurus management has its own specific requirements that are irrelevant for users who have to index and retrieve documents or that are not even compatible with the requirements of these users. Information systems, including thesaurus management modules that allow direct (interactive) modifications to the thesaurus, are designated as *integrated tools*. When using these tools, it is more difficult, or even impossible, for thesaurus managers to perform certain operations such as:

- introducing modifications on an experimental basis
- reviewing previously introduced modifications and systematizing the collection of update data designed to inform users
- revising indexations resulting from previous modifications as a separate task
- producing printed editions on small scale-systems during normal use time or launching any other memory-intensive computer operations that are likely to overload the system—thesauri are especially memory-hungry

However, integrated tools have advantages. They provide:

- immediate availability of corrections and updates

- immediate availability of usage statistics
- identical functions and interface for both managers and users

specialized tools

Nevertheless, a *specialized tool*, i.e., a separate software module for thesaurus management, is recommended where there is more than one user group or target information system, for such cases usually require the isolation of update work in order to:

- control the management of successive versions and reserve equal treatment for all user groups
- formalize updating data to be transferred to target systems, i.e., separate user systems that may demand different input formats

avoiding drawbacks

The possible drawbacks of using specialized tools run parallel to the advantages listed above for integrated tools. They can be avoided by:

- ensuring a straightforward update process from the management system to the target user system(s)
- ensuring straightforward dataflow for use-related statistics from user systems to the management system in cases involving completely separate systems
- using the same or compatible software for management and for user functions

1.4.2 *Data Modeling*

common data structures

Defining a computerized system includes describing the data to be handled and their structure, as well as defining the different functions (creation, update, report generation) the system must perform on these data in order to achieve its intended management task. As far as data and data structures are concerned, all thesauri share common characteristics. For instance, all thesauri include preferred and non-preferred terms to express concepts and define semantic relations between concepts, but many thesauri also have features of their own. For example, a given thesaurus may distinguish between two types of hierarchical relations—namely both specific-generic relations and partitive relations—where others do not include such a refinement. This chapter takes as a starting point the elements and relations described for multilingual thesauri in the relevant international standards (ISO 5964), with the exception of "facets".

1.4.3 *Functions*

As explained above, users and managers require two sets of overlapping functions. Figure 1 illustrates this overlap and provides references to the corresponding parts of this chapter that deal with each group of functions. Indexing and retrieval are discussed only briefly.

overlapping functions

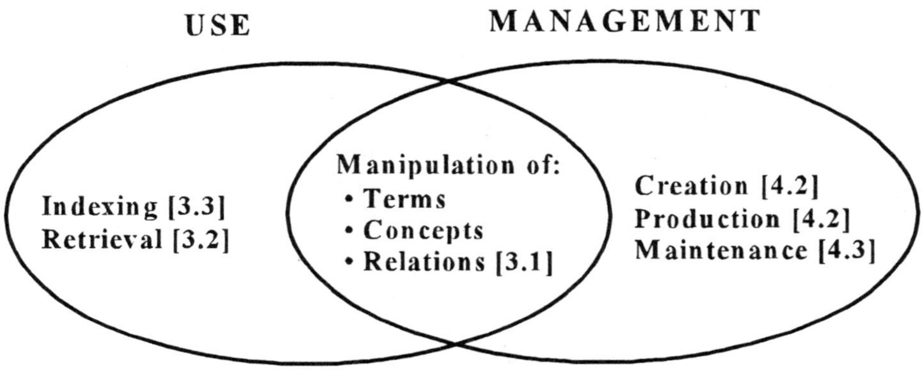

Figure 1: *Thesaurus Use and Management Functions*

In some cases, especially when creating custom-tailored information systems, additional modalities are possible. Not only are thesaurus manipulation (consultation) and management treated as separate functions. The operation that links selected concepts to documents (indexing), query formulation, and retrieval tools are also separate functions. Quality is the most important issue in this regard, notably the reliability and the speed of the interface between the thesaurus consultation module and each of the other modules (thesaurus management, document indexing and document retrieval).

customized systems

1.4.4 *User-Friendliness*

Although user-friendliness is often a hidden dimension in the lists of software evaluation criteria, it is a major factor in deciding to select any given system. This chapter does not treat user-friendliness in detail, but appropriate references will be made whenever this dimension is relevant. In some cases, the corresponding requirements are mandatory ones. To cover all the other cases, it would be beneficial to isolate this criterion in the list of selection

system specifications

specifications in order to give it appropriate priority, especially when customizing a system, where clearly stated guidelines will help when negotiating with the developers.

2 DATA AND DATA STRUCTURE

2.1 Conceptual Data Modelling

2.1.1 *Introduction*

data modelling levels

Software development techniques distinguish among these successive levels used to describe the data to be manipulated by an automated system according to the degree of conceptualization inherent in these descriptions. The *conceptual level* most nearly approximates the way users look at the problem: data are described in terms of concrete or abstract objects that are manipulated in the domain under consideration according to the relations that exist among these objects. The *logical level* uses a particular class of software tools to integrate constraints proper to computerized data management, e.g., the description of data to be handled with a relational database management system will have to be made compliant with the various "normal forms". The *physical level* most closely reflects the actual requirements of a given system in a given hardware and software environment. Performance is then the major concern of the system developer, who has to ensure that the planned system will be able to manipulate the data efficiently in order to achieve user objectives. Transforming a conceptual data model into a description of files to be implemented at the physical level is sometimes designated as a *degradation* process. For instance, a typical element of this degradation involves allowing for some data redundancy.

degradation

conceptual level

In this section, data to be manipulated for thesaurus management are described at the highest conceptual level in order to provide the reader with a description that remains valid whatever tool is used and that allows comparisons between different degradation approaches (see 2.2.1 below). The following section describes some typical implementation options and draws attention to the corresponding potential drawbacks for the thesaurus manager. The last section of this chapter is devoted to practical considerations with respect to the data as they have to be actually manipulated by the computer.

2.1.2 *Data Elements and Relations*

Thesauri can be described as systems made up of sets of data elements and of the relations established among those elements. These relations constitute a structural description of a conceptual view of the system in which significant subsystems are progressively delineated.

descriptors

2.1.2.1 *The Concept Subsystem.* The concept subsystem is made of the three basic elements—descriptors, non-descriptors, and scope notes—and of the relations that link them. The *descriptors* are the authorized indexing and retrieval terms. They consist of single or compound terms, generally in the singular form. Adjectives and verbs are excluded. *Non-descriptors* are entry terms that cannot be used as indexing terms. They are synonyms or quasi-synonyms of the terms chosen as descriptors and are provided to help the user find the most appropriate authorized indexing by referencing descriptors.

non-descriptors

By linking each non-descriptor to the descriptor to be used instead, the semantic equivalence relations form bridges between the vocabulary of natural language and that of the artificial language. Multi-equivalence relations may also be used to link one non-descriptor to two descriptors either to be used together (post-coordination guidance) or between which the user is invited to choose (in the case of polysemes).

multi-equivalence relations

Scope notes are short texts that either give a definition for, or state restrictions and give advice on the use of, given descriptors. Application relations link each scope note to the descriptor to which it applies. Some thesauri distinguish between scope notes, definitions, and "history notes", which contain information on how the concept was expressed in previous versions of the thesaurus.

scope notes

In a monolingual thesaurus (or in a given language version of a multilingual thesaurus), the intended *concept* can only be grasped correctly by considering together the chosen descriptor, its non-descriptors, and any scope note that might have been attached to it.

Figure 2 shows that some elements belong to "classes" (namely *terms* and *texts*), that is to say, they share common formal characteristics. These two classes are especially important in a multilingual thesaurus: language identifier relations link each term and each text to a single language, which allows for the distinction

multilingual thesauri

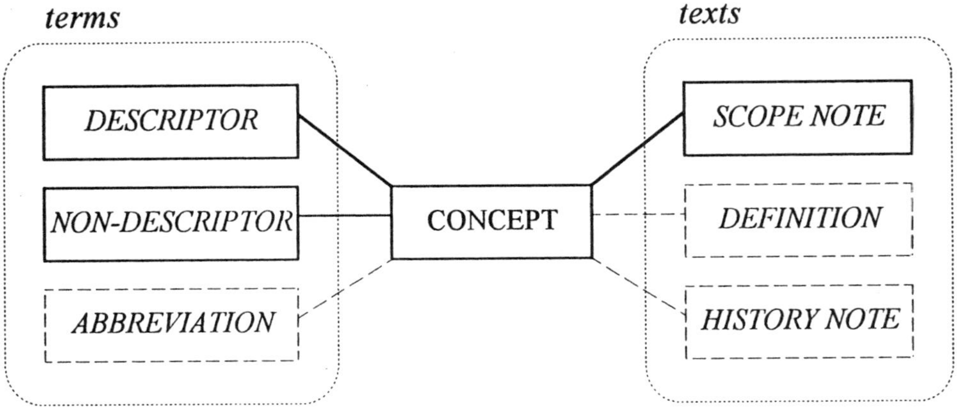

Figure 2: *Element Classes*

among the different *language versions* as subsystems. Typical attributes of terms and texts are their type, language, and wording. For management purposes, it is also necessary to record and maintain information on proposals and decisions taken related to each element.

2.1.2.2 *Structuring relations and subsystems. Hierarchical relations* are established between pairs of concepts where one term (referred to as the *narrower term*) can be considered as subordinate to the other (*broader term*). The distinction between specific-to-generic and partitive relations is usually not made in thesauri. Some thesauri allow polyhierarchical relations in which one concept can be associated with more than one generic concept.

hierarchical relations

The purpose of this essential means of structuring a thesaurus is to help users explore the vocabulary in a systematic way. It can also be exploited in retrieval mechanisms. *Hierarchical chains* are subsystems made of concepts linked together by hierarchical relations. *Associative relations* are established between concepts designated as possible alternative indexing terms, which form complementary paths for exploring the vocabulary.

hierarchical chains

associative relations

Inclusive relations link each hierarchical chain—and, as a consequence, each concept—to a super-set that is generally defined according to a classification-like approach. These relations are sometimes organized into several levels. The *domains* subsystems (sometimes designated as "fields" or "microthesauri" and

inclusive relations

identified by codes or titles) thus form the highest levels of the semantic structure of the vocabulary.

Figure 3 emphasizes the relations (hierarchy, association and inclusion) because, for management purposes, it must be possible to manipulate them as such and to record and maintain information about them, such as proposals and decisions. The "hierarchy/order" relation is used to manage domains and sub-domains and to order them according to a logic different from mere alphabetical order.

2.2 IMPLEMENTATION OPTIONS

2.2.1 Degrading the Model

One customary approach to implementation is to consider the concept as a single object and to manage it in a single file made of records with fields to encode descriptors, non-descriptors and scope-notes. This option leads to the following problems:

single-object concepts

- It is necessary to limit the number of non-descriptors and to determine their maximum number beforehand.
- It is impossible to attach a given term to another concept without re-typing it.
- It is necessary to scan all "concept" records to perform functions involving all terms or certain types of terms, such as tracking duplicates among descriptors and non-descriptors or extracting all modified descriptors.

managing relations

Managing relations is an important aspect, especially in non-relational database management systems. Ways of implementing "one to many" and "many to many" relations will determine how easy it will be to perform operations such as linking a given concept and all its narrower terms to a new generic concept, removing all relations for a deleted concept, and restoring a previous situation.

degradation options

For the sake of performance, some degradation options may be chosen for thesaurus use. For example, in some cases it can be difficult to follow the line from a given concept to the domain(s) to which it belongs via the hierarchical relations to the most generic concept(s) and then from this (these) top-term(s) to the domain(s). If this is the case, the code or title of the domain(s) can

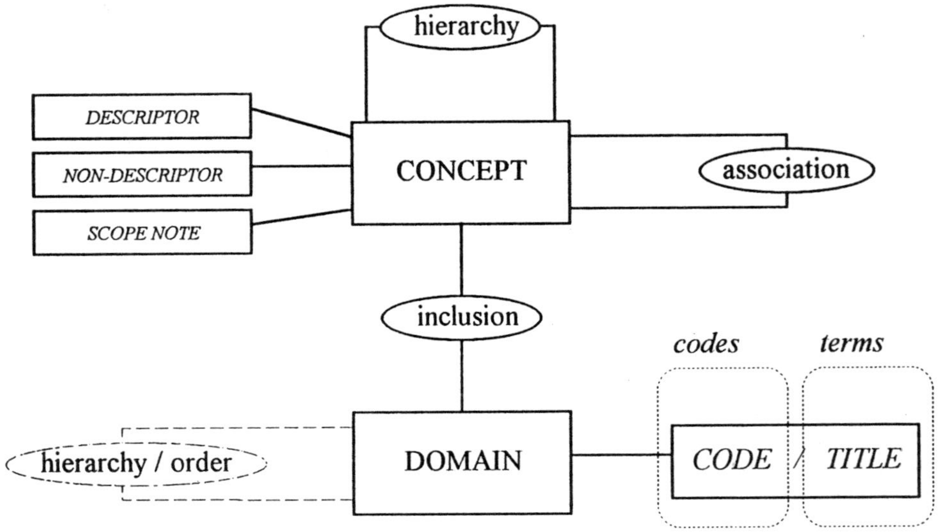

Figure 3: *Thesaurus Relations*

be encoded in the concept record so that it is unnecessary to have to perform this search every time the information is needed. This approach can, however, lead to significant redundancy that will become problematic if a domain title or code is modified. Nevertheless, thesaurus management will be more efficient if all the elements and relations handled are implemented as separate objects. In other words, the closer the implementation is to the conceptual model, the safer. The first step of computer-assisted thesaurus management involves establishing the conceptual data model that corresponds to the thesaurus being designed.

conceptual models

2.2.2 *Off-the-shelf vs. Custom-tailored Systems*

Numerous software programs designed for managing fully standard-compliant thesauri have been coming on the market. Selecting an existing product requires a carefully conducted assessment process. The more numerous the specific characteristics required for the application, the more attractive a custom-tailored approach begins to look. Needless to say, a customized product will inevitably be more expensive. The corresponding cost-benefit analysis should include the relative *costs* of:

software selection

- using an available software development team that would be capable of creating a tool from scratch
- taking advantage of parameter configuration options offered in an existing software program (Note that the question of who does what at what price always has to be answered very precisely.)
- benefiting from custom adaptations that could be proposed by the provider of an existing tool

cost-benefit analysis

This analysis should also include purely *qualitative* considerations such as the ability of the envisaged software development team to deal correctly with textual data and, more generally, with the complexity of the thesaurus management environment, which should not be underestimated. Another option is to outsource the whole thesaurus management function (see Section 5).

qualitative considerations

2.3 Data Management

2.3.1 *Data Elements*

2.3.1.1 *Length of Textual Data.* Some software systems limit the length of terms, which is quite efficient if the users are required to type in the descriptors when indexing documents or formulating queries, but which is irrelevant if this is not the case. Ideally, it should be possible for terms to be as long as necessary to take full advantage of one of the basic characteristics of thesauri as documentary languages, which is the ability to use words instead of codes to express concepts. Depending on the thematic field covered by the thesaurus, the typical limitation to between 30 and 40 characters may be unacceptable. This holds true for texts as well, e.g., in the scope notes. When textual data longer than terms are stored in special text-fields, the retrieval aspect also has to be especially examined, e.g., is it possible to perform quick searches on words in these data?

limitations

2.3.1.2 *Character Sets.* The ability of a given software program to satisfactorily handle textual data in a given language or group of languages is essential, especially in the case of multilingual thesauri, but also in the case of a thesaurus in one "rare" language, i.e., in any language for which no fully integrated solution for textual data management is as yet widely available. Beyond the thesaurus management system itself, the software and hardware

multilingual character management

environment in which it is to be used (database management system or programming language, operating system, keyboard, display, and printing devices) will determine if and how efficiently textual data can be typed in, recorded (using one or more than one byte per character), displayed on the screen, searched, sorted alphabetically, or printed. (Each element of this list is equally important). Solutions for one language at a time exist for most languages. Problems appear, however, when it is desirable, for instance, to display several character sets on screen at the same time. In this case, compromises are likely to be necessary. Before accepting any given solution, one should consider:

INFO

Unicode INFB 20:861

- which form of the thesaurus is likely to be the most used (the printed or the on-screen version)
- whether the use of upper and lower cases—together with accented letters—is important, which may differ considerably according to the language
- whether different formats should be necessary, e. g., a complete character set for the printed version, but upper case only for the target information system

2.3.1.3 *Numerical and Coded Data.* A thesaurus management system has to allow for the manipulation of numerical data as well. The thesaurus manager needs statistical data that give information on the characteristics of the thesaurus as such and on the frequency with which specific concepts are used in indexing and retrieval. As far as coded data are concerned, a thesaurus management system should offer all functions generally available for such data, such as choosing from a pull-down menu that presents the limited set of permissible values (e.g., language and type of a term, or the status of a modification).

statistial data

2.3.2 *Relations*

Whatever the method of managing relations at the physical level, a thesaurus management system must provide the manager with all the necessary functions required to take and keep track of decisions on relations, to easily manipulate reciprocal and transitive relations, and to perform consistency checks. *Modifications of relations* include the addition, deletion, and modification of linked objects. Quite often, a single modification has consequences that affect more than two objects, which means that the thesaurus

consistency checks

modifying relations

manager needs a system that requires minimum typing and provides maximum capability for automated processing and data control.

reciprocal relations

Thesaurus managers also need to be able to change their minds, which supports a preference for approaches that isolate relations as objects, allow contradictory relations to co-exist, and so enable precise control of relation status. As far as *reciprocal relations* are concerned, it is mandatory that the system take them into account automatically, i.e., if *A* is said to be a narrower term of *B*, the relation indicating that *B* is a broader term of *A* must be automatically set.

transitive relations

Transitive relations include, for instance, the "inclusion" relation whereby if a "top-term" is attached to a domain, then all its (direct and indirect) narrower terms also belong to that domain (sometimes called inheritance). The main concern of the thesaurus manager is that the consequences of a given modification must be taken into account so as to make all the consecutive modifications anywhere else in the thesaurus available immediately, and, if the case arises, to generate a report of inconsistency problems before any further modification is introduced.

rules

Rules are applicable to keep the structure of a thesaurus, which is sometimes referred to as a "semantic network", consistent. For example, it is admitted that a given term cannot be at the same time, directly or indirectly, both a narrower and a broader term of the same term, which would lead to unmanageable endless loops in the network.

polyhierarchy

A distinction has to be made here between standard rules and rules specific to a given thesaurus, and also between "mandatory" rules, the violation of which will not be accepted by the system and which will immediately generate an error message, and rules whose violation will only make the system generate a warning to the manager, possibly not online but in the form of a printed report. For example, polyhierarchy might be admitted in some thesauri and not in others. If it is admitted, it might be recommended to limit its use to concepts that are at the lowest level in the hierarchical chains, i.e., concepts to which no further narrower concepts are attached.

Thus, thesaurus management functions—and this is especially true for ready-made systems that incorporate *a priori* consistency rules—should include options for:

- tailoring the mandatory level associated with each rule that is implemented
- implementing new rules

3 User Functions

3.1 Searching for and Manipulating Concepts

3.1.1 *Retrieving Terms*

3.1.1.1 *Formulating Queries.* All data types should be searchable, including management data types such as dates. Since the basic elements of a thesaurus are terms or are made up of terms, the search functions of a thesaurus management system should include options to formulate queries with search criteria based on words and phrases, as well as the ability to:

thesaurus search functions

- search for exact wording (which is definitely enhanced by having access to the computerized inverted file or index in the form of a "word wheel") or for the occurrence of word(s). In case of words, the system should allow for the indication of definite order (or not) and of proximity operators.
- right, left, and medial truncation(s) (any number of characters) and wild cards (definite number of characters) on any word

Moreover, it should be possible to formulate queries that are limited to either:

- a given type of element ("search descriptors that ... [search criterion]" or "search non-descriptors ..." or "search scope notes ...") or not ("search elements of any type that ... [search criterion]")
- a language version (or not), taking into account that accents are essential in some languages and not in others

3.1.1.2 *Query Results.* The first result expected by users is to know whether the search was successful or not. If it has been, they want to know the number of hits, and if the query has been made on more than one type of element, they need to know the distribution of hits among the different types so that they can choose which element(s) they want to manipulate next. What the end user (i.e., the one who indexes and retrieves documents) needs is the concept. From the word specified, it is not necessarily useful to

hits

user needs

stop on the element found (term or text) before accessing the desired concept.

manager needs

The situation is different for thesaurus managers, for they look at all elements as distinct objects. That is why, if the terms and texts are implemented as distinct objects for both the end user and the manager, it must be possible to disconnect the mechanism that automatically proceeds to the next step (e.g., from the retrieved scope note or non-descriptor to the corresponding descriptor or concept, or from the domain title to the top-terms).

user expectations

3.1.2 *Displaying Concepts*

data display and access

Here again, needs may differ for the end user and for the thesaurus manager, but in case of a multilingual thesaurus, both will first want to choose a working language that will determine which language version is used with respect to display options (without preventing access to all data). End users primarily need to identify the concept presently available for further manipulation (the current concept). In most situations, providing them with the corresponding descriptor in the working language will be sufficient. In other cases, they will want to consult all the information available about the current concept, which may be partly available directly in the current record and partly distributed over linked objects. This information should be displayed in a manner that respects the following principles:

data principles

- All data in a record should be shown, or accessible at a single keystroke, and coded data should be made explicit (e.g., "English" rather than "language 2").
- All existing relations should be shown and made explicit (e.g., "A and B are the narrower terms" rather than "there are two narrower terms"), taking into account the working language.
- Format should be based on that of the "reference" presentation of the printed thesaurus, even if screen and paper are not comparable. Computer interactivity can be used to compensate for the limited screen size. The printed version is a good reference because it is familiar to users and has probably been designed to be as user-friendly as possible (see 4.2.)

graphical presentation

As far as graphical presentation is concerned, a graphical display of hierarchical chains parallel to the one provided on paper will be much appreciated, provided that computer screen generation time is not excessive, and especially if all actions normally possible remain available in this context. The thesaurus manager's specific needs may be derived *mutatis mutandis* from the preceding paragraphs taking into consideration that they are interested in:

- not only concepts but all elements and relations as distinct objects, each one having its own display format
- management data not necessarily available to end users
- possibilities for tailoring display formats, including options for explicitating relations, such as showing status, e.g., "A and B are existing narrower terms, C is a new one, D is a proposed one".

3.1.3 *Exploiting Relations*

network navigation

Relations are essential to thesauri, and computers offer the possibility of navigating through the relational network and exploitation techniques that go far beyond what is possible on paper. Desirable functions can be described in a few words, but user-friendliness, which may be the most important aspect, is certainly the most difficult to describe. It must be possible to go from any element to any other element linked to it by any type of relation with a single keystroke or mouse click. (This operation is so frequent that having to press two keys is one too many.) It should also be possible to keep track of the path that has been followed and so to return to any previous position in the network.

3.1.4 *Manipulating Sets*

accumulating sets of objects

What end users and managers need when consulting their thesaurus is not always a single object (such as one concept) but quite often a set of objects (e.g., a set of concepts) that is of interest for their current task. Such objects are encountered in the course of performing searches on terms and navigating along relational links. One way to manipulate sets involves the possibility of picking up these objects as one works along without stopping the thesaurus consultation and losing the context.

identifying sets

A second way to manipulate sets involves selecting sets of elements linked to the current one, for example, all narrower concepts of the current concept (on one, several, or all levels of specificity) or all top-terms included in the current domain. An interesting, more sophisticated form of this mechanism can be applied not only from a single element but from a set of selected elements (the "current set"), for example by selecting all concepts related to the ones in the current set. In using such a mechanism, users should be queried so that they can specify which logical operation shall be applied. In the example, they might be asked if the selection of concepts is related to *any* concept or to *all* concepts in the current set.

Boolean operators

When selected sets are available, the third way to manipulate them is to perform Boolean operations using techniques like those offered in documentary systems where a history of the previous searches is available, for example, selecting all concepts having "vitamin*" in the wording of the descriptor AND those having this word in the scope note (but) NOT those belonging to the domain "chemical products".

3.2 *From Concepts to Documents: Retrieval*

document records

One type of element among those that can be linked to concepts could be document records, in which case the generic mechanisms described in the previous section can be used directly for document retrieval purposes. If thesaurus consultation remains a separate module, it may be useful to automate some of the concept manipulations that are needed for formulating and tailoring "search strategies" and that imply the semantic relations linking concepts in the thesaurus.

posting

Such manipulations include "posting", which in this context means collecting concepts that are linked to a given concept by a specified type of semantic relation:

- "Generic posting" involves selecting the broader concept(s) of the current one.
- "Specific posting" involves selecting the narrower concepts, in which case it is necessary to be able to specify the number of levels of specificity to be taken into account.
- "Associative posting" involves selecting related concepts. Another form of this kind of posting involves selecting concepts that are "related via hierarchy" (or "siblings"), that is to say, concepts that have the same broader concept as the current concept.

3.3 *From Documents to Concepts: Indexing*

Indexing can be seen as "translating" the natural language of documents into a documentary language. Some automated systems offer textual data analysis tools that tend to transform (reduce) the natural language of titles, abstracts, or full texts into sets of single and compound terms that can be compared to those appearing in the thesaurus used. Such a transformation implies *normalizing* the terms encountered in natural language by applying the same rules as those used during thesaurus elaboration, such as restricting terms to nouns, to the singular form, to a limited character set, etc. This process may also include the use of weighting mechanisms.

documentary language

Thesauri can also be enriched to favor comparisons with natural language by introducing more numerous non-descriptors as "lead-in" terms to fill the gap between intellectual and automated analysis of text content. Such non-descriptors range from synonyms that may be useful for guiding indexers to mere plural forms or spelling variants that are useful for the computer only. A more basic, though practical, function allows indexers to interactively select terms in available textual data and to request a check of their selection against thesaurus contents. In any case, the result of automated processes must be proposals for concepts and indexing statements that will be validated by the human indexer.

automated processes

human text analysis

4 Management Functions

4.1 Thesaurus Creation

4.1.1 *Data Collection*

Computers can certainly assist thesaurus management from the very early stages. If the corpus of documents that the corpus is designed to index (or a representative sample of this corpus) is available in electronic format, systems especially designed for this purpose can be used to extract terms from this source of "raw" data. In this case, attention should be paid to:

SEE *Term Extraction 8.4.1:725*

- the nature of quantitative information provided (Basic information units include the number of occurrences for a term, and further ranking can take into account the position of occurrences in the analyzed texts, such as "in title", "in abstract", "in conclusion".)
- the nature of extracted terms, which should not just be single terms, but compound terms as well, and could be made available in a "normalized" form (see 3.3)

- the availability of information on co-occurrences of extracted terms
- the possibilities for customizing or fine-tuning the tool (regarding the extent of the corpus, extraction strategies, ranking techniques, etc.) so as not to be overwhelmed by a large amount of unmanageable data

structured resources

There might also be an opportunity to take advantage of existing sources of processed data, i.e., structured or controlled vocabularies such as dictionaries, glossaries, or any type of lists of terms, the contents of which include added semantic value relevant to the scope covered by the planned thesaurus and to the context envisaged for its use. Existing documentary languages—including classification schemes—are a particular example for such resources and may provide a starting point for building a new thesaurus based on portions of an existing thesaurus (or thesauri).

data importation and migration

In all these cases, the main issue is the interface or the import tool: how will available or extracted data be loaded into the thesaurus management system where it has to be used? This question is also relevant to the case of migrating from one thesaurus management system to another: how and at what cost will the data be loaded into the new system? This issue involves two major considerations:

exchange formats

- There is a need to agree on data exchange format(s) (see 4.2.2.).
- The thesaurus management system must be able to deal with data that might still be (partly) unstructured (see 4.1.2).

projecting costs

Even when some automatic data collection is possible (as described above), and certainly when this is impossible, thesaurus creators have to face the problem of manually encoding a large amount of data. Concern for the ease of (massive) input may look like addressing a "one-off" problem but, given staff-related costs, it might be a critical question to answer how the large number of words that will make up the thesaurus will be encoded. One option may be to use a customized, simplified encoding module with the appropriate structuring functions in order to achieve the right balance between input speed and the completeness of data required for management purposes.

4.1.2 *Structuring*

manipulating relations

It must be possible and even easy to manipulate relations (see 2.3.2). The early stages of thesaurus creation involve more specific requirements with regard to manipulating sets of elements and performing related actions on them. Thesaurus creation may require intermediate development stages where the nature of some terms (terms which are neither descriptors nor non-descriptors yet) or the nature of some relations (e.g., "grouping" of terms into a domain without proper inclusion relations) are still provisional. Structuring functions could include mechanisms such as:

structuring functions

- linking a grouping of terms to a "pseudo-concept" (a concept to which no attributes have yet been assigned)
- choosing a preferred term from among grouped terms and automatically assigning the appropriate thesaurus status (the chosen term is the descriptor, all the others are nondescriptors)
- grouping concepts to be precisely structured later

ongoing management

Structuring functions that are also useful during ongoing thesaurus management include:

- in case a concept with broader and narrower concepts is deleted, automatically modifying relations by hierarchically linking the "orphans" to the concept that serves as their "grandfather"
- selecting or listing all polyhierarchies, multi-equivalences, and duplicates

4.1.3 *Enrichment*

checking mechanisms

Once the structure of the thesaurus has been stabilized, its manager concentrates more on completing and enriching thesaurus contents. At this stage, manual input is inevitable, but intellectual work becomes much more important and time-consuming than mere encoding. In the case of a multilingual thesaurus in particular, managers will want to automatically check the completeness of contents, for instance by identifying concepts with missing descriptors (in a given language or in just any language).

processed data

For establishing language equivalents or generating scope notes (or definitions), managers might also benefit from the availability of multilingual sources of "processed" data for automatically loading such terms and texts (see 4.1.1). This phase of thesaurus creation is also when the manager starts using available statistics on thesaurus characteristics (see 4.3.5).

4.2 Thesaurus Production

4.2.1 *Printed Formats*

4.2.1.1 *Presentation Types*. ISO 5964 provides very extensive specifications for the possible design(s) of printed editions of thesauri. They distinguish between alphabetical, systematic, and graphical presentations, and state that if there is more than one type of presentation, one of them must contain all available information about each concept (the "reference" presentation). A *structured alphabetical presentation* is usually established as this reference presentation. It lists all descriptors and nondescriptors in alphabetical order, with each term accompanied by the following data:

structured alphabetical presentation

- nondescriptors: the descriptor to be used instead
- descriptors: domain, language equivalents, scope note, nondescriptors, broader terms (on all levels), narrower terms (on all levels), related terms

This type of presentation is not practical when users want to look at the contents of the thesaurus according to its semantic structure. This is why the structured alphabetical presentation is usually supplemented by a *systematic presentation* or by a *graphical presentation*, both of which are more "use-oriented" in that they are organized according to domains and display complete hierarchical chains (sometimes with associative relations) and usually contain descriptors only. If it is impossible to incorporate language equivalents in the "reference" presentation, they can be given in a separate *multilingual alphabetical presentation*. Some thesauri also provide users with a means to search for terms on the basis of words as they appear in the form of a *permuted alphabetical presentation*.

systematic presentation

graphical presentation

A thesaurus management system must allow for producing the printed editions that are most desirable for users of the thesaurus, i.e., the system must not impose or limit the print formats. Different systems offer different solutions for printing:

print formats

- Some systems have a given number of available print formats that cannot be changed.
- Some systems offer options for reconfiguring print format parameters.
- Some systems include or give access to a report generator that allows for full reconfiguration of existing format parameters and for creation of new ones.

print output

Viewed from a technical perspective, actual printing options range from direct control of the printer by the thesaurus management system to production of tagged text files to be processed by word-processing or publishing software. These variations depend on the programming language or database management system used, which may include more or less sophisticated publishing features. When selecting a system, planners need to consider the availability and flexibility of these options, along with some of the specific questions discussed below.

alphabetical presentation

4.2.1.2 *Alphabetical and Systematic Presentations*. To produce alphabetical presentations, it is necessary to be able to *sort* a large number of terms and, in some multilingual thesauri, to do so according to sorting orders that differ from one language to another. Given that it might be impossible to rely on default sorting options implemented in low-level software layers, it is recommended that a thesaurus management system have its own sorting tools or that it allow for customizing the sorting tools used, or both. The question of the character set(s) handled has to be examined in connection with this point (see 2.3.1).

permuted presentation

KWIC Index

Producing a permuted alphabetical presentation introduces specific problems: the terms listed have to be sorted not only according to their complete wording (i.e., their spoken order), but also based on each "non-empty" word that appears in the term. This implies maintaining a (language-specific) list of "empty words" (or "stop words"). In some agglutinate languages, this also means configuring the permuted presentation so that it is possible to split some words into their meaningful components and to produce the appropriate page layout to display these lexical components (keyword in context or KWIC; see Figure 4).

	Audiovisuelles Programm
Ausbildung der	**Ausbilder**
	Ausbildung der Ausbilder
	Ausbildungsprogramm
	Ausbildungsrichtung
	USE Studienrichtung
	...
Audiovisuelles	**Programm**
Ausbildungs~	**programm**
	...
Ausbildungs~	***richtung***
	USE Studienrichtung
Studien~	**richtung**
	...
	Studienrichtung

Figure 4: *Keyword in Context Representation*

For the production of structured and systematic presentations, some systems limit the number of levels of specificity that can be listed recursively (together with alphabetical order between terms of the same level). Depending on the thesaurus, such a limitation may be unacceptable. It is especially useful in multilingual thesauri in particular to be able to customize parameters for relation indicators.

4.2.1.3 *Graphical Presentations.* Off-the-shelf software solutions that provide integrated graphics capability are rare. What would be useful for thesaurus management would be a system where modifications introduced in a graphic would be taken into account by the system and "translated" into modifications in the corresponding relations. The simplest form of graphical presentation is the *tree structure* (see example in ISO 5964). Tree structures can be produced manually, but some systems generate them automatically. *tree structures*

A more sophisticated form is the *arrow graph* presentation (see 5), which is produced using computer-assisted design software. Textual data can be automatically loaded into such systems, but the actual position of the terms, arrows, and lines that illustrate relations has to be designed manually. In some thesauri, the coordinates of the descriptors in the "arrow graph" are given in other types of presentation. Consequently, it makes sense to integrate *arrow graphs*

Figure 5: *Arrow Graph*

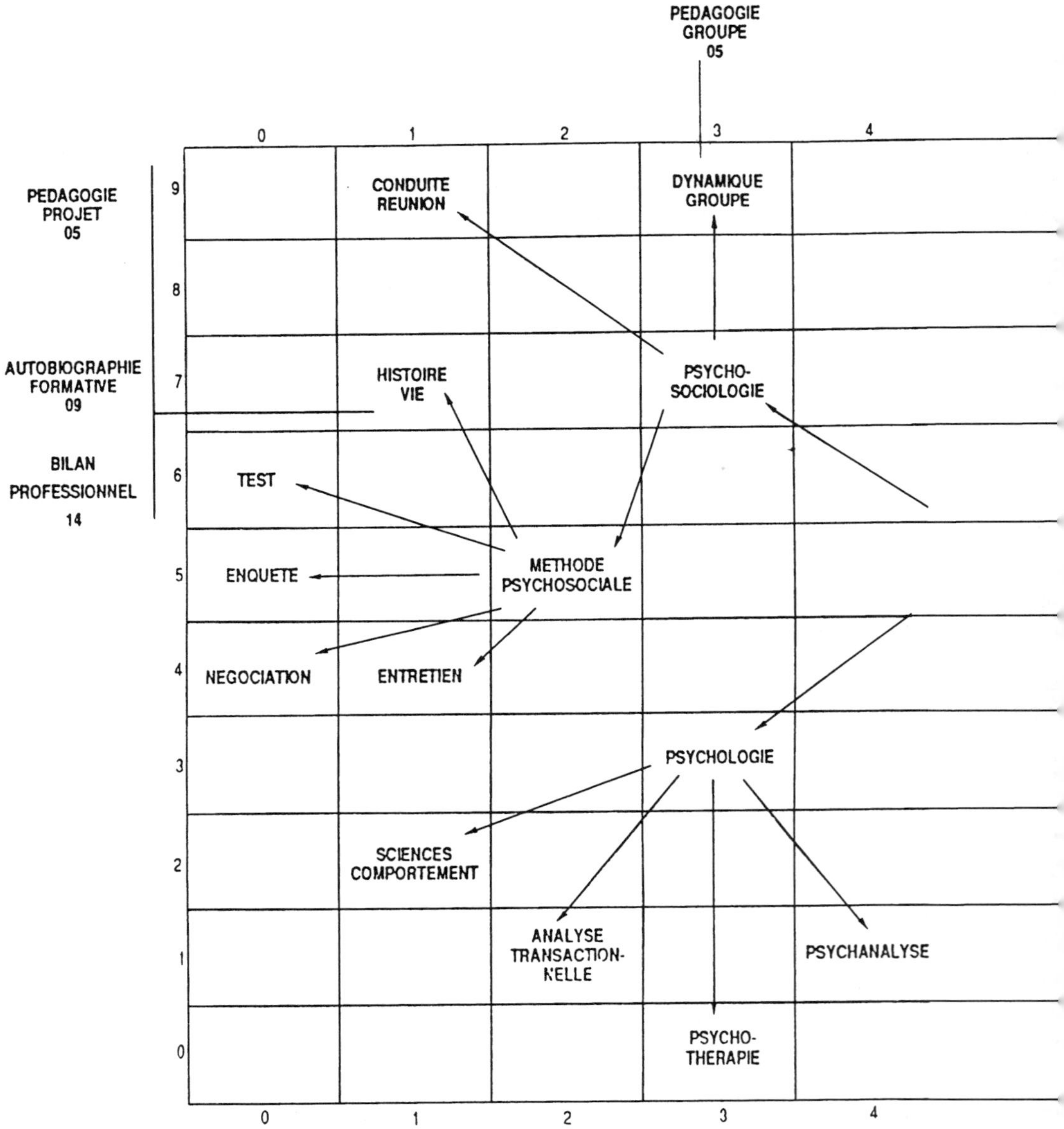

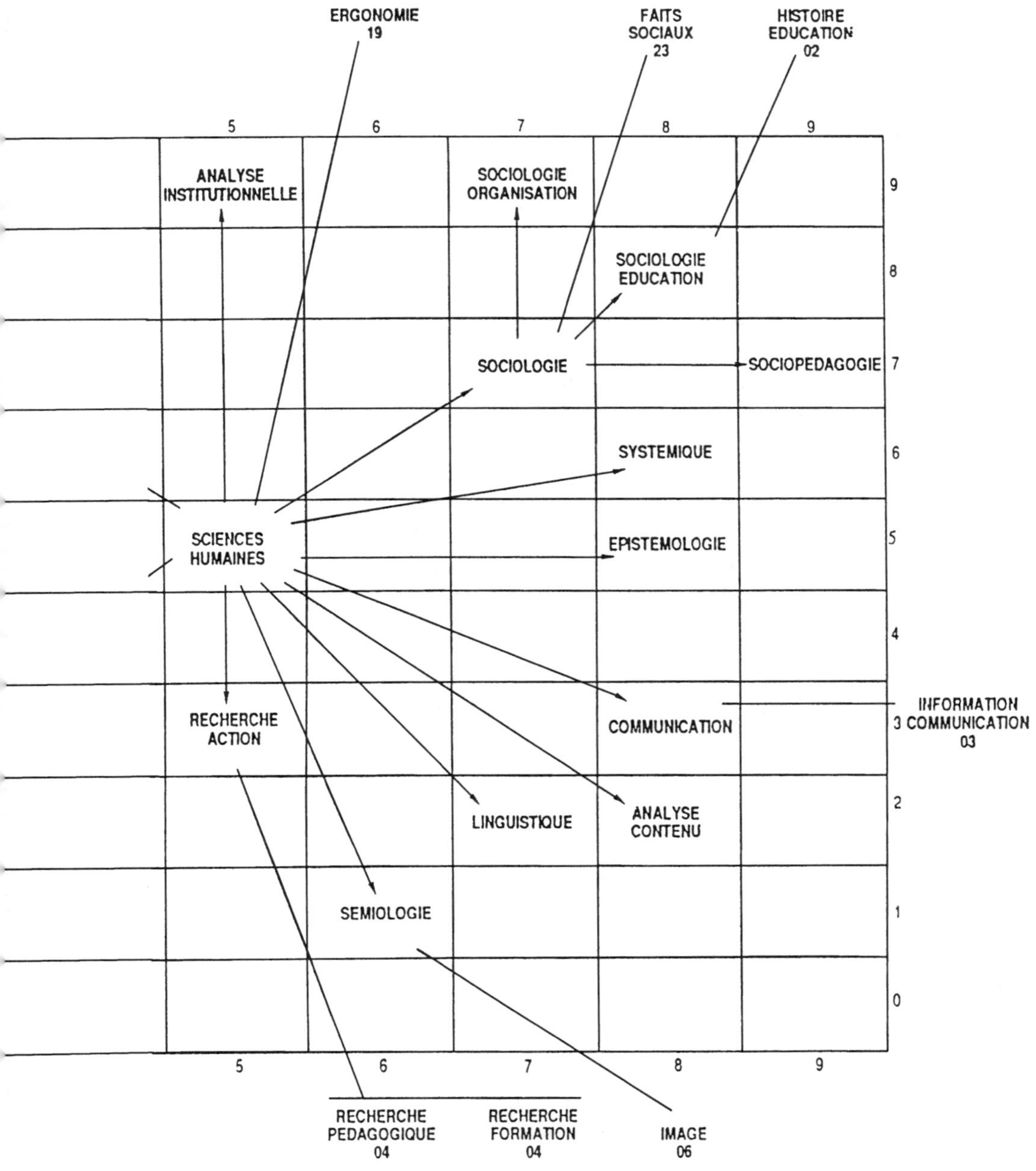
ERGONOMIE 19
FAITS SOCIAUX 23
HISTOIRE EDUCATION 02
5 6 7 8 9
ANALYSE INSTITUTIONNELLE
SOCIOLOGIE ORGANISATION
9
SOCIOLOGIE EDUCATION
8
SOCIOLOGIE
SOCIOPEDAGOGIE
7
SYSTEMIQUE
6
SCIENCES HUMAINES
EPISTEMOLOGIE
5
4
RECHERCHE ACTION
COMMUNICATION
INFORMATION
3 COMMUNICATION
03
LINGUISTIQUE
ANALYSE CONTENU
2
SEMIOLOGIE
1
0
5 6 7 8 9
RECHERCHE PEDAGOGIQUE 04
RECHERCHE FORMATION 04
IMAGE 06

graphics creation and manipulation capability with general thesaurus management functions. In practice, however, such data are encoded manually in the thesaurus system after the graphical display has been stabilized.

terminographs

Terminographs offer another form of graphical presentation (see Figure 6). The example given has actually been produced semi-automatically using a sophisticated, but proprietary system. Such a presentation can also be produced manually using a computer-assisted design package.

4.2.2 *Electronic Formats*

data interchange

Electronic formats are especially relevant in situations where the thesaurus is managed and used in different systems. Where both systems are built on the same database management system or can handle the same database file format, and if both are under total control of the same body, any "internal" format may be used. For external purposes, standardized formats, such as ISO 2709 or FORMEX, can be used for communicating data to targets systems. In such cases, the ability of addressees to correctly handle these very sophisticated formats can limit operational effectiveness. Partners may also agree on any specific format, in which case:

- either the thesaurus manager takes responsibility for providing a format that will allow straightforward loading of data into identified target systems ("specific target systems formats")
- or the thesaurus manager provides a single, highly readable, "flat" text file with "tags", together with the explanation of the tag names used, possibly introducing some data redundancy to facilitate easier data loading; such files assume no responsibility for further use of the data provided beyond the guarantee of completeness and internal consistency ("tagged text files").

4.3 THESAURUS MAINTENANCE

4.3.1 *Recording Modification Requests*

modification request forms

Thesaurus managers generally use special forms for collecting or recording user requests for modification. Any forms circulated among the user community must be designed to achieve the right balance between the point of view of the users (who want to be able to report easily any problem they encounter) and that of the thesaurus manager (who looks at modifications in a more technical way and usually needs more complete data).

automating request management

Obviously, automating this process increases efficiency, and the number of proposals to be processed will determine to what extent such automation is useful. Automating the management of modification requests may range:

- from providing a specific encoding module, which "translates" requests as received from users into sets of data that can be handled by the system
- to automated pre-processing of envisaged modifications, which generates special reports with informative and warning messages
- to automation of clerical work, which features automatic production receipt acknowledgement and reports on the state of processing

4.3.2 *Recording Decisions*

modification history

Decision records should document which modifications have to be taken into account by the system, but decisions on modifications include decisions *not* to modify the thesaurus, and these decisions should also be recorded because simple deletion of related data would prevent the manager from:

- informing proposers about the processing of their requests
- relying on previous decisions when taking new ones

content management strategies

This second aspect is especially important. The quality of the recorded information will determine to what extent the thesaurus management system will contribute to the long-term coherence of strategies for managing the thesaurus content. This information should not be limited to the decisions taken (which can easily be coded) but should also include their motivation (which implies recording textual data). Management efficiency can be improved by making information available such as: "this term has already been proposed as a descriptor but has been rejected for the following reason ...".

Figure 6: *Terminograph*

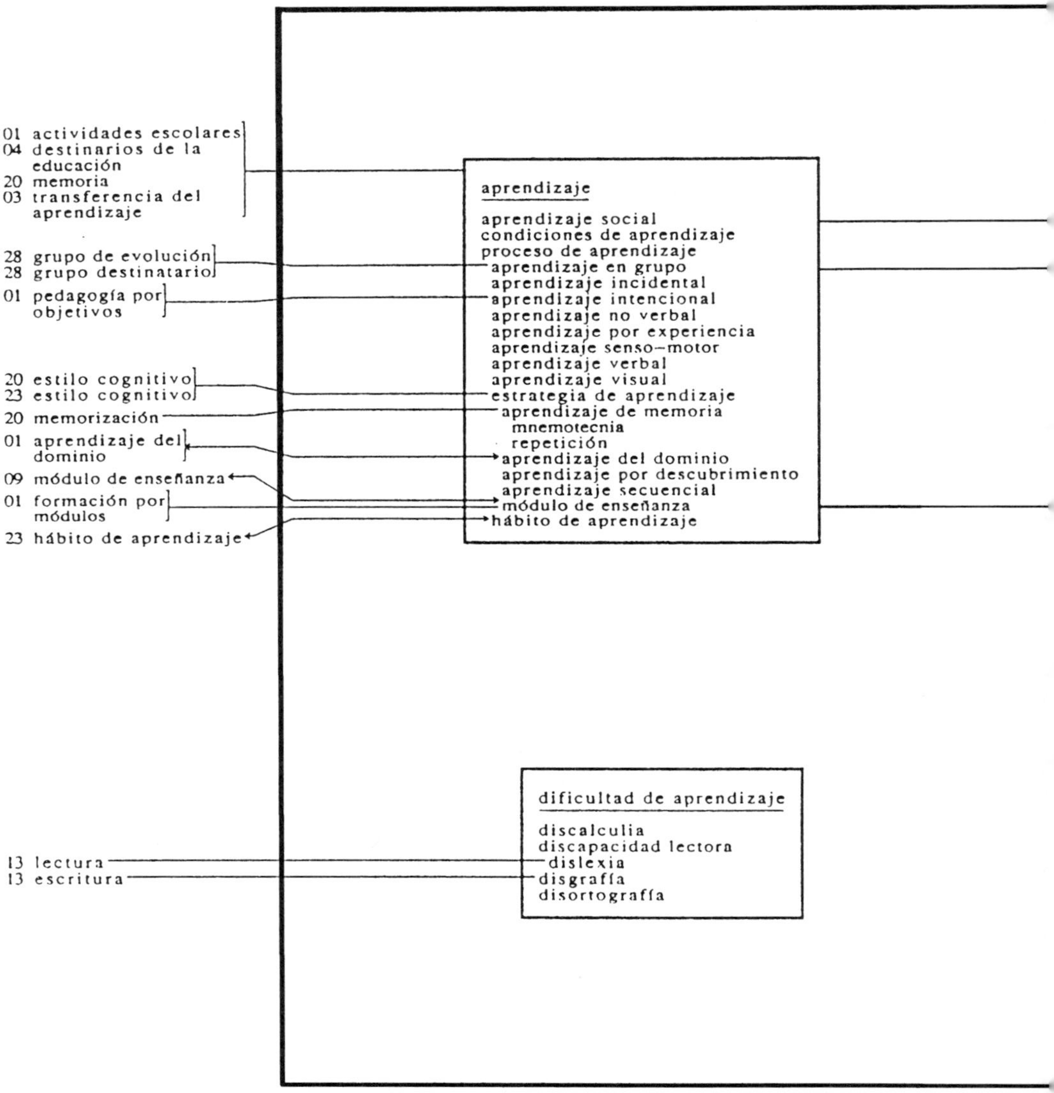

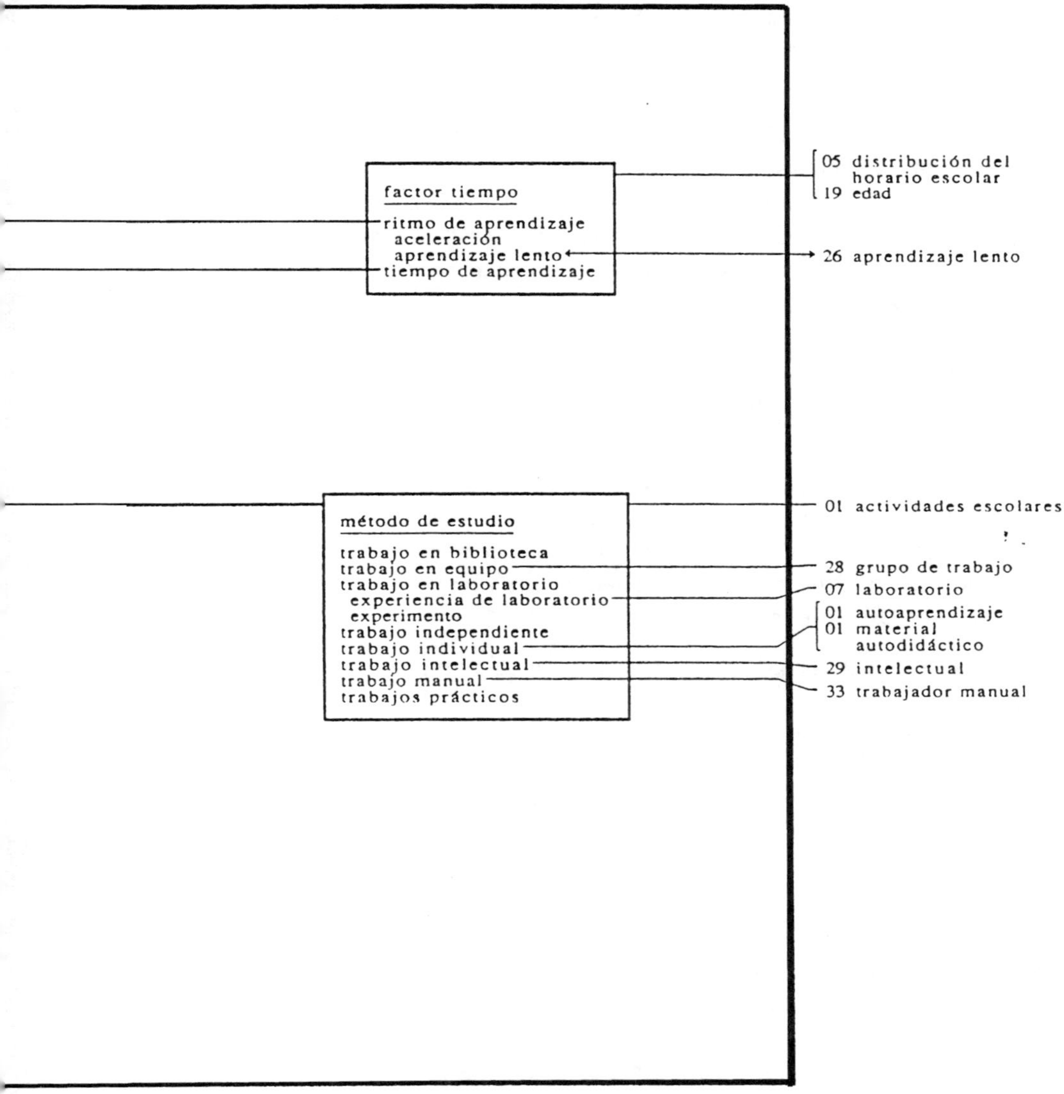
factor tiempo
ritmo de aprendizaje
aceleración
aprendizaje lento
tiempo de aprendizaje
05 distribución del horario escolar
19 edad
26 aprendizaje lento
método de estudio
trabajo en biblioteca
trabajo en equipo
trabajo en laboratorio
experiencia de laboratorio
experimento
trabajo independiente
trabajo individual
trabajo intelectual
trabajo manual
trabajos prácticos
01 actividades escolares
28 grupo de trabajo
07 laboratorio
01 autoaprendizaje
01 material autodidáctico
29 intelectual
33 trabajador manual

4.3.3 *Issuing Update Products*

Section 4.2 discussed the production of complete editions of a thesaurus for the users and target systems; this section examines update-process products that are necessary:

- for managers and management bodies to help them in the course of the update process ("update-process working documents")
- for users and target systems
  - to allow them to use updated data even if complete editions cannot be produced as often as managers want to provide users with new data (i.e., as often as managers stabilize intermediary updated versions of the thesaurus) ("update products")
  - to allow them to take necessary actions for migrating to a new version even if they are provided with complete updated editions ("migration products")

decision strategies

4.3.3.1 *Update-process Working Documents*. Such documents are prepared for the use of those responsible for decisions on content modifications and have to be designed to fit management needs. This means that data have to be presented according to two different types of distinctions:

- the distinction between the nature of the various modifications envisaged
- the distribution among domains, because managers need to have a comprehensive view of the context of the decisions to be taken and of the possible interactions of successive decisions in the same semantic field so that they can organize their work accordingly

Types of possible modifications include:

managing modifications

- adding and deleting concepts
- adding and deleting hierarchical, inclusive and associative relations
- creating and deleting non-descriptors and scope notes
- modifying the wording of existing descriptors, non-descriptors and scope notes

In the management of a multilingual thesaurus, it is useful to distinguish between the following kinds of modifications. Some changes that concern all language versions, such as the addition

modifying multilingual thesauri

and deletion of concepts and modifications to structural relations, have to be examined collectively. Other changes, typically modifications to terms and texts, apply to one language version at a time and can be examined by language-version managers only if the management system has been organized to accommodate such a distinction. However, this capability cannot be based on a purely technical definition, such as limiting it to specific types of elements to be modified.

language version interaction

Some modifications envisaged for only one language version, such as the modification of a descriptor or the introduction of a scope note, may nevertheless be of interest for, or have important consequences in, the other language versions because they tend to modify the intended concept. Thus, the manager must have the option to assign the appropriate status to each modification still to be examined or completed. In this case, it is also necessary to produce documents that combine data from all language versions, even if they do not necessarily appear together in any printed format. Such a document might consist of a "multilingual concept form" incorporating not only all language equivalents but also non-descriptors and scopes notes from all language versions.

update management data

Update-process working documents also need to incorporate data that are relevant solely for management purposes, such as:

- modification status (documents will incorporate previously approved, pending, and newly proposed modifications)
- definitions that aid the understanding of the proposed concept but that will not necessarily be kept in the thesaurus itself if the concept is actually introduced
- comments made by users, managers, or consulted experts
- sources for proposals, terms, texts, and comments

It is beyond the scope of this chapter to discuss possible ways of actually presenting these data.

summary information

Finally, managers need summary information on the consequences of the decisions they have taken:

- Good thesaurus management systems allow for some early feedback during management meetings or soon thereafter.
- If prepared for management purposes, the kinds of possible comprehensive update documents described below may also take into account modifications still pending.

supplements and amendments

4.3.3.2 *Update Products*. The thesaurus manager may want to provide users with update data in the intervals between two complete editions. For end users using the printed thesaurus, update documents that modify available printed presentations of the original version can be categorized as follows:

- documents that have to be used in addition to the original version ("addenda", e.g., that complement the alphabetically structured presentation, including new descriptors and non-descriptors)
- documents that "cancel and replace" parts of the original version ("localized" update documents, e.g., a complete new version of the graphical presentation of a domain where numerous modifications have been introduced)
- documents describing modifications introduced to manually amend the original version ("descriptive" update documents, e.g., lists of modified terms or lists of deleted concepts)

target systems

If more than one update takes place between two complete editions, the successive versions of update documents should be cumulative, i.e., each new version should cancel and replace the corresponding previous versions, but a "descriptive" update document should incorporate a means of distinguishing what is new as compared to the previous version of the same update document in order to allow progressive as well as global updating. In computerized documentary systems where the updated thesaurus is implemented, implementing update data is a two-step process involving both updating the electronic version of the thesaurus and updating the index (see 3.3). These issues are relevant only if the thesaurus management system is distinct from the system(s) where the thesaurus is used ("target system(s)").

update scenarios

Unlike the printed edition of a thesaurus, the electronic version can (and should) always be updated integrally in target systems. Three different situations may be distinguished:

- The target system is able to load a complete new version provided in an agreed format.
- The target system is able to update its own version of the thesaurus using update-related descriptive data.
- No automated updating procedures have been agreed upon.

As far as the first case is concerned, data exchange format issues have been discussed in 4.2.2. As for the remaining two cases, exhaustive and detailed update-related descriptive data have to be prepared and provided to target system managers either in the agreed format (second case) or in a format that is as generic and self-explicit as possible in order to allow for automatic or manual update work, which remains the final responsibility of target-system managers (third case) (see also 4.2.2).

4.3.3.3 *Migration Products*. Updating thesaurus contents is not the final step. Depending on the update strategies chosen in the documentary systems where the updated thesaurus is used, these target systems may need more or less supplementary data to allow "re-indexing", i.e., revision of existing "indexings", of both indexing statements (groups of concepts attached to indexed documents) and retrieval statements (such as search strategies recorded for "selective dissemination of information" (SDI) purposes).

reindexing

Re-indexing involves "translating" indexings made using an earlier version of a thesaurus into indexings made using a newer version. Re-indexing is the safest procedure in the long term in order to preserve options for taking full advantage of the availability of successive, improved versions of the thesaurus so that the older version of the thesaurus may be completely removed after the operation. Failure to re-index requires the retention of previous versions of the thesaurus. Thesaurus managers have to let users and target systems managers know which strategy they are recommending because these managers take it into account in their decisions on modifying thesaurus contents.

migration data requirements

No detailed description of data specifically needed for re-indexing purposes is proposed here, for too many different situations may be encountered in real life because of the number of options that can be chosen in implementing thesaurus use in a documentary system. One example of a bad implementation option is hard-coding the descriptor wording in the indexing field of document records. Such cases require a great deal of migration data. In cases where the thesaurus system shares the same structure, fully interacts with target systems, and thus ensures their management, coded identification of present and past concepts can remain unique over time, independent of language or thesaurus contents and structure, which keeps migration data requirements to a minimum.

supplementary data

Supplementary data related to re-indexing may include lists of deleted concepts, of new nondescriptors corresponding to deleted concepts, of cases where manual re-indexing is necessary, of new and modified descriptors, etc. Modifications to the structure of the thesaurus are less relevant because concepts are usually considered as isolated in indexings, with the exception of those situations where "posting" techniques have been used in indexing (these techniques are recommended for retrieval only—see 3.2).

reindexing multilingual thesauri

In the case of multilingual thesauri, special attention has to be paid to "monolingual exceptions", i.e., cases where actual re-indexing may be different according to the language version used when indexing. For instance, if the preferred term assigned in a given language is incorrect as compared to other language versions, it can cause complex problems. If indexer 1 uses an incorrect term in a given language version when creating an indexing statement, he or she may understand the concept treated by the entry differently from the concept that is understood by indexers using any of the other language versions. From the moment when such a mistake has been corrected in the thesaurus by choosing a new preferred term in the problematic language version, all indexers will most likely understand the same concept, but it would be wrong to assume that this applies to the past as well. A decision must be taken on how to re-index, i.e., translate old indexings in order to preserve the understanding of the concept that indexer 1 had originally. This is especially important if the original preferred term remains in the new version of the thesaurus, but is then attached to another concept.

candidate concepts

Documentary systems quite often include options to index documents using "candidate descriptors", i.e., usually language-dependent terms rather than language-independent candidate concepts. Hence it may be desirable to take them into account when re-indexing, but the language-dependent nature of candidate descriptors introduces difficulties that will often limit the options for automated processing.

4.3.4 *Managing Successive Versions*

Specialized management tools significantly simplify the difficult task of managing successive versions of a thesaurus (see also 1.4).

If there are more than one user group or target system and different update frequencies are involved (and moreover, if printed editions are produced at yet another frequency), the corresponding management functions have to allow for producing printed and electronic editions of complete new versions and of update-process products corresponding to the current specific situation at any given time. *multiple user groups*

Such a possibility may be described, theoretically, as follows: the thesaurus management system must allow consultation of any version of the thesaurus (i.e., state of the thesaurus at a given time) and it must be possible to request that the system perform certain operations (*not* update operations) "as if" the specified version were the current one. Such situations are rare, however. In practice, it is recommended to limit the number of different versions sent out "into the field", and to allow deletion of some of the corresponding management data in the thesaurus management system when a "major" updated version is stabilized and circulated to all users in all available formats. The corresponding minimum (mandatory) requirement is the possibility of managing two versions. *limiting versions*

4.3.5 *Statistics*

4.3.5.1 *Management Statistics.* Thesaurus managers are interested in figures that provide a global picture or "profile" of their thesaurus. Standards only recommend providing users with the number of terms included, with breakdowns by preferred and non-preferred terms and by language versions. Supplementary figures are useful for managers who want to make comparisons with other thesauri as well as between successive versions of their own thesauri as a guide for content management strategies. Such figures include the following, where those items marked with an asterisk are language version-dependent: *supplemental statistics*

- the number of concepts (i.e., of descriptors), non-descriptors* and scope notes*
- the number of mandatory and optional multi-equivalences*
- flexibility, i.e., the ratio between the number of non-empty words appearing in compound terms and the number of these terms that also appear as single terms*
- the maximum and mean length of terms*

- the number of hierarchical and associative relations, the number of concepts without hierarchical relations, and the number of concepts without associative relations
- the number of top-terms, i.e., of hierarchical chains
- the number of polyhierarchies and the number of cross-domain polyhierarchies
- the distribution of concepts according to levels of specificity

It should be possible to have a distribution by domains (and sub-domains) for all data. Most of these figures can be used as input data for further computations of ratios and indicators that are not useful to implement in the thesaurus system itself. It should be possible to download data into a spreadsheet to take advantage of sorting, computing, and graphical presentation features.

4.3.5.2 *Usage Statistics*. The number of times each concept has actually been used in indexing (and, if possible, in retrieval) is of great interest to the thesaurus manager. The frequencies provided have to be used carefully as indicators and must take into account the size of the indexed corpus, which also implies the date when concepts are introduced. Some thesauri include this information in their printed edition, which is useful for users if the thesaurus is used in only one or a very restricted number of major documentary systems and if the printed edition is updated frequently. Breakdowns of frequencies according to some of the management statistics listed above and co-occurrence data may also be useful, but they are difficult to systematize in a practical way.

frequency data

5 Choosing the Right Solution

The purpose of the preceding sections has been to outline potential problems and difficulties related to using a computer for managing a thesaurus. The following checklist outlines the key points to consider when making final decisions on thesaurus design and particularly on which thesaurus management system to use.

5.1 Outsourcing vs. In-house Thesaurus Management

One way to ensure good thesaurus management is to let external specialists take care of it. This is especially true for projects involving very specific needs or that are of very ambitious size with

large-scale projects

respect to the number of languages treated or the number and size of user groups served, but the appropriate technical means are not available in-house. It is not, however, the purpose of this section to go into detail with respect to this alternative. Every consideration cited in the text and every technical aspect has to be examined, but if the decision is made to outsource, negotiations for a service contract can concentrate more on results than on process.

relative advantages

The main advantages of in-house thesaurus management are control and flexibility, especially if it is feasible to supervise teams directly, control the costs related to their work, and to take advantage of this flexibility as needed. Outsourcing thesaurus management has the following advantages:

- It provides a clear picture of all costs and of all technical and organizational constraints.
- It benefits from up-to-date and potentially very sophisticated software tools without having to install them in-house or learning how to use them.
- It benefits from the experience gained by specialists who have been previously involved in managing several thesauri.

contingency planning

When envisaging subcontracting thesaurus management, special attention should be paid to preserving long term interests. Contingency planning should provide options for the event that something might go wrong and it could become advisable to change service providers. On the technical side, contractual agreements should stipulate that the subcontractor will turn over all necessary management data should the contract be terminated. This is especially important if deliverables are mainly use-oriented—even if they are in electronic format, such deliverables do not contain all available information and are not organized in the most practical way for management purposes.

5.2 Decision Checklist

limiting versions

- *List and weigh all specifications.* This chapter has discussed selection criteria, but actual experience in thesaurus management is certainly the best guide for ranking these criteria. If such experience is lacking, bringing in an expert consultant is a good idea.

- *Evaluate different products.* If mandatory hardware requirements (operating system, memory size, etc.) are not too restrictive, the market is mature enough to provide some choice. It is unwise, however, to rely on demonstrations only. One should ask for names of existing users and arrange to see operational systems in their actual use environment.
- *Test before deciding*. This is especially true if a thesaurus already exists and it is desirable for the new software to be able to reproduce the existing printout. If no thesaurus exists, it is important to test creation and maintenance functions on the intended equipment. This kind of prove-out testing can be on a small scale, but as far as a real performance review is concerned, some (large scale) test data should be prepared so as to actually see, for example, how quickly the system being tested can find a given word from a few thousand entries.
  Performing a test means preparing a set of test data including all particular cases that can be identified and specifying a test protocol including a weighted list of criteria and taking the results into account. Tests are always time-consuming, but they are worth the money because thesauri are time-consuming anyway and adequate testing can prevent costly waste of time down the line.
- *Plan for further development.* Account for software maintenance and upgrades, installation, parameter configuration, training, back-up, and periodic evaluation procedures.

REFERENCES

Aitchison, Jean; Gilchrist, Alan; and Bowden, David. 1997. *Thesaurus Construction: A Practical Manual*, 3rd ed. London: Aslib.

International Classification. 1990. "Special Issue on Thesaurus Software." *International Classification*, 17, No. 3/4.

ISO 5964:1985. *Guidelines for the establishment and development of multilingual thesauri*. Geneva: International Organization for Standardization.

ISO 2709:1981. Format for Bibliographic Information Interchange on Magnetic Tape. Geneva: International Organization for Standardization.

ISO 2788:1986. *Guidelines for the establishment and development of monolingual thesauri*. Geneva: International Organization for Standardization.

Van Slype, G. 1987. *Les langages d'indexation: conception, construction et utilisation dans les systèmes documentaires*. Paris: Les Editions d'Organisation.

Section 6.2

Terminology and Indexing

RICHARD A. STREHLOW

1 KEYWORDS AND INDICES—DIFFERENT FUNCTIONS IN RETRIEVAL

document retrieval

Keywords, the essential technical terms found in a document, are typically used to retrieve documents from a corpus or group of documents that might contain desired information. Authors are frequently asked to list the keywords that they believe might be used to retrieve their document. Keyword lists and concordances are also commonly used.

information access

An index, on the other hand, is an aid to access the information found in a document with greater precision. It represents the results of a systematic analysis of the topics discussed or treated in a document.

term selection

The intellectual content of a technical document is represented by the terminology used in the title, abstract, and body of the document. Selection of properly useful terms is therefore central to the writing of any representation of the document and in the document itself. The use of standard terminology or a controlled vocabulary by authors adds some assurance that the target readership will be able to find documents that may contain desired information.

The intellectual content of a technical document may also be represented by an index. The goal for users of an index is the retrieval not of a document, but rather of the information contained in it. Rapid and efficient access to information in a document often requires an index. An index is terminologically rich, but is not a mere listing of keywords and locations. It provides pointers to the precise information presented in the document. The entries in an index will frequently include keywords from the document, but they will generally also include contextual information dealing with the *aboutness* or topicality of the concepts represented.

For example, a document may contain several references to *quinoline* (an organic compound containing a heterocyclic ring structure of carbon atoms and a nitrogen atom). A keyword search on *quinoline* would retrieve this document and probably many others. An index to the document might list *quinoline*, but add topical information, such as its:

topical information

- definition
- natural occurrence
- physical properties
- purification
- reactions
- synthesis

It is possible that none of the terms in this list might occur in the text, but they would probably occur in an index.

The need for an index increases with the size of a document or a collection of related documents because it minimizes scanning time. It is well known that when using systems that retrieve documents electronically based on keyword searches, it takes time to page through the multiple occurrences of the term. An index is different in that it permits random access to specific information.

abstract

Whereas an abstract is generally a narrative that can in a sense stand in place of the document itself or other knowledge record, an index is a list of terms or phrases that were selected by an indexer to systematically present a topical analysis of the document. It is, therefore, an organized reader's guide for locating information that may be present in the document. Phrases that are used by an indexer are typically very short, and the organization is typically alphabetical.

2 THE INDEXING PROCESS

resources

Detailed treatments of the indexing process may be found in Mulvany (1994), Bonura (1994), Lancaster (1991), Sharp (1965) and Cleveland and Cleveland (1983). The Cleveland and Cleveland and the Lancaster references also address the issue of abstracting. All of these sources are commended as useful for a prospective indexer. The Mulvany book is an especially comprehensive presentation of the many details of writing indexes for books. Bonura provides a concise presentation of indexing with an emphasis on technical works.

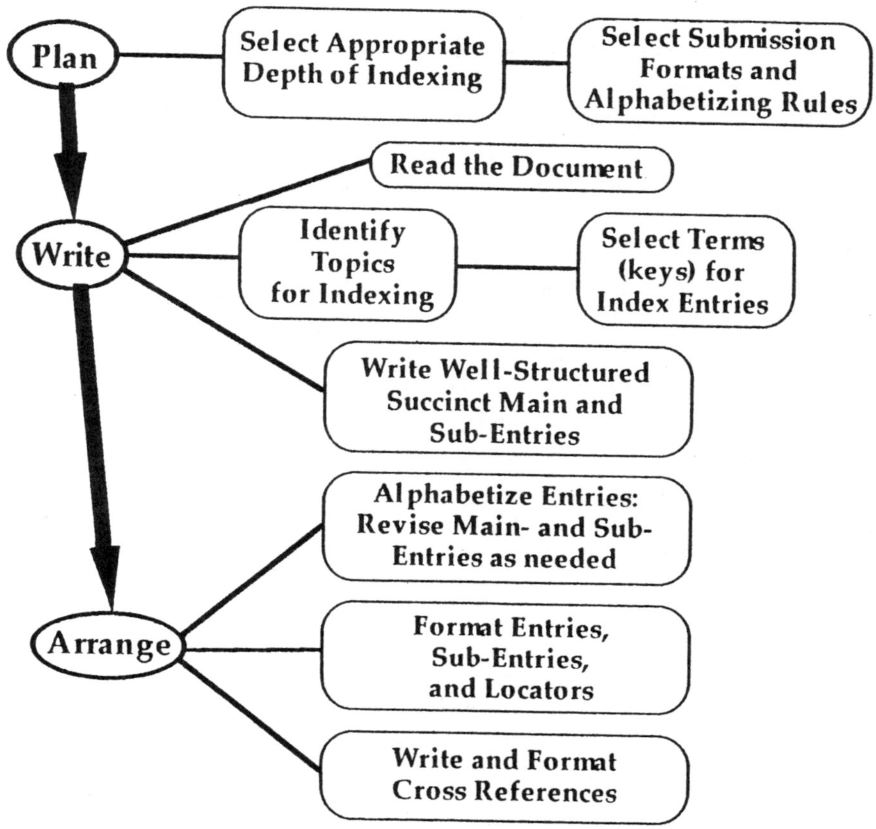

Figure 1: *Work Flow for Indexing Documents*

Although there are several distinct steps in writing an index, it is basically a two step process involving first a conceptual analysis of the document and second a translation of the concepts into appropriate terms. An index entry, the fundamental entity in an index, consists of a heading with at least one reference locating that item in the document. The more detailed process of writing an index involves more than this basic two-step process, and is shown in Figure 1. *index entry*

The process of indexing requires the indexer to guess what questions a future user of a book or database may pose and to assume plausible associations in order to add additional terms or phrases to the index. Because the purpose of an index is to *conciseness* decrease the time that a prospective reader needs to spend to locate information, an index should be concise.

index depth

An index is characterized by its depth—the degree to which a topic is represented in an index (ISO 5127-3a: 1981). Its depth reflects a combination of exhaustivity and specificity in the index entry. The index writer balances these factors to produce an index of desired depth. Indexing of serial publications and text databases poses unique problems of balancing exhaustivity and specificity. Use of a standard or authoritative thesaurus or a specific controlled vocabulary is frequently required.

An index for a technical document may require 5-10 entries per page of the document (one entry per 100 words for a 1% representation). A minimal index may be much less than this. Technical papers are generally not indexed unless they appear in compilations or transactions of symposia, in which case the Table of Contents is deemed to be an insufficient access guide.

The conceptual analysis of the document begins with the selection of topics. A topic in a document suitable for indexing may be a:

- word or name
- phrase
- sentence
- paragraph
- chapter
- definition
- equation
- section
- procedure
- premise
- table or table entry
- example
- figure or diagram
- concept
- idea
- conclusion

user questions

The indexer generates these topics from a comprehending examination of the document and lists them. The task is then to describe the topics in the way a reader might try to find them. Here the indexer does not ask the question, "What is this about?" Rather, "What question would a typical user have that is answered by this topic?" The focus is on questions. Automatic indexing methods do not readily simulate this required mind-set, but are often useful aids to an indexer. Because the required mind-set is different from that of the author, professional indexing specialists are frequently employed.

To illustrate this, Bonura suggested several questions that a user of computer software documentation might have:

- What does the escape key do?
- Where do I start?
- When is my data saved?
- Why doesn't the printer print?
- How do I install the software?
- When is my data saved?
- Whom can I contact for help?

rapid access

The goal for an indexer is to provide speedy, effective access not to the full intellectual content of the document, but to possible highly focused needs of the users. The difference in required mind-set from that of an author makes it sometimes difficult for an author to compose a high grade index.

3 Term Selection

controlled vocabulary

The terms used to key the index entries may be the keywords from the document, but the indexer may need to set contexts and to identify preferred terms or more common spellings that the author did not use. A controlled vocabulary, if available, is often helpful. Three very important characteristics of a controlled vocabulary are:

1. Preferred synonyms are to be specified.
2. Homographs are to be distinguished.
3. Links should be provided between hierarchical and associative relations.

specific terms

Controlled vocabularies may be found as thesauri or standard terminologies. These sources may be most useful to the indexer, especially if they were not used by an author. A venerable principle of indexing is that a concept should be indexed using the most specific term that covers the topic (Cutter 1876 [sic!]). Multiple specific terms are preferred to a smaller number of general terms. This yields enhanced precision. A standard terminology that displays the conceptual structure of a discipline aids an author in ensuring that the terms used to provide access points to the document are appropriate and useful. (Authorities, however, caution against incorporating a particular classification scheme in an index.) Standard terminologies are also useful for indexers, but for a different reason.

standard terminology

Definitions may assist an indexer because a keyword may be too specific to be a successful index entry (Strehlow 1994). A standard terminology with well-drafted definitions, can offer the indexer a single more general term useful for the reader. Frequently the genus of a defined term will be at a desired level of abstraction. Unfortunately, many definitions use genuses that are too broad to serve as useful index terms. Here the indexer may need to rely on carefully written phrases incorporating the specific term.

document scope

The scope of a document is always a factor in term selection. For example, in a general book about organic compounds, an entry about *quinoline* might be placed as a sub-entry under the broader entry for *heterocyclic compounds*. In a specialized book about *quinoline*, there could be many entries under *quinoline* itself to permit adequate treatment. The question-answer model for the indexing process should provide guidance to the indexer in listing the aspects to be presented.

general guidelines

Terms found in the title, abstract, contents list, introduction, opening phrases of sections or paragraphs, illustrations, diagrams, tables and captions, underlined terms or phrases, and references are likely to be especially significant in writing index entries. Of course, well-written documents using consistent terminology are easier to index effectively. Although indexing conventions vary depending on language and national traditions, some general suggestions offered by Cleveland and Cleveland are valuable for writing and organizing an index:

- Use plural form.
- Use subheadings to differentiate different subtopics.
- Use scope notes in entry.
- Don't post up to a broader term.
- Don't scatter similar entries (place kinds of government under government, instead of city, state, etc.).

grouping entries

Bonura offers many specific recommendations that deal more directly with terminology related problems. He shows the value of appropriately grouping entries to clarify meaning and uses the example of "operator" that has different meanings for the entry "operator, messages to" and "operator, binary". He also recommends that all definitions be indexed.

A key point in indexing is that an index should be conceptual and provide direct access to the important ideas in the document and not be simply a list of terms or names, which could be found by a computer search.

information access

Retrieval of technical documents is based on the terminology of the field, the keywords found in the document, the presence or absence of related terms, and other factors. The purpose of an index is to access precise information. An index is terminologically richer than text, abstracts, and most titles. It is an organized systematic analysis of the concepts and topics to be found and profits from a consideration of the terminology used in the document, but provides significantly more for the reader. Indexers can profit from careful use of controlled vocabularies and standard terminologies.

REFERENCES

Bonura, Larry S. 1994. *The Art of Indexing.* New York: John Wiley and Sons.

Cleveland, Donald B. and Cleveland, Amy D. 1983. *Introduction to Indexing and Abstracting.* Littleton, Colorado: Libraries Unlimited, Inc.

Cutter, Charles A. 1876. *Rules for a Dictionary Catalog.* Washington: U. S. Government Printing Office.

Lancaster, Frederick Wilfred. 1991. *Indexing and Abstracting in Theory and Practice.* Chanpaign, Illinois: University of Illinois Graduate School of Library and Information Science.

Mulvany, Nancy C. 1994. *Indexing Books.* Chicago and London: University of Chicago Press.

Sharp, John R. 1965. *Some Fundamentals of Information Retrieval.* London: Andre Deutsch.

Strehlow, Richard A. 1994. "Keywords, Index Terms, and Thesauri." *ASTM Standardization News*, January, 17.

ISO 5127-3a:1988. *Documentation and Information — Vocabulary — Section 3a: Acquisition, Identification, and Analysis of documents and Data.* Geneva: International Organization for Standardization.

Section 6.3

The Role of Terminology in Retrieving Information

RICHARD A. STREHLOW

1 THE PROBLEMS OF RETRIEVAL

document classification

Access to documents and data is achieved using language. The "library problem" of how a collection of documents is best organized, classified, or arranged for easy access by users has long been a primary concern of classificationists, indexers, and librarians.

information growth

However, the problems of retrieval have become increasingly severe because of the growth of computerized information, which in this chapter is superordinate to data and to documents. Not only is the number of documents (both in hardcopy and electronic formats) related to a subject area increasing—so is the number of subject fields. In addition, full texts of documents have become widely available in computerized form. These changes have caused a qualitative change in the retrieval process.

describing information

The basic linguistic issue in retrieval is the representation of documents and data. A user of an information database must describe the desired information. How the information is described requires the inquirer to know whether specific data or more general information is needed. There is a fundamental difference between the retrieval of data and the retrieval of documents. These differences are described in Table 1.

representing documents

The representation of documents is a many-faceted, longstanding process and is accomplished in a variety of ways. Authors, publishers, indexers, abstractors, classifiers, reviewers, and educators have all been part of the traditional system of representation. Representation can be provided by titles, authors' names and institutional affiliations, abstracts, summaries, subject descriptions, index terms, and many other types of contextual data. The means used for representation include virtually any characteristic of a document. In the past, libraries relied on subject classification and

Table 1: *Comparison of the Retrieval of Data and Documents* (Blair 1990:7)

Document Retrieval	**Data Retrieval**
Produces or refers to one or more documents that might contain the desired information, e.g., "What documents or references are there in a collection that refer to the physical properties of benzoic acid?"	Produces the answer to a specific question, e.g., "What is the melting point of benzoic acid?"
Probabilistic	**Deterministic**
Speed of retrieval is based on the time for the inquirer to think and make logical decisions.	Speed of retrieval is based on the characteristics of the retrieval system.

cross-referenced author lists as the primary access routes. With the increasing number of specialties, a multiplicity of specialized journals led to frequent representation of a paper by author and journal name alone.

term-based searching

All of these types of data as well as metadata, i.e., data about data, constitute a distinctly higher order of abstraction and can provide access to documents. Although a general consideration of information retrieval must require a consideration of all these means of document representation, this discussion is focused on term-based searching. The fundamental questions associated with representation of documents by terms are discussed below.

terminology control

The representation of computerized data for retrieval typically uses highly structured indices and keys. Rigorous terminology control is required for retrieval from databases. The terminological aspects of data elements have received additional treatment (in Strehlow and Wright 1993: Forrey 152-157; Westbrook and Grattidge: 158-170; Grattidge et al.: 217-232; Strehlow et al.: 129- 139). The focus here is on classic retrieval mechanisms, although many aspects are also mirrored in Internet search practice.

retrieval as dialogue

The retrieval process involves a kind of dialogue. Just as dialogue is central to the negotiation and determination of meaning in conversation, it is also crucial in querying a computerized information collection. It involves participants at a physical and temporal distance, however, and differs profoundly from common conversation. It also differs for documents and for data.

controlled vocabulary

When searching for documents, an inquirer guesses those terms that might have been used by an author or indexer of a document that might be useful. When searching for and retrieving data from a database, it is usually necessary to learn or refer to a highly controlled vocabulary for each database that is queried.

2 Retrieval Effectiveness and Its Evaluation

user objectives

The effectiveness of retrieval is a function of two primary factors:

- The degree of exhaustivity desired or needed by an inquirer
- The qualitative differences for retrieval from large collections

The relation between the desired degree of exhaustivity and the size of the collection may be understood by comparing some characteristics or kinds of objectives a user might have (see Table 2).

precision & recall

Measuring or estimating retrieval effectiveness is not simple. It actually presents a major problem to those who wish to evaluate and compare various search strategies. Retrieval effectiveness is usually described by the *precision* and *recall* that are associated with a retrieval operation. *Recall* is the fraction of relevant documents retrieved from a collection, and *precision* is the fraction of documents that are relevant in a retrieved set of documents.

These concepts are expressed mathematically by the equations (Salton 1988:248):

Recall, $R = a/(a+d)$

Precision, $P = a/(a+b)$

where,

a = the number of relevant retrieved documents

b = the number of non-relevant retrieved documents

c = the number of non-relevant not-retrieved documents

d = the number of relevant not-retrieved documents

fallout

A second measure of effectiveness, *fallout*, is discussed by Blair (1990:Note 10:116). Fallout is the conditional probability that a non-relevant document will be retrieved, the fraction of the non-relevant documents retrieved to all non-relevant documents, (Fallout = b/c). Other measures, such as expected search length, have also been used.

Table 2: *Aspects of Exhaustive and Non-exhaustive Searching of Large and Small Document Collections* (Blair 1990:164-175)

	Small Collection	**Large Collection or a Database**
Exhaustive Search	Generally feasible only when the inquirer can examine every document in a collection Examples: Examining an issue of a magazine for articles relating to home ownership	Required where significant economic or social consequences may be involved. Examples: Patent search or search for precedents in a legal corpus; medical data
Non-exhaustive Search	The most common type of search where the objective is to find any document that might pertain to a subject Examples: • A common library search • Search where the search mechanism depends on a subject classification system	Initial scoping of literature to focus on areas for further, more detailed study Examples: Search designed to obtain a small number of relevant documents that enable a more traditional search of older (non-computerized) literature

relevance

Both concepts—precision and recall—present problems in their quantitative evaluation. The culprit is the concept of *relevance*, which is ill-defined because it involves the judgment of the inquirer. Hence it varies with the individual and over time. For a single user, however, it appears to be possible to evaluate relevance, at least at any given time. User variability will depend on the knowledge of the user at the time of evaluation and may change significantly as a result of examining the retrieved set of documents.

A second problem develops when the inquirer confuses retrieved documents that are merely useful or illuminating with those that were specifically relevant to the inquiry. An even more serious problem affecting the application of the concepts of recall and precision lies in evaluating the number of relevant documents in a collection (Blair 1990:69-120). In general, recall and precision are found to be inversely related, i.e., an increase of one results in a decrease of the other as depicted in Figure 1.

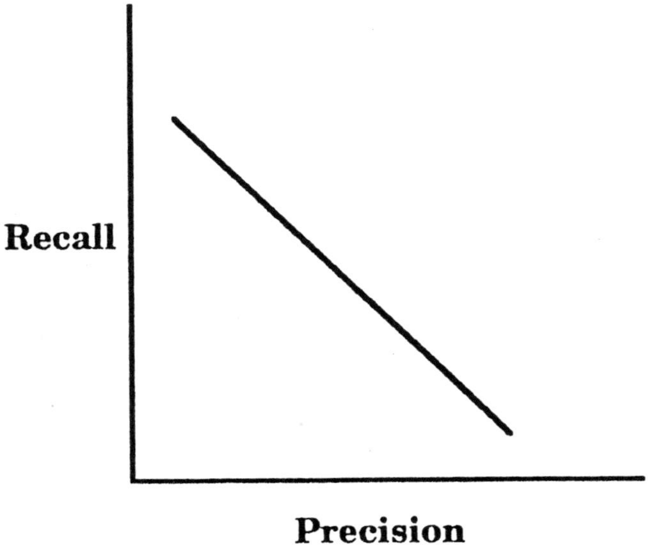

Figure 1: *Indicated Relation between Recall and Precision*

effectiveness

Blair conducted an experimental determination of effectiveness on a collection of 40,000 documents (Blair: 88-112). Fifty-one queries were made on a legal document collection and 40 were evaluated (11 others were used for statistical validation). The mean value for precision was 79.0% and for recall was 20%. Four-fifths of the relevant documents were not retrieved. Figure 2 depicts a possible relation between the size of a collection and the recall for a given query. The absence of data in the literature makes it difficult to estimate the functional relation between these variables.

One principal reason that recall was low was due to the fact that it is impossibly difficult for inquirers to predict the exact words, word combinations, and phrases that are used in all (or even most) relevant documents and primarily in only those documents (Blair 1990:101). This example is discussed further below.

prediction criterion

The effectiveness of a search is not merely a matter of the quantitative results. It involves the satisfaction of the inquirer's need (Blair 1990:9-13). Here, many subjective factors are involved. Principal among these are the *prediction criterion* and the *futility point criterion*. The prediction criterion is the standard employed

Figure 2: *Indicated Relation Between Recall and Collection Size*

by an inquirer to judge if a term selected for use in a retrieval exercise correctly yields the desired documents or data. The futility point criterion is the standard employed by an inquirer to determine whether the number of documents retrieved is less than the maximum number of documents that the inquirer would be willing to examine. If the response to a query yields too many documents to be evaluated effectively, the inquirer may accept the consequences of a less-than-exhaustive search.

futility point criterion

In general, it is difficult to assess the quantity and value of retrieved non-relevant information. Any decrease in this number can result in a loss of relevant *hits*, i.e., retrieved references or documents. An early comment on the phenomenon of inquirer satisfaction was by D. R. Swanson, as quoted in Blair (1990:8-9).

non-relevant information

non-retrieved information

> Although library searches probably seem more often than not to be successful simply because a relatively satisfying amount of material is exhumed, such success may be illusory, since the requester cannot assess the quantity and value of relevant information which he fails to discover.

3 Information Retrieval—A Principal Use of a Terminology

common words

Terminologies or special languages frequently are used as documentary languages. This is because that in attempting to retrieve information, the inquirer seeks access to the intellectual content of a document or the specifically named data in a database. Common words are not useful in representing the intellectual content of a document simply because they are common.

word frequency

Zipf explored the frequency of word usage (as cited by Blair: 1990:229). He developed a model of communication that applies to the dialogue of retrieval. The model describes two competing economies of term usage. The economy of the author attempts to minimize the numbers of terms used to allow for varied intents. The economy of the inquirer would maximize the number of terms in order to gain a maximum precision.

Zipf's rule

Zipf's rule states that a compromise between these two economies results in a mathematical relation between the frequency of word usage and its rank. Although the relation has been tested and verified on many samples, I am not aware of any test that involves multiword terms. However, it is plausible that the relation will still be valid for these cases.

rank

Frequency of word usage is the number of times a word is used in a document. *Rank* is obtained by listing the words in order of their frequency and then numbering them in that order, with the number assigned to a word representing its rank. The most common word in a document might occur 340 times and would be rank 1. There would be several words used only once. According to Zipf, the product of the rank and the frequency is (approximately) a constant. The equation for this is:

(Rank)(Frequency) = Constant, or

$$(R)(n/N) = A$$

where: R is the rank

N is the number of words in the text

n is the number of times a given word of rank r is used

A is a constant

Experimental values of the constant are about 0.1 according to Salton (1988: 104-108).

variant usage

A less familiar result of Zipf's considerations involves variant usages. In general text that is not subjected to a principle of controlled vocabulary use, the number of different ways a word can be used, i.e., the number of meanings attributed to a given term, increases as the square-root of the term's frequency (Blair 1990: 247). This kind of polysemy diminishes the precision of a search.

In addition, different terms may be used to refer to the same concept. In one of the queries from the legal database reported above, the term *accident* was used. The referent was represented by a variety of terms in the unretrieved documents as listed in Table 3. Such polynymy (synonymy) resulted in decreased recall, possibly paralleling Zipf's conclusion on variant usages.

standard vocabularies

Standard vocabularies are a central component of the retrieval process. The significance for the author is that precise terms should be used in the title, abstract, and text, as well as in keyword lists. It is necessary for the inquirer to be equally precise. These two conditions can often be met if standard vocabularies are used.

Some of the examples that occur appear to be innovative or even unexpected. Some relevant documents used in the retrieval study involving this legal matter discussed events prior to the accident or mentioned only technical aspects of the reasons for the accident. Thus the concept sought was referred to by multiple terms.

search effectiveness & collection size

These theoretically based considerations as evidenced by the example lead to the conclusion that search effectiveness will always decrease with increase in the size of a collection. Term-based retrieval from full-text sources will tend to be less effective than retrieval from titles, indices, and abstracts. In other words, a qualitative difference exists for full-text retrieval as compared with retrieval from abstracts, index terms, or titles. The reason for this is that writers (authors, indexers, and abstractors) are more likely to exercise careful control of monosemic and mononymic terminology in abstracts, indexes, and titles, whereas discourse conventions frequently result in rich variation in running text segments.

variants

Other variant usages include spelling variations, synonyms, quasi-synonyms, related terms, acronyms, etc. (Buchan 1993). All of these varieties of usage complicate term-based retrieval and decrease its effectiveness. Search strategies should allow for maximum possible variety with respect to multiple term meanings and multiple term-concept assignments (see Section 5).

Table 3: *Terms Used to Refer to an Accident in a 40,000 Document Collection* (Blair 1990:101)

Terms representing the same concept as *accident*
event
difficulty
unfortunate situation
situation
problem
incident
Oblique references to *accident*
the subject of your last letter (in a letter)
what happened last week (in a letter)
Mr. A: We all know why we are here. (in the minutes of a meeting)

4 THE DIALOGUE OF RETRIEVAL: A DIALOGUE AMONG AUTHOR, INQUIRER, AND OTHERS

term-based search strategies

The retrieval dialogue for maximized term-based search effectiveness involves both the author-publishing team and the inquirer-information system team. The authoring-publishing team includes authors, referees, pre-publication reviewers, editors, abstractors, and indexers. All of these can help to increase the probability that the document will be found when information relating to its intellectual content is sought.

special language

The author's dialogue begins in the use of a subject-oriented terminology. One assumes that the author wants his document to be read. It is necessary for authors to know and use the special language characteristic of their fields. For fields that are young and active, the terms of the special language may be changing. Here synonym use is essential in order to ensure that the document may be found when desired.

terminology standards

If a normative or standard usage exists, the author has the primary obligation to know and use it appropriately. Published terminology standards issued by a normative standards organization are significant sources for choice of vocabulary. A subject thesaurus may be used to ensure that a representative number of superordinate concepts are referenced.

controlled vocabulary

Abstractors and reviewers are a significant part of the information publishing team and have long provided access to documents by supplying expert analysis of the intellectual content of documents and usually publishing these derivative documents. The same considerations hold for these writings as for the other authors. The same probabilistic effects of term usage frequency exist for them. Where exhaustive retrieval may be required, it is advisable to conduct a structured review using a fixed key word list as a highly controlled vocabulary (Interrante *et al.*1993).

Indexers also have long focused on the need for a controlled vocabulary (Ellis 1993). Indexers usually make more extensive use of thesauri, subject classifications, standard terminologies, and other related tools than do authors.

subject-field knowledge

The inquirer's approach to the dialogue of retrieval has received more attention from retrieval specialists than has that of authors because a collection of documents is usually taken as the given parameter or target for a search. Knowledge of the subject field and its terminology is critical. In the dialogue the inquirer first acts as speaker. The system response is to produce a number of hits. The inquirer evaluates the results and modifies the search parameters interactively to produce an acceptable result.

online sources

The consequence of these considerations is that one should expect to reiterate and modify each query used. If speed is essential, online sources should be employed. If the inquirer is not fluent with the mechanics of database operation (query entry, command structures, etc.), the search can be conducted with the aid of an experienced expert, although more and more individuals are now called upon to master search strategies..

relevance feedback

Some data on the effectiveness of using iteration of relevance feedback indicates that a 30% to 60% increase in precision for given recall may be achieved (Salton 1988:308). Although this provides some useful guidance to the inquirer, application to large systems is problematic. A genetic model for retrieval has been proposed to optimize retrieval effectiveness by use of relevance feedback (Blair 1990:249-270).

5 Techniques of Constructing a Search Strategy

Various models for retrieval have been proposed. Blair (1990: 27-67) alone discusses a number of logically distinct models plus some hybrids. The basic structure of a retrieval system is shown in Figure 3. Retrieval models may use:

retrieval models

1. A single descriptor (term or other characteristic)
2. A set of descriptors logically combined (a complete Boolean retrieval system)
3. A set of descriptors with cut-off values
4. Weighted descriptors or weighted indexing
5. Cosine rule
6. Specialized full-text methods

Most searches are conducted following the first two models listed above and often involve indexed collections.

cosine rule

Most inquirers do not use the more complex retrieval systems that may be implemented in certain systems. Cut-off values for some descriptors may be added as in the third model. (A set of five terms might be specified with a rule that if any three are found in a document, it will be retrieved.) Weighted descriptors can be used to list documents in the order of likelihood of their being relevant.

vector model

The cosine rule is based on the vector model of a document (Salton 1988:313-326). Each document may be looked at as being representable by a list of the terms used in it. A search vector is also a list of terms (albeit with many fewer terms). These lists are considered as vectors. The cosine rule is a mathematical representation of the closeness or similarity of these two lists. The set of document vectors (lists of terms) can be presented and mathematically treated as a matrix representing a document space for a specified collection.

full-text retrieval

Full-text retrieval systems use methods similar to those used for indexed systems and have become more common. Because of the kinds of considerations presented above, the indexing or abstracting process may be absent. This requires that the full range of the author's term selections be considered.

SEE

Thesaurus Management 6.1:381

Automated thesaurus systems are under active development. Both normative and descriptive thesauri have been examined (Blair 1990:53-61; Salton 1988:299-303). Blair considers automating both normative and descriptive thesauri (based on co-occurrence

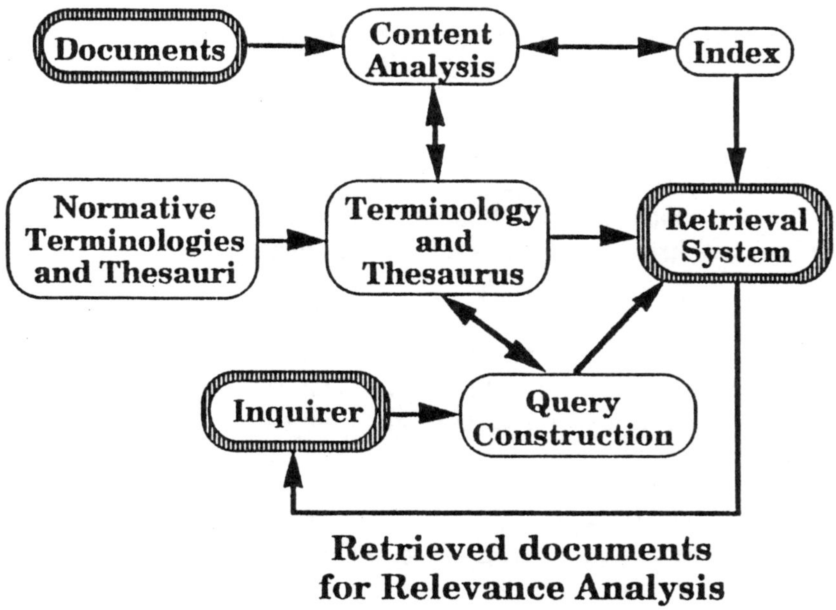

Figure 3: *Basic Structure of a Retrieval System and Process*

terms) and makes a strong case for use of descriptive thesauri. This is based on the assumption that authors may not be knowledgeable about the normative sources and would tend to use personally based associations.

search process

A search for information is usually conducted by composing one or more logical queries. A strategy typically includes several aspects and should be written down for reference during the search. A well-formed question is the first requirement. It is desirable to perform this step in writing. The inquirer should describe the question in terms of the information sought, the desired satisfaction level, and the number of documents that the inquirer is willing to examine.

satisfaction level

The information sought may be general information about a topic, specific data, etc. It may involve highly technical terms or more general ones. The satisfaction level may be described as shown in Table I, or, for example, in terms of:

- any document
- most recent documents (to obtain references, as an entry to the literature)
- several documents
- most documents

The number of documents the inquirer is willing to examine is fundamentally an economic question. It is affected by both speed and costs of the search and by other demands on the inquirer's time (See 2.4).

term selection

The selection of terms is subject-specific. Term selection thesauri are published by retrieval systems managers. Subject thesauri are useful in suggesting related terms. It is not clear that generally changing to a superordinate term will necessarily increase the number of retrieved documents because the text may use the most precise terms the author desires. In the case of an indexed document or one with a keyword list, the document would presumably be more likely retrieved by a query using a superordinate term. Of course, compound terms of more than one word may be best handled by Boolean combinations of the multiple descriptors because the head word and modifiers are usually separated in the text. Retrieval systems frequently allow for proximity searching that permits combinations within specified numbers of words.

Boolean searches

proximity searching

truncated retrieval terms

Truncation is commonly permitted by retrieval systems so that a glosseme (i.e., an appropriately truncated term) can be used to trigger retrieval with any of its common affixes. Typically this strategy involves specifying the glosseme and appending a question mark. For instance:

-instrument?	retrieves	instrument
		instrumental
		instrumentally
		instrumentation
-?symmet?	retrieves	symmetry
		symmetric
		asymmetry
		asymmetrical

affixes

A comprehensive list of common affixes is found in the *Oxford Advanced Learner's Dictionary* (Oxford 1995). Frequent users of retrieval systems are advised to generate their own preferred lists of affixes because many are customary in particular subject fields. In English and other Indo-European languages, inflectional and derivative forms are common. Constructing such a search tests the linguistic competence of the inquirer.

Varying the Boolean query will generally permit regular change in recall and precision. Data that indicate the range of results that might be obtained are shown in Table 4.

Table 4: *Example of a Broadened Boolean Search and the Effect on Retrieval* (Salton 1988)

Query (Narrow to broad)	**Number Retrieved**	**Relevant Number Retrieved**
(A and B and C and D)	2	2
(A and B and C) or (A and B and D) or (A and C and D) or (B and C and D)	6	5
(A and B) or (A and C) or (A and D) or (B and C) or (B and D) or (C and D)	23	15
A or B or C or D	86	25

When automated, this type of broadening is called quorum level searching. Systems that provide this capability may speed a search.

6 Enhanced Retrieval Techniques

Entities other than terms generally are also used in constructing a search. Some of these items were mentioned above in 5.1. Most practical searches, however, are term-driven because terms most clearly relate to the particular concepts that are part of the intellectual content of a document. Combination queries using terminology along with other data (e.g., authors, etc.) as listed in Section 1 are common.

More recently, document structure itself has been proposed as an appropriate enhancement for retrieval (Strehlow *et al.* 1991). With the development and application of Standard Generalized

Markup Language (SGML) documents increasingly can contain structural and contextual data. This data may be incorporated into future enhanced retrieval techniques. SGML is an international standard that provides a mechanism for describing information structures (ISO 8879).

INFO
SGML
INFB 21:863

Automated systems are under active development (Salton 1991). The use of automated Extensions of Boolean queries, thesaurus broadening, semantic nets, syntactic analysis, and other methods offer hope that enhanced information management methods will continue to be developed, especially in the framwork of ISO 23950:1995 (ANSI/NISO Z39.50)..

Now, however, term-driven retrieval coupled with terminological awareness and logical analysis by a subject specialist is still central to the art of retrieval.

REFERENCES

Blair, David C. 1990. *Language and Representation in Information Retrieval*. Amsterdam: Elsevier Science Publishers, Amsterdam.

Buchan, Ronald L. 1993. "Variant Terminology." Strehlow, Richard, and Wright, Sue Ellen, eds. *Standardizing Terminology for Better Communication: Practice, Applied Theory, and Results*. ASTM STP 1166. Philadelphia: ASTM, 95-105.

Ellis, Wayne. "Identification and Selection of Keywords for Indexing Voluntary Consensus Standards." Strehlow, Richard, and Wright, Sue Ellen, eds. *Standardizing Terminology for Better Communication: Practice, Applied Theory, and Results*. ASTM STP 1166. Philadelphia: ASTM, 255-269.

Forrey, Arden W. 1993. "Standardization of Names for Clinical Laboratory Tests and Resulting Analysts." Strehlow, Richard, and Wright, Sue Ellen, eds. *Standardizing Terminology for Better Communication: Practice, Applied Theory, and Results*. ASTM STP 1166. Philadelphia: ASTM.

Grattidge, Walter, Lund, W. Bruce, and Westbrook, Jack H. 1993. "Developing a Database for Use with Materials Property Data." Strehlow, Richard, and Wright, Sue Ellen, eds. *Standardizing Terminology for Better Communication: Practice, Applied Theory, and Results*. ASTM STP 1166. Philadelphia: ASTM, 217-232.

Grattidge, Walter, and Westbrook, Jack H. 1993. "Committee E49's Response to Terminological Challenges from Computerized Materials Property Data." Strehlow, Richard, and Wright, Sue Ellen, eds. *Standardizing Terminology for Better Communication: Practice, Applied Theory, and Results*. ASTM STP 1166. Philadelphia: ASTM, 152-157.

Interrante, Charles G., Mesina, Carla A., and Fraker, Anna C. 1993. "Information Retrieval Using Key Words and a Structured Review." Strehlow, Richard, and Wright, Sue Ellen, eds. *Standardizing Terminology for Better Communication: Practice, Applied Theory, and Results.* ASTM STP 1166. Philadelphia: ASTM, 242-251.

ISO 8879:1986. *Information processing — Text and office systems — Standard Generalized Markup Language (SGML).* Geneva: International Organization for Standardization.

ISO 23950:1998. *Information Retrieval (Z39.50) — Application Service Definition and Protocol Specification.*

Salton, Gerard. 1988. "Automatic Text Processing, the Transformation Analysis, and Retrieval of Information by Computer." Reading, Massachusetts: Addison-Wesley Publishing Company.

Sidney, Albert. 1995. *Oxford Advanced Learner's Dictionary.* Oxford: Oxford University Press.

Strehlow, Richard A., and Wright, Sue Ellen, eds. 1993. Standardizing Terminology for Better Communication." Philadelphia: American Society for Testing and Materials, STP 1166.

Strehlow, Richard A., Tallant, T.O., Mason, James D. and Kienlen, P. L. 1992. "Text Encoding for Retrieving Scientific and Technical Information." Galinski, Christian, and Williams, Malcom, eds. *Proceedings form the International Symposium on Terminology and Documentation in Specialized Communication, T&D '91.* Ottawa: Minister of Supply and Services Canada, 188-197.

Strehlow, Richard A. 1983. "The Varieties of Compound Terms and Their Treatment." Heymann, F.J. and Interrante, C.G., eds. *Standardization of Technical Terminology: Principles and Practices*, ASTM STP 806. Philadelphia: American Society for Testing and Materials, 26-33.

Strehlow, Richard A., Kenworthey, William H., and Schuldt, Ronald. 1993. "Terminological Aspects of Data Elements in Databases." Strehlow, Richard, and Wright, Sue Ellen, eds. *Standardizing Terminology for Better Communication: Practice, Applied Theory, and Results.* ASTM STP 1166. Philadelphia: ASTM, 129-139.

Section 7

Commercial and Industrial Applications

Section 7.1

Terminology in Technical Writing

TANGUY WETTENGEL AND AIDAN VAN DE WEYER

1 WHAT IS TECHNICAL WRITING?

definition

Technical writing can be defined either through its products, that is, the output of technical writing activities, or by the skills required for the activities themselves. The most concrete and obvious products of technical writers are documents, such as catalogs, user manuals, reports, and design specifications. However, not to be ignored are information management tools, such as classification methods, document design protocols, information modeling strategies and formalisms, and document quality assessment guides.

technical writing skills

The skill of a technical writer is to enable, through an appropriate strategy of information design, the quick retrieval of relevant data in a given situation. This task requires an optimal match between, on the one hand, the reader's expectations and previous knowledge and, on the other, the document's content and its organization. Because it is based on theoretical knowledge, this skill can be improved and even taught, although the research fields concerned (ergonomics, artificial intelligence, semiotics, linguistics, cognitive science, and terminology) are not often familiar to technical writers.

target audience analysis

Much of the theoretical literature on technical documentation amounts to little more than recipe-like instructions, recommending a crisp style, the avoidance of passives, and the use of short sentences and many nouns. More significantly, a correct emphasis on target audience analysis is often left unexplained, giving the technical writer no tools to set up a useful picture of the extent and nature of the readers' knowledge and of the types of information they might find useful in particular situations. Successful tuning to the target audience's needs and background knowledge requires the use of devices that are designed to express knowledge and to highlight this knowledge from different points of view and for different uses, in other words devices to express incomplete and biased information.

Models of devices and, above all, of the distorted picture of a device that an operator is forced into because his proficiency in performing a familiar and repeated task relies on this very distortion, are, as the cognitive science literature has shown over many years, the best (and possibly the only) way to take the readers' culture and expectations into account.

models of devices

1.1 Model-driven Information Design

Modeling has been undertaken systematically in artificial intelligence contexts because it is the only way to organize computer knowledge. It has often been left out of ordinary target audience analysis because it simply does not fit into the standard culture of technical writers. We ardently hope that these stigmata of the dark ages of technical writing will soon be healed.

The model appearing in Figure 1, adapted from Morin (1996), illustrates the representation an electrical engineer might have of an industrial installation. As may be noticed, only the aspects relevant to his design task appear in the impoverished system he has described to the technical writer. A terminology attuned to such a pragmatic view is particularly useful for this type of reader because it restricts itself to the information required by his task.

model-based terminology

The reader might wonder why models of this kind are relevant to technical writers and how they may enter the landscape of terminology. Let us point out that they enable the following target-shaped definitions:

target-shaped definitions

actuator
Device regulated by a controller that drives a machine.

sensor
Device that transfers the data supplied by a machine to a controller.

target-sensitive terminology

When comparing these definitions to the ones given by an authorized dictionary (IEC 1992), the relevance of target-sensitive terminologies can easily be discerned:

actuator (electric)
An electric transducer that converts an electrical signal into a signal of any kind, such as mechanical displacement.

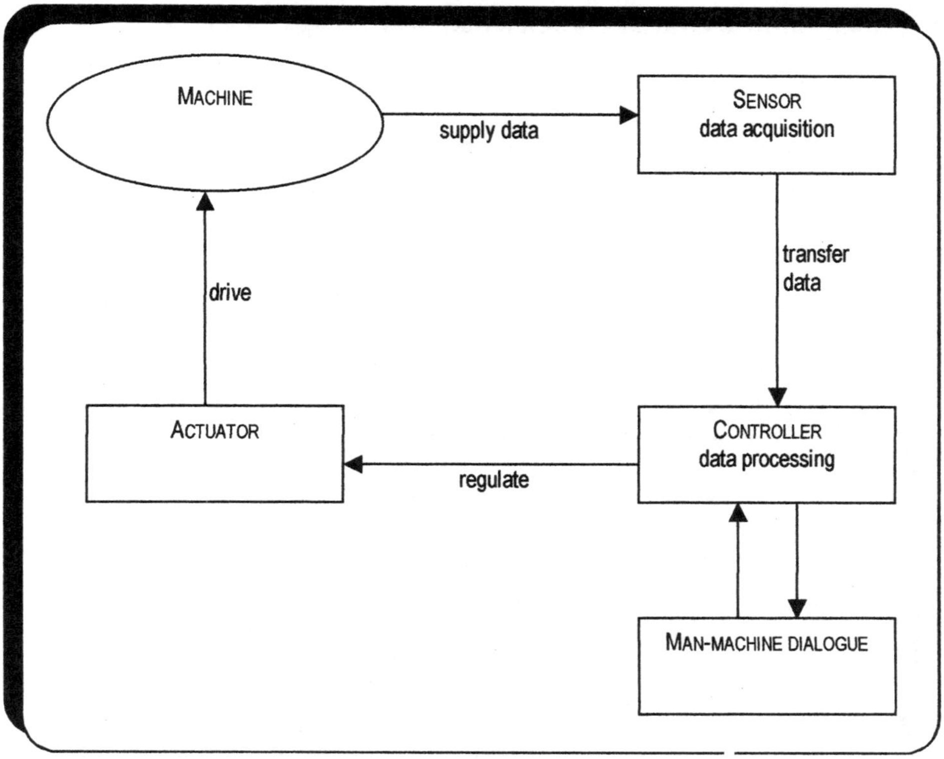

Figure 1: *An Electrical Engineer's Image of an Installation*

sensor
That part of a measuring transducer which converts the input signal into a form suitable for measurement

conceptual domain structure

From the document design point of view, the model provides the conceptual domain structure that can be mapped onto a target-oriented segmentation of specifications, as has been shown elsewhere (Morin 1996). In user manuals, the impact of model-driven design is even more impressive because the distortion of the device that has to be manipulated is much greater due to the naive perspective users often have of technically complex systems and to the lack of interest they may show for anything within the device that is not strictly bound to their tasks.

2 TERMINOLOGIES FOR TECHNICAL WRITERS

One of the main areas of technical writing is the production of task-oriented texts (e.g., user manuals), that is texts designed to instruct the reader in performing purposive activities. The illustration in Figure 2 is a representation of the major contexts of monolingual terminologies. Technical writers deal mostly with the highlighted areas. *task-oriented texts*

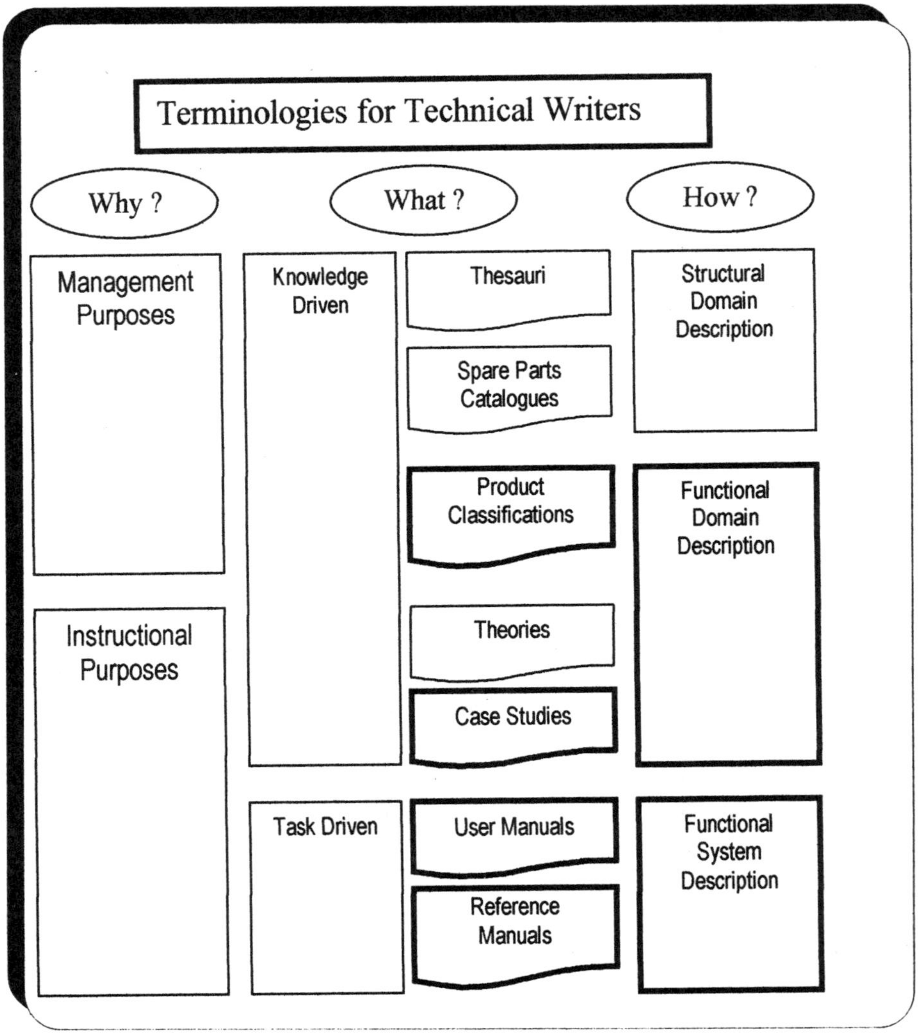

Figure 2: *The Domain of Technical Writing*

instructional intention

As can be seen, technical writing is concerned with using terminologies for instructional rather than for information management purposes (e.g., the creation of thesauri). Processes (such as the functioning of a nuclear reactor) need a peculiar concept definition strategy because only some aspects of the documented object may be relevant to the process. The major one is the object's functional identity, both in a broad and in a narrow sense (the role it plays within the global system and with respect to neighboring components).

sequential & pragmatic links

It should be stressed that the widely used partitive and generic concept links are less crucial to functional identification of objects than sequential and pragmatic ones. Moreover, process-oriented terminologies make a parsimonious use of them because such terminologies are not intended to be mere classification devices, but rather are working tools in dynamic situations (process control, machine operating, etc.).

process-oriented terminology

A distinguished feature of process-oriented terminologies is their commitment to action. This follows from the fact that devices are operated and that processes are often monitored by *people*. The particular bias that action-orientation introduces in concept systems is that the functional identity of an object is task-driven, which means that only those features relevant to the operator's activity (to his potential tasks) are to be documented. Sparse and critical information may secure understanding in these contexts.

domain analysis

These process-oriented terminologies are suited for comprehensive domain analysis (such as a study of the relations among the components of the World Wide Web—see Section 5) as well as to describe only the elements necessary to perform a particular set of tasks (such as a study of the ignition system of a combustion engine for the purpose of fault diagnosis—see 4.2).

3 Functional Definitions

SEE

Definitions 1.3.1:64

Definitions are frequently needed in technical writing, as technical documents often introduce new knowledge. The effectiveness of definitions in conveying fresh knowledge is not only a matter of perspective (functional rather than structural for technical writing contexts), but also depends on their architecture and on their consistency throughout the text in which they are embedded.

3.1 Semantic Structure of Definitions

As with any other language expression, a definition can be split into different chunks. For example, the following definition of an airbrush for ceramics can be divided into five chunks:

aérographe				
Pulvérisateur	à air comprimé	utilisé	pour déposer de l'émail	sur les pièces.*
Chunk 1	Chunk 2	Chunk 3	Chunk 4	Chunk 5

****airbrush***: *Compressed air pulverizer used to deposit enamel on pieces.*
Note: The French examples have been given only approximate English translations as they are taken from actual monolingual terminologies.

In the context of an utterance, these chunks have a semantic value. The chunks of the preceding example could be given the following values:

INFO
Text Designations INFB 22:867

Semantic Value	Chunk
Identifier	pulvérisateur
Identifier's property	à air comprimé
Action pointer	utilisé
Goal	pour déposer de l'émail
Location (goal)	sur les pièces

The semantic value of a chunk can be determined through such factors as position (e.g., pre-verbal), part of speech (e.g., noun) and phrase structure (e.g., *with* + noun or *to* + infinitive). The various semantic values of these chunks can be called *semantic primitives* because they classify the informational content of the chunks within an utterance. Such semantic primitives may include *identifier* (for object definitions), *instrument* (for definitions of actions), *goal*, *action pointer*, *composition*, *location*, *property*, *state*, *time reference*. (The idea of semantic primitives can be found in Fillmore (1968) and in Schank and Abelson (1977).)

semantic primatives

The *identifier* is a major component of a definition because it conveys information about what the defined object is. The *identifier* classifies the object in terms of some more general category to which the object belongs, e.g., *airbrush* belongs to the class *pulverizer*, which also includes other objects. The choice of this general category should be strongly dependent on the reader's knowledge and intended task. With respect to this knowledge and

INFO
Standard Definitions INFB 12:340

this task, there are *base-level* (see Rosch 1976) general categories, which should be used when choosing an *identifier*. For example, in a definition of *cup*, the super-class *identifier* could be *crockery, container*, or *object*, but for a reader concerned with ceramics manufacturing, *crockery* is the most appropriate and, thus, the base-level.

When building definitions, the semantic primitives can be ordered in different ways, as in the following definitions of an electrical switch:

switch		
A component	fitted with an actuator and contacts	to make and break a connection.
Identifier	Properties	Goal

or,

switch		
A component	to make and break a connection,	fitted with an actuator and contacts.
Identifier	Goal	Properties

semantic structure

The ordering of semantic primitives is known as *sequencing*, and a sequence of semantic primitives is the *semantic structure*. In different contexts, some semantic structures are more effective, because they relate to the information needs of the reader. For example, if the reading is purposive and related to action (i.e., if a reader wants to know how to perform an action), then the *goal* primitive should be given prominence, which can be done by putting it in an extreme position (i.e., first or last). In the following example, from a terminology of bread-making, the *goal* is in the last position of the sequence:

mist: *Water vapor released into the oven to give a luster to the bread.*

buée			
Vapeur d'eau	libérée	dans la four	pour donner de la brillance au pain.
Identifier	Action Pointer	Location	Goal

order of information

Consistent sequencing of semantic primitives throughout a document allows the reader to expect a specific order of information in a given context and makes the text easier to use.

This aim may take precedence over the advantages of giving prominence to certain primitives as outlined above.

3.2 Grammatical Structure of Semantic Primitives

grammatical structures

Similar ergonomic concerns reward the use of a consistent grammatical structure when phrasing semantic primitives across a set of definitions. For example, if the *goal* primitive is a verb, it can be expressed either in a *to* + infinitive form, or in a *for* + gerund form (again taken from a terminology of ceramics manufacturing):

stirrer
Machine used to keep a paste in suspension.

or,

stirrer
Machine used for keeping a paste in suspension.

systematic correspondence

Whichever grammatical form is chosen for a semantic primitive should be used throughout a document in order to ease the reader's translation of grammatical into conceptual structure by means of a systematic correspondence.

4 Task-driven Approaches for Skill Acquisition

4.1 Action-centered Concept Systems

terminological use

As pointed out above, terminologies might be used for different purposes, such as translation, information management, writing instructional text, or self-teaching. Depending on the particular use for which a terminology is intended, the focus of definitions and the choice of concept-to-concept relations should differ.

terminological target orientation

Target-orientation in terminology can be defined as the match between the reader's perspective of some domain and the semantic structure of definitions. If readers are using the terminology to support self-teaching of a skill within a domain that is new to them, their interest may be mainly geared to the actions they might have to perform. This case can arise in organizations where quality assessment documents are intended to provide learning facilities to newcomers. A French company dealing with porcelain transfers explicitly required action-oriented term definitions on this basis.

action-oriented definitions

The method used to introduce such a bias into definitions strongly depends on the design of the underlying concept system. Following the principle that definitions can be read out of concept systems in a straightforward way, these particular terminologies should refer objects to actions in such a way that the definition of an object systematically relates it to some action. In this type of terminology, actions will occupy higher levels than objects, as shown in Figure 3.

tokens and types

The representation of the given concept system fragment includes labeled links (*action – method*, for instance). They identify predicates, such as *action* and *object*, and their arguments, such as *method*, *support* or *instrument* for *action*, and *property* for *object*. *Objects* linked to one another enter generic (*G*) or partitive (*P*) relations, a special case of the latter being *P**. This signifies that the parts of some whole (e.g., *table service*) are all tokens of the same type (*piece*). To enhance readability, *P** appears next to the partitive concept (*piece* as linked to *table service*). Generic relations may also link category tokens to category types, as is the case between *large* and *small*, and *property: size*. This facility provides an economic device for the representation of facets.

The following examples show an action and an object definition drawn out of the proposed concept system:

habiller
Modifier l'aspect d'une pièce au moyen d'un décor posé par application.

to decorate
To change the appearance of a piece using a design placed by application.

pièce
Élément d'un service de table tel qu'une tasse ou une assiette, de dimensions et forme variables, pouvant être habillé.

piece
Part of a table service, such as a cup or a plate, of variable size and shape, that may be decorated.

The effect of this structure is twofold: if, by means of a browser, the learner inspects the concept system in order to get acquainted with new concepts, he will enter it by the action level, which is naturally bound to step-by-step skill learning. When looking up the definition of some unknown object, he will be informed of the role

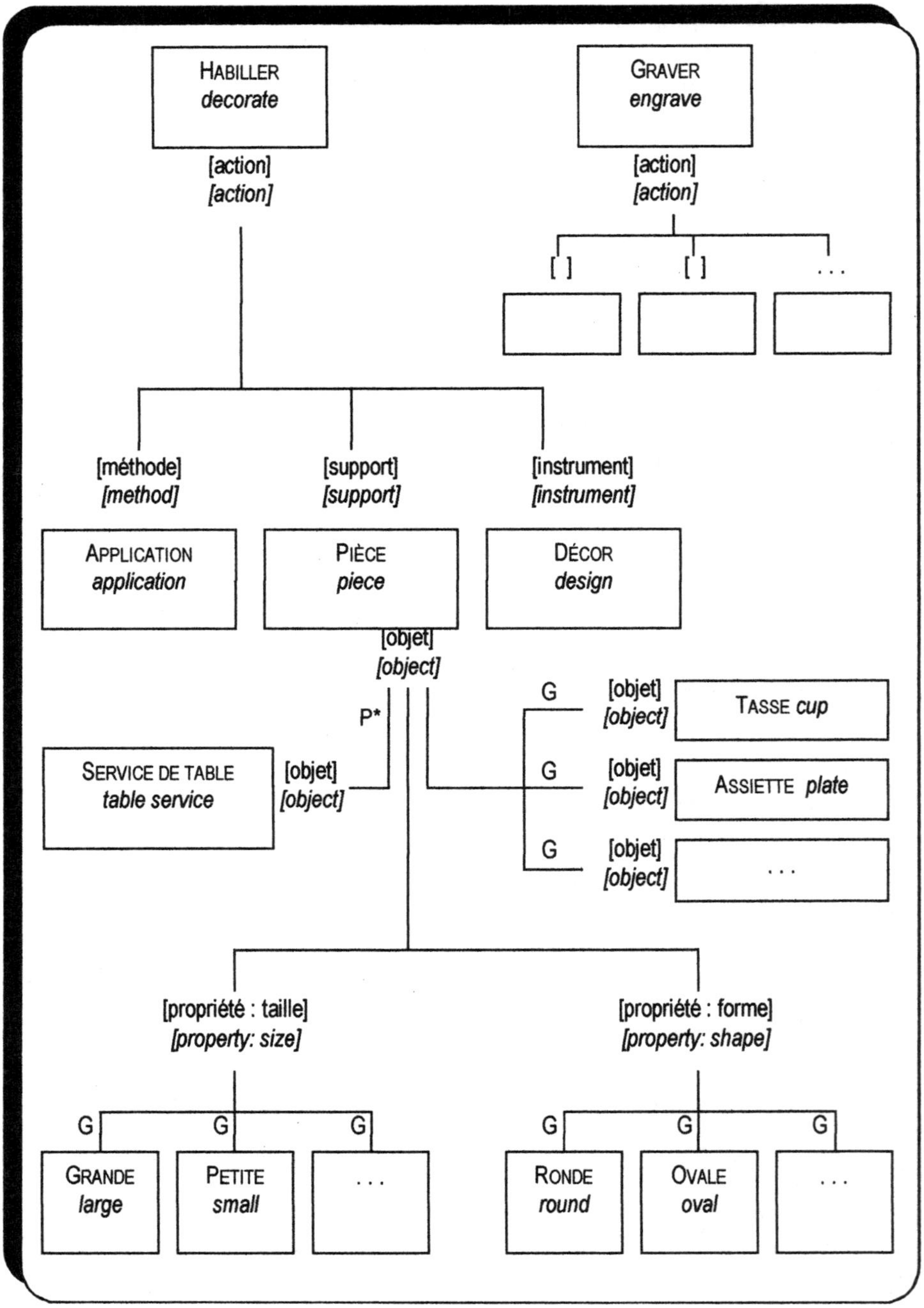

Figure 3: *An Action-Centered Concept System*

the object plays in actions (as shown in Figure 3) that he will probably perform after having acquired the skill.

4.2 Object-Centered Concept Systems

task-driven object identity

Even though a skill may not be explicitly addressed by the structure of the concept system (e.g., with actions as its focus), a terminology can capture object identity in a task-driven fashion and thus enable learning. A restricted functional system description is illustrated by the example in Figure 4, which highlights a subsystem of an internal combustion engine from the perspective of a mechanic dealing with the ignition system in diagnosis and repair situations.

task-related subsystem

The selection of such a subsystem is generally dependent on a task (diagnosis and repair of ignition systems), which explains why the reported information is dramatically reduced to the knowledge relevant to that task. The model has been set up entirely with the needs of mechanics in mind and reflects the image that they have of the system. The process has been broken down in order to understand the role of each part.

The boxes with solid borders represent the relevant engine parts, which are to be defined in the terminology. The boxes with dotted borders show phenomena that take place within the ignition process. These are not to be defined as they form part of the background knowledge of the reader and are used to integrate the other concepts into the system. The arrows show the links between concepts, with the arrow head indicating the direction of process-flow during ignition, and the arrow's label describing the nature of this link. For example, the following relationship is shown: "la bougie génère l'étincelle" (*the spark-plug generates the spark*).

The definition of a concept can be read out by following its links, e.g.:

bougie
Dispositif alimenté par la bobine servant à founir l'étincelle qui provoque l'explosion.

spark plug
Device powered by the coil, serving to produce the spark that creates the explosion.

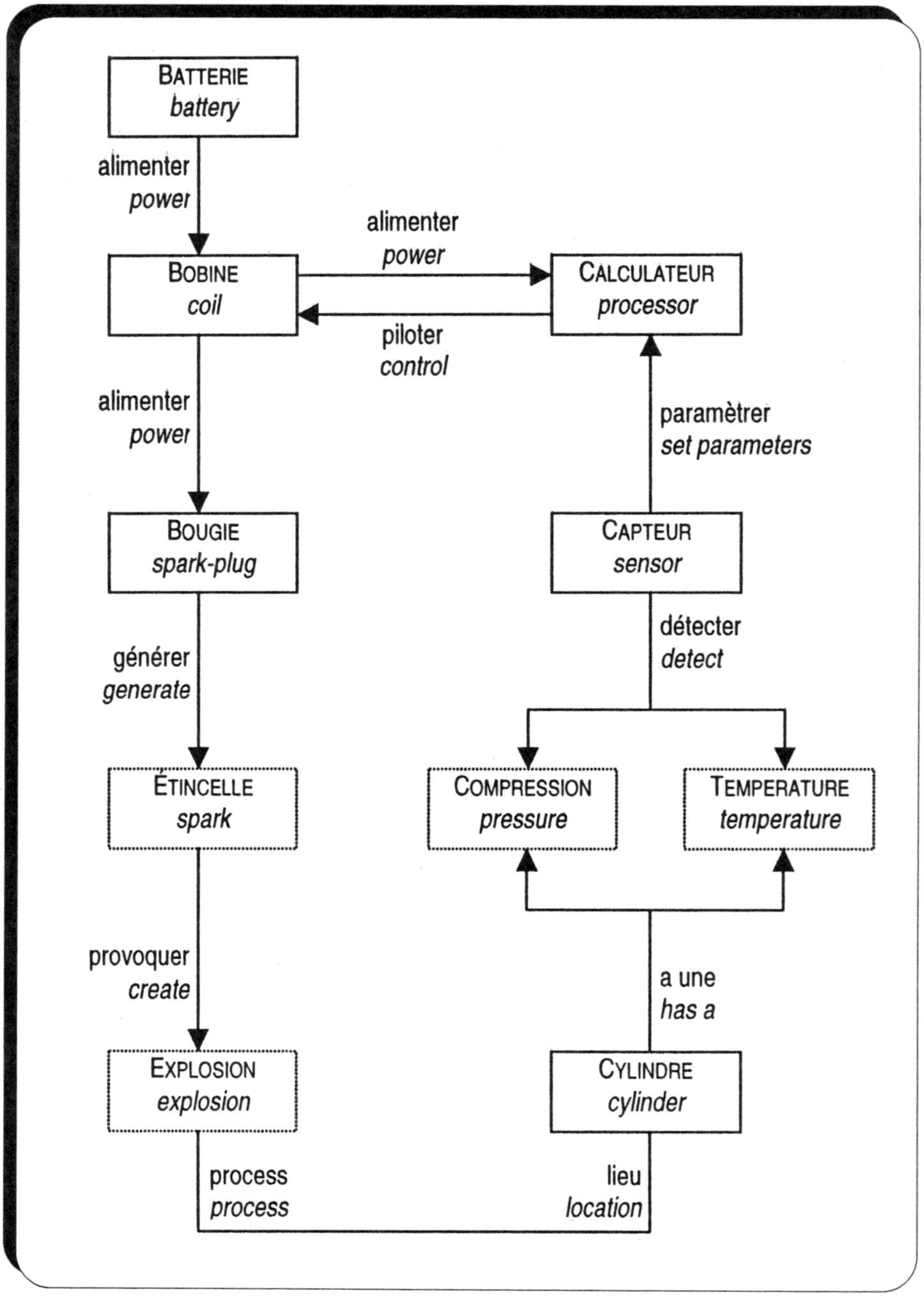

Figure 4: *An Object-Centered Concept System*

5 Knowledge-driven Approaches for Domain Overview

An object-centered terminology may also be designed to provide general reference about some domain. Figure 5 shows an extract from the concept system underlying a terminology of the World Wide Web, where objects are functionally defined with respect to each other.

As an alternative representation for relations within concept systems, we have introduced here a facility borrowed from KL-ONE (Brachman and Schmolze 1985). Individual concepts, as opposed to generic ones, are shown in shaded ellipses. Circles within squares stand for a *role*, identified by a label, such as *format* with respect to *page*. The relation is read out in the following way: "A page has a format that is a mark-up language". *Mark-up language* is understood as a constraint on every filler of the *format* role (such as *HTML*).

We consider such a concept system to be a comprehensive functional domain description because it encompasses the entire system from a functional point of view. Such descriptions enable task-independent information acquisition rather than skill learning and are therefore thought of as being knowledge-driven.

functional domain description

Definitions read out of the concept system carry the following information:

HTML
Markup language used as the format of a page, that has standard WWW code, which is HTTP, as its transmission protocol.

page
Resource having a mark-up language as its format, text as its content and Java programs as possible resources.

The incompleteness of the preceding definitions may be accounted for by the fact that they only take the pictured part of the concept system into consideration.

6 Knowledge-driven Approaches for Product Classification

Design techniques are not only relevant to documents where information serves instructional purposes. Management of complex data

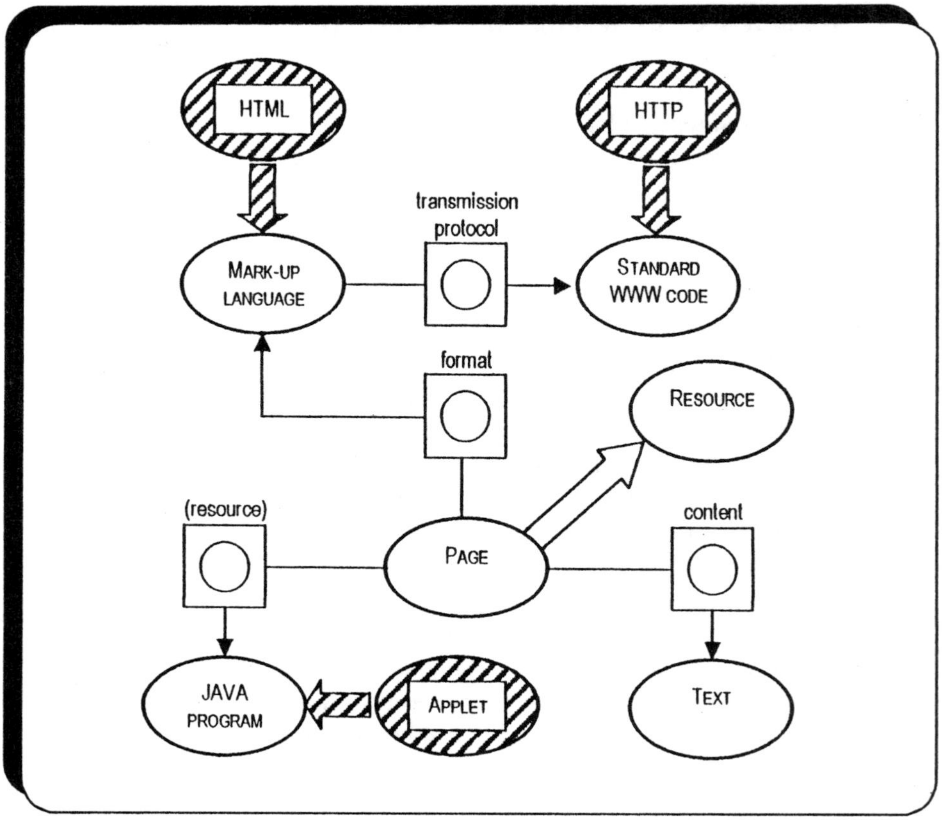

Figure 5: *An Object-Centered Concept System for Domain Reference*

often requires sophisticated information design, as is the case in convoluted product classification contexts. The reason why these contexts demand terminological solutions is that misleading names of product types, synonymy, and polysemy are frequent causes of confusion in classification systems.

commercial logic

But the main reason for confusion lies in the secret belief of domain experts that names of products refer straightforwardly to concepts. This is rarely the case because names often arise in an uncontrolled way as the production scope of a firm grows. Moreover, they may be subjected to commercial logic rather than to consistency in the management of a classification system.

catalog inconsistency

The case we would like to discuss below occurred in a company producing electrical devices. At some point in the firm's development, the sales catalog became difficult to understand because different products had nearly the same name, and the same product, when intended for different usage, had different names. Products were also grouped in *families* (product types) based on different criteria (same usage contexts, same market sector, etc.). One of the side effects of this slightly chaotic situation was that the firm's overview of its market share was blurred.

technically coherent classification

In order to rectify this situation, the firm decided to change its product classification method. The first step taken was to free it from any kind of marketing constraints, that is, to build up a technically coherent classification of products to which marketing family features could be attached without becoming themselves classification criteria. Since taking into account different usage contexts for the same product had contributed to making the former classification inconsistent, these contexts have been equally excluded from the technical product view, which consequently became restricted to function, i.e., to the effect of the classified device on an installation (*to make and break a connection*, for example), and which further specified it by features such as operating technology (for example, *manual* or *automatic*). The intended effect was to allow a clear overview of those functions that were actually being served by the range of products.

function cluster

Product (type) names were originally thought of as terms designating a single concept comprising complex and over-detailed information (function, usage context, capacity, accessories, etc.). The new classification system featured instead concepts corresponding to clusters of function(s) and constraints (see Figure 6).

generic conceptual content

Directly linking these clusters to product names would have implied that only one product corresponded to the concept, which was obviously false, or would have referred several names (and thus products) to the same concept, which would have been unsatisfactory for classification purposes. What needed to be shown was that several products related to the cluster's *generic* conceptual content (*to make and break a connection* [*function*] + *operating technology* [*constraint*]), which remained constant over the changes in the constraint value (from *manual* to *automatic*, in the case of *operating technology*).

Figure 7 shows the structure of the *switch* concept, which is built up by constraining a function (*to make and break a connection*) by means of an operating technology (*manual*). However, the addressed function concept only refers to *switch* if it is

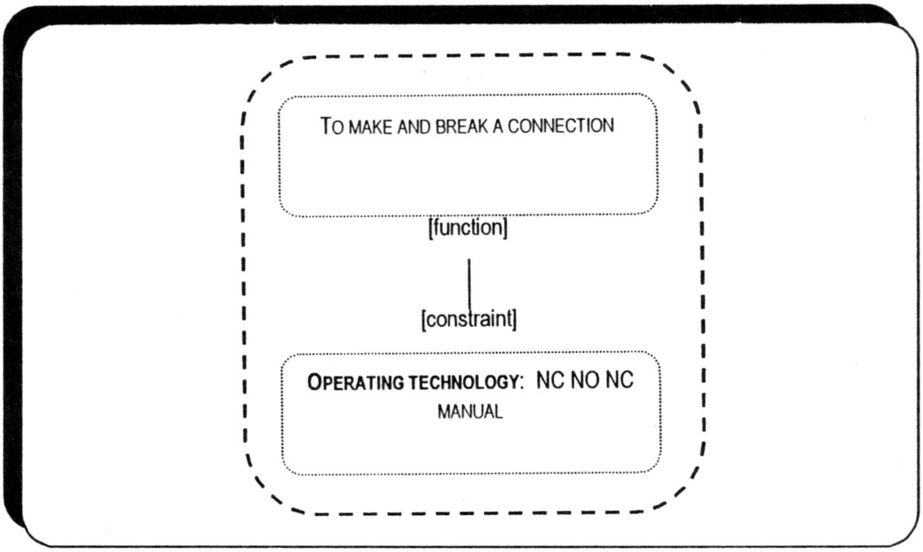

Figure 6: *A Concept Clustering for a Function and a Constraint*

constrained by the operating technology cited above. If the constraint's value were changed (*automatic* instead of *manual*, for example), the same function would refer to *relay*, as shown below.

constrained functions

The cluster's generic conceptual content is thus a fixed element of the classification system that does not refer to any particular product. By contrast, as soon as the cluster's constraint is tagged to some value (such as *NC NO NC manual*), the cluster relates to a specific type of products, namely switches. We call these clusters *constrained functions*. In Figure 7, *switch* is a concept defined (see below) by the above constrained function.

functional principle

As is well known, switches may have different capacities, usually expressed in Amperes (A) and Volts (V). The actual products of our firm range from 20A to 200A and from 250V to 600V. In order to express the common features of these products, they must have a single anchoring point, e.g., the *switch* concept. As this concept does not stand for any real product, it only represents a constrained function to which products of different capacities are linked, that is, an intermediate level of abstraction between the constrained function(s) and some concrete products. We call this level the *functional principle* and distinguish the corresponding concept labels with the suffix *FP*.

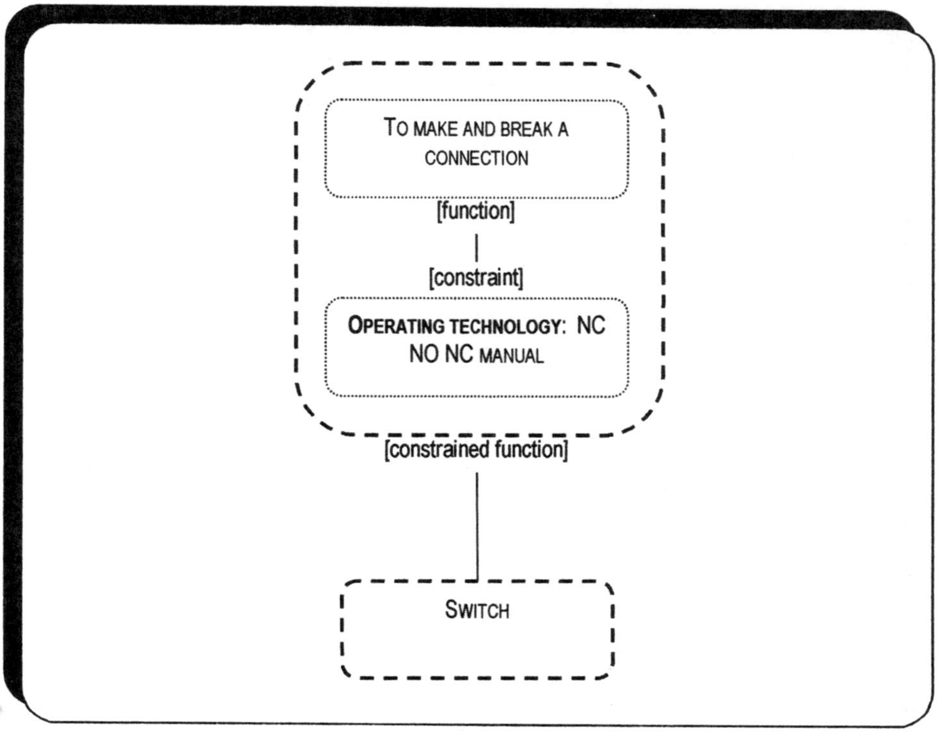

Figure 7: *Functional Meaning of the Switch Concept*

In Figure 8, the contrast between concepts standing for products and concepts representing functional abstractions is pictured. As opposed to solid lines, dotted lines mean that the reported concept does not refer to an actual object. Figure 9 represents the difference between two instances of the same function cluster with different values for the constraint.

One of the major advantages of including the functional principle level in the classification system is that it accounts for combinations of functions in the same product set. Figure 10 highlights this type of situation by showing the *thermal magnetic circuit breaker functional principle*, which combines evaluated versions of the *to make and break a connection* and the *overcurrent detection* function clusters.

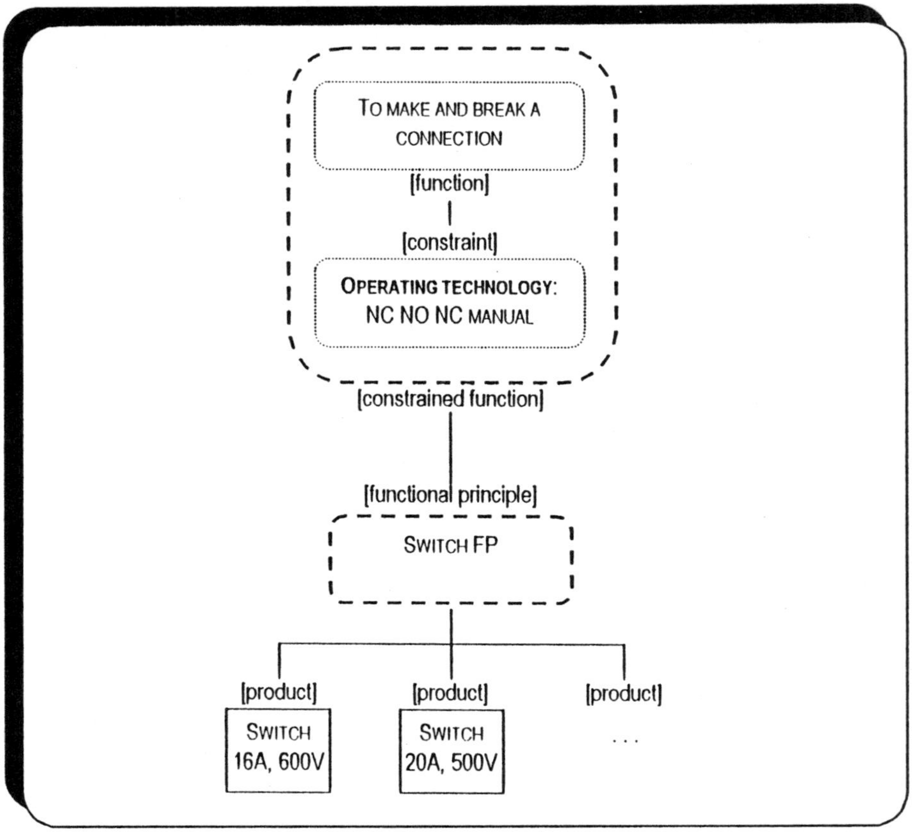

Figure 8: *Constrained Functions, Functional Principles, and Products*

defining factors

Another interesting feature of functional principles is that they consistently define product types (switches, thermal magnetic circuit breakers, etc.) allowing the proposer of the classification system to fix the level of product types themselves. If, in the example given by Figure 10, the function cluster had been considered as unconstrained (instead of as a constrained function by evaluating the *detection technology* to *thermal magnetic*), the resulting functional principle would have been *circuit-breaker*. This case is examined in Figure 11.

product types

Functional principles correspond to product types and may be defined by reading the function name *(to make and break a connection*) and the constraint value (*manual* for *operating technology*) out of the concept system (see Figure 9):

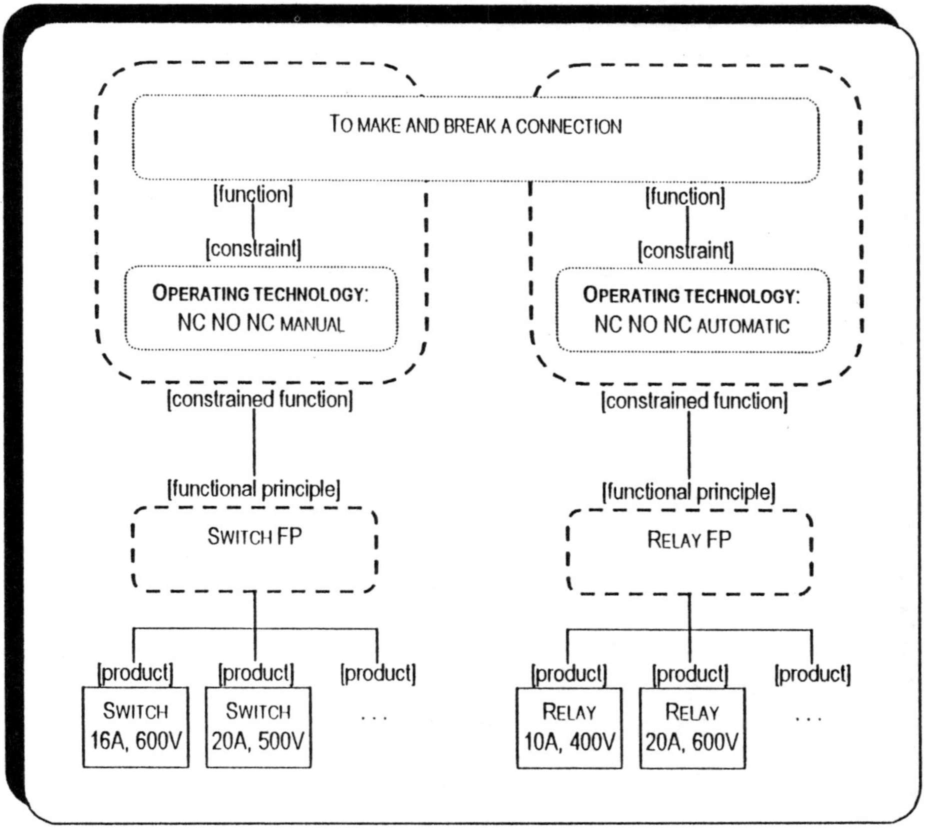

Figure 9: *Functional Principles Derived from the Same Function*

switch FP
Product type used to make and break a connection manually.

The fact that a functional principle may refer to more than one constrained function cluster (Figure 10) is reflected in definitions in the following way:

thermal magnetic circuit breaker FP
Product type used to make and automatically break a connection including a thermal magnetic overcurrent detector.

functionally appropriate definitions

It should be noted that the link between *detection* and *automatic circuit breaking* does not explicitly appear in the definition. The reason for this is that the intended terminology is not designed to provide a theory of electrical devices. Doing so would be a failure

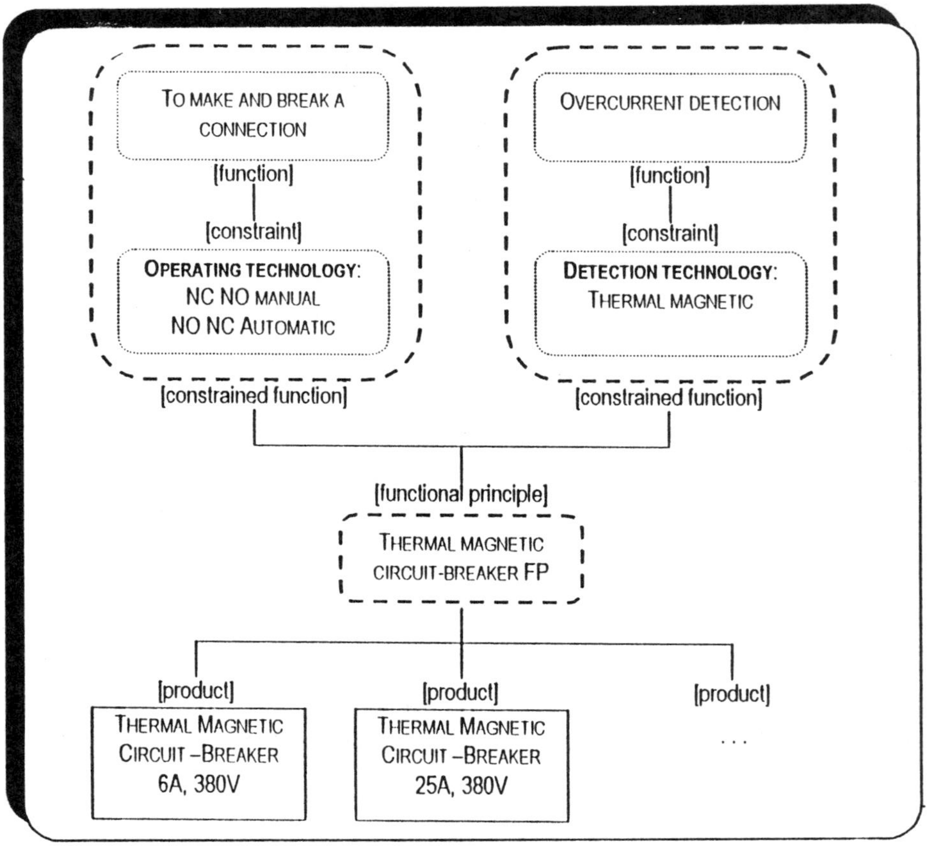

Figure 10: *A Functional Principle Combining Two Different Function Clusters*

with respect to the target audience's expectations. The user of such a classification system, an electrician, knows that some kind of detection is a necessary cause of automatic circuit breaking. Moreover, because of this knowledge, he will expect the device description to include information about the detection domain (which is provided by the specification *overcurrent*) and the technology implementing the detection process (here, *thermal magnetic*). The current definition is therefore as complete as is needed.

In the case pictured by Figure 11, the fact that the constraint has purposively been left unevaluated is mirrored by the following definition of *circuit breaker FP*:

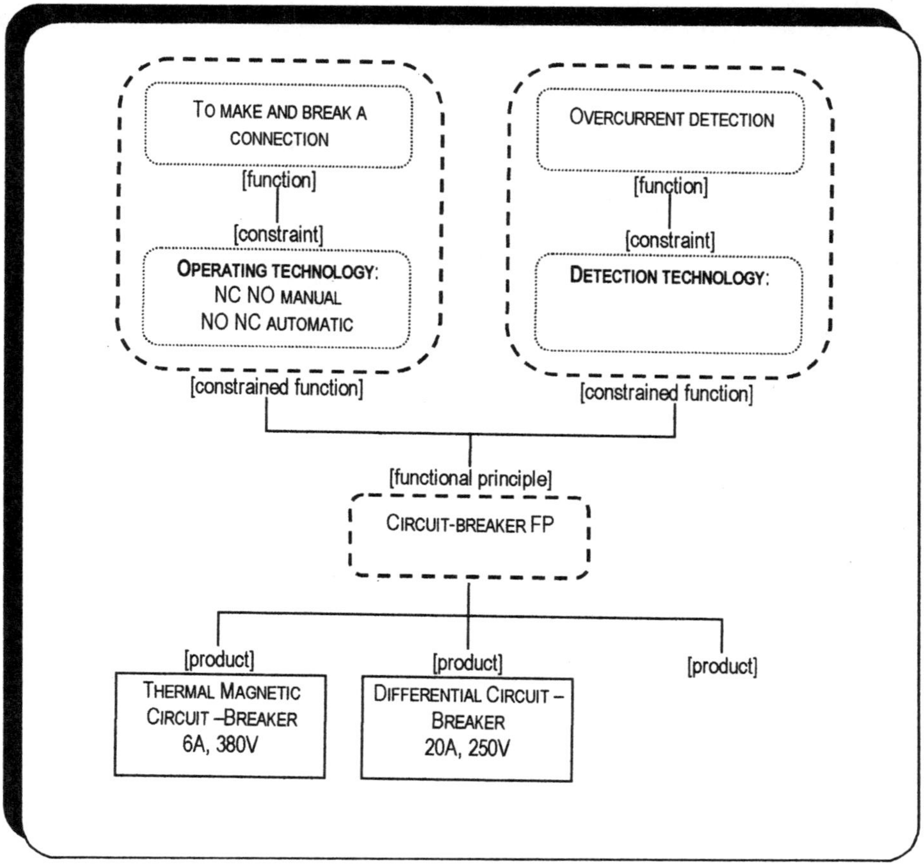

Figure 11: *A Functional Principle Related to an Unconstrained Function Cluster*

circuit-breaker FP
Product type used to make and automatically break a connection including an overcurrent detector.

Products themselves are not addressed by the terminology because each is systematically linked to a defined functional principle, that is, a product type, and they are only distinguished in the classification system by their capacity.

product type

REFERENCES

Brachman, Ronald J., and Schmolze, James G. 1985. "An Overview of the KL-ONE Knowledge Representation System." *Cognitive Science 9,* 171-216.

Fillmore, Charles. 1968. "The case for case." *Universals in Linguistic Theory*. Bach, E. and Harms, T., eds. New York: Rinehart & Winston, 1-88.

IEC. 1992. *Electricity, Electronics and Telecommunications: Multilingual Dictionary*. Vol. 1. Amsterdam: Elsevier.

Morin, Gervais. 1996. "Identifier les Représentations d'Installations Electriques Industrielles chez les Destinataires d'un Cahier des Charges: Une Application de la Théorie des Modèles Mentaux." M.A. Thesis, Limoges: University of Limoges.

Rosch, E., et al. 1976. "Basic Objects in Natural Categories". *Cognitive Psychology,* No. 8, 382-439.

Schank, R.C. and Abelson, R.P. 1977. *Scripts, Plans, Goals and Understanding*. Hilldale: Erlbaum.

Section 7.2 Industrial Management

7.2.1 Terminology as an Organizational Principle in CIM Environments

SUE ELLEN WRIGHT

1 INTRODUCTION

high-quality document management

The extensive recent growth of terminology management has been spurred by mounting pressure for government agencies, companies and other institutions to produce high-quality documentation pertaining to virtually all aspects of their operations. As a result of early association between terminology and document production, the relationship between *quality assurance* and *productivity* within the framework of terminology management has generally been defined in terms of efficient, high-quality *document production*, both in the source language and in target-language translated texts. Furthermore, because of this emphasis, terminology management within enterprises has been centered in language service or technical writing departments and, even more specifically, has focused on multilingual document generation and translation.

terminology management

Important as these factors are, limitation of terminology compilation to the translation area represents a narrow view of the significance of terminology management for government, industry, and the economy in general. The following discussion will explore the extent to which terminology management is intimately linked to productivity and efficient work flow throughout a variety of enterprises and agencies, with special emphasis on the role of terminology in Computer Integrated Manufacturing (CIM) and Computer Integrated Enterprise Management. CIM in this context will reflect the definition offered by Hannam, i.e., "the integration of business, engineering, manufacturing and management information that spans company functions from marketing to product distribution" (Hannam 1997:2). Although some authors (notably Scheer 1994) view CIM as a somewhat narrower concept within the environment of business process engineering, the CIM concept seems to have held the day against various competitor terms such as CIE (computer integrated enterprise) and CIME (computer integrated manufacturing enterprise). See also Wright 1994. This article is intended as an introduction for terminologists unfamiliar with these environments and will not provide the degree of detail required by data element administration specialists.

2 Terminology and Information Management

2.1 Terminology Applications in Enterprises

Because of the traditional relegation of terminology compilation to a subordinate role within the overall scheme of organizational activity, this work has been conducted in isolated corners of the corporate structure. The traditional compartmentalization of activities militated against holistic approaches to knowledge and information management. This trend was characterized by a concomitant inability to integrate knowledge and practice across departmental boundaries and a failure to provide mechanisms for adequate information retrieval that take into account the passage of time as well as physical and cultural distances between individual units within an organization (see Gaylord 1987:1-2).

holistic information management

SEE

TQM 7.2.3:488

The relative isolation of translation-oriented terminology also meant that terminological products were primarily used (and sometimes even generated) at the very end of the production process, such that the shipping department inserted terminology-laden translated documentation immediately following end-item inspection and just before applying the shipping label. In contrast, forward-thinking enterprises have now integrated terminology generation and documentation into the document and the product generation process, a development that has resulted in increased document quality and general reliability.

SEE

Technical Writing 7.1:445

Technical Translation 2.1.1:147

Further attention to the terminological aspects of the overall know-how residing in enterprises has resulted in terminology being used for the efficient retrieval of information from knowledge bases at all stages of the document life cycle. The critical importance of retrieval strategies has become increasingly apparent as vast numbers of World Wide Web users clamor for efficient tools in the form of Internet search engines and directories.

SEE

Information Retrieval 6.3:426

2.2 Terminology within the Management Structure

broader terminology applications

The gradual movement of terminology compilation out of the narrow confines of translation-oriented terminology management into the broader focus of technical documentation on the one hand and information retrieval on the other is accompanied by increased involvement by terminologists in technical communications, information management, and data processing.

vertical information structures

The expansion of terminological activities into other venues, however, only represents an initial step toward the recognition of the importance of sound integrated terminological practice. Galinski (1989) projects the image of terminology management escaping from its encapsulated isolation in traditional management structures to assume a top-down, bottom-up function that flows throughout corporate information structures.

TQM

CA systems

This development runs parallel to the altered role played by data processing and quality assurance in modern organizational structures: instead of remaining the concern of individual departments and work groups, Total Quality Management (TQM) and Computer Aided (CA) systems are now firmly established as essential components of virtually every corporate department, from Product Design to Manufacturing Engineering to Production to Customer Service. Although they retain some elements of their earlier specialized structure, Quality Control (QC) has grown into Quality Assurance (QA) and evolved from there to Total Quality Management. In like manner, Data Processing (DP) has given way to Information Management. Together they have come to function very much as in-house consultants to other departments in their application of quality and information management procedures in order to implement new procedures associated with Total Enterprise Management. Terminology applications today exhibit the same kind of cross-disciplinary profile as these more well-known, but not necessarily more important, production-support functions. Indeed, both QA and Information Management are dependent on the elaboration of high-quality terminological resources.

Figs. 1 and 2 illustrate Galinski's representation of the difference between the traditional segregation of support services (QC, DP, technical communication, translation, as well as terminology management) and modern information-management-oriented organizational structures. This vision of terminology management within

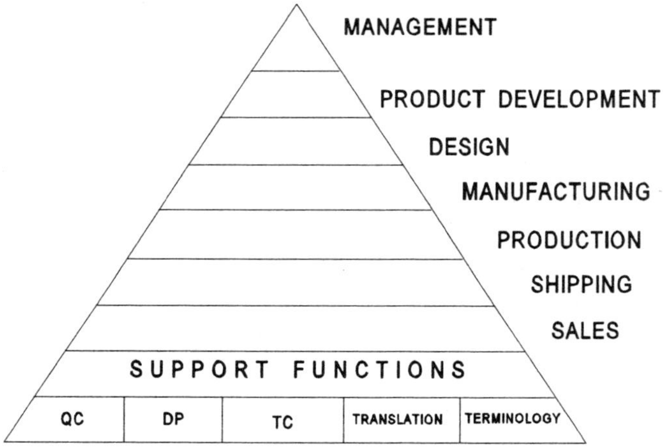

Figure 1: *Compartmentalization of Information Services in Traditional Enterprise Structures*

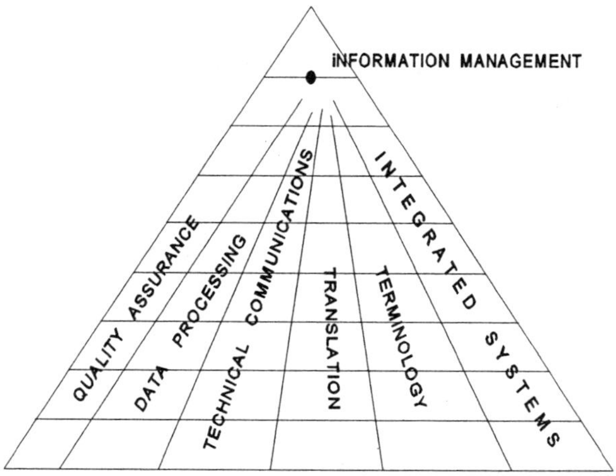

Figure 2: *Top-down, Bottom-up Information Flow in Integrated Systems*

the framework of the corporate structure mirrors a similar triangle used to illustrate integrated information systems in general (Scheer: 1994:5).

3 The Role of Terminology in Computer Integrated Environments

CIM

The most revealing view of terminological applications in the corporate environment can be gained by examining the role that terminology plays in the integration of computer applications on an enterprise-wide scale.

3.1 Master Data Files and Data Dictionaries

item master

The core of an integrated information system is the *master data file* (frequently called the *item master*). In an ideal system, a systematically structured master data file takes the form of a data (element) dictionary documenting the data element concepts included in the system. Because data element names may vary from application to application or department to department throughout an organization, the concepts are likely to be associated with unique ID numbers or codes for all virtual objects managed by the system. For instance, in a manufacturing system, this might include, among other items, all:

- parts and assembly-related items
- raw and production-support materials
- systems, tools and equipment, gages, etc.
- information for master process charts and routings
- factory-data-entry items (cost-accounting data)
- personnel-related information
- supplier and customer-related items
- billing and accounting items

data dictionaries

Virtually every object that can be recorded in a computer-support system belongs in the master data file. (Of course, in reality, this may not be a single physical file because many systems will organize different categories of data into separate, interactive files.) As noted, more sophisticated systems go beyond the *item master* concept to create *data dictionaries*, which are, "database(s) about data and databases. [They hold] the name, type, range of values, source, and authorization for access for each data element in the organization's files and databases" (Freedman 1989:176). Although Freedman's definition is widely accepted, different authors have viewed the data dictionary from various perspectives (Gaylord 1987:73-75; Weatherall 1988:84; Bray 1988:73-74; Hannam 1996:196). As noted below, data dictionaries possess many terminological features, such as definitions and the delineation of each object's position within the hierarchy of the system architecture.

3.2 Islands in the Stream

islands of information

Early computerized information systems (and many that still exist) were—and still are—plagued by structural inadequacies. The creation of numerous stand-alone application centers resulted in the development of *islands of information*, which increase costs and contribute to inefficiencies, in terms of both practical manufacturing processes and overall information flow (Hannam 1996:9, 15 ff.; Weatherall 1988:156-157; Gaylord 1997:190-191). Indeed, initial movement toward CIM was inspired by the realization that only in overcoming this insularity would manufacturers and information specialists achieve effective, fully *rationalized* process flow.

integrated systems

The implementation of integrated systems and the elimination of isolated sub-systems involve linking all units to a network backbone. Often visualized as a single trunk line running throughout the enterprise, the backbone is more likely to be an interconnected network designed to facilitate communication between the master control system and individual sub-systems, as well as intercommunication among all the nodes on the net. Of course, in optimum cases this process is more efficiently achieved by replacing isolated sub-systems with harmonized master systems. Even in such cases, however, the importation of legacy data in the form of one-time snapshot conversions is highly desirable and poses the same kind of incompatibility associated with autonomous systems..

standards for data exchange

Regardless of the communications solution proposed for any one enterprise or for exchanging data among closely associated enterprises, developers and standardizers have responded to the need for compatible data resources by creating a number of industry standards for facilitating data sharing and exchange inside and between enterprises, across hardware platforms, and among various software environments. On the technical side, *open systems interconnection* (OSI) solutions have evolved, including MAP (*manufacturing automation protocol*), IGES (*initial graphics exchange specification*), and STEP (*standard for the exchange of product model data*) for the purpose of integrating design and manufacturing information within and across enterprise boundaries (Hannam 1996:210-223).

data element principles

More closely allied to terminological practice, however, are the standards developed by Sub-Committee 32 of ISO/IEC Joint Technical Committee (JTC 1) for Data Element Principles. This six-part suite of standards treats data definition and administration, i.e., "the human intellectual activity required to exploit data management technology" (Graves and Gillman 1996; ISO 11179, Parts 1-6). These standards are designed to provide uniform, essentially terminological principles for the naming, definition, and classification of data elements. This activity takes place in conjunction with the creation of metadata registries designed to adequately define data element concepts in such a way as to enable the interchange of data among different systems. A logical extension of the effort to normalize data element naming practice is the movement to establish metadata registries (Metadata Registry Implementation Coalition, http://hmrha.hirs.osd.mil/mrc/).

3.3 Linear Development Structures

linear project planning

In traditional development structures, the project cycle progressed in a linear fashion. Figure 3 illustrates a simplified overview of these activities. Projects were conceived in Product Development (often in conjunction with Marketing or Sales Engineering) and carried through at least to the testing stage without significant involvement from downstream Manufacturing and Production Departments. Manufacturing Engineering entered the picture later to develop actual product manufacturing facilities after much of the design had been established, a process not infrequently involving problematic re-design actions in cases where Product Development had not adequately taken the manufacturability of the new product into account. Depending on the relative efficiency or inefficiency of the system, Quality Planning and Materials Sourcing (Materials Management) entered the picture later, along with Production Planning.

Due to conflicting corporate cultures and lapses in communication among different management islands, polysemous and synonymous terminology evolved that further contributed to a

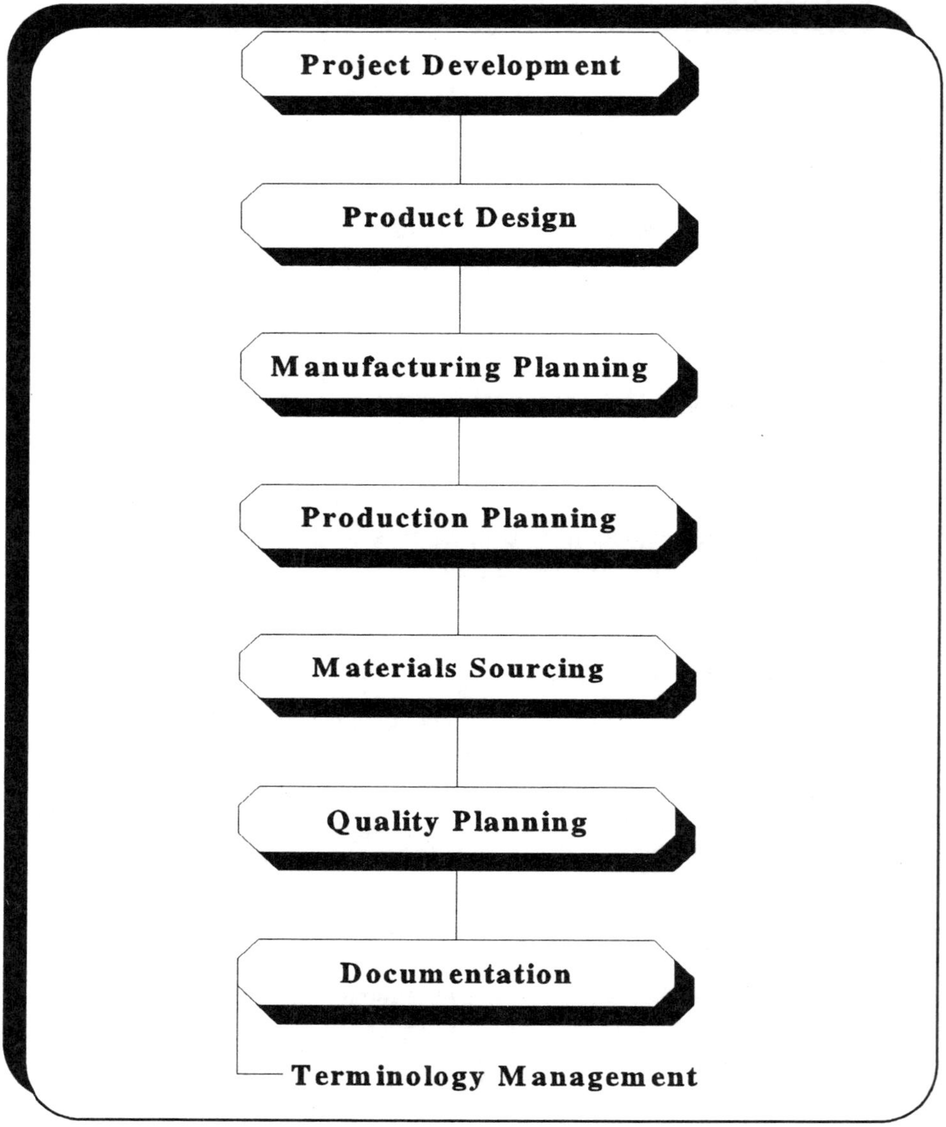

Figure 3: *Traditional Sequential Project Development*

"Tower-of-Babel" syndrome, impeding efficient organization and communication. Under these conditions, documentation—the generation of both monolingual and bilingual support literature—most frequently began during the end stage, with technical communicators and translators scrambling under highly stressful conditions to sort out conflicting terminology and to generate reliable, user-friendly language and information products. Under these conditions, terminology management is an afterthought grafted unnaturally onto the project management structure.

crisis-oriented terminology management

3.4 Simultaneous Project Management

In response to market demand for cost savings and increased QA, industry has been motivated to achieve unbroken information flow throughout the life cycle of any given product or process. The current goal is to achieve optimum efficiency at all stages of development and to shorten project development time through parallel development procedures in all affected production-support sectors.

project life cycle

Instead of progressing in a linear fashion, with one activity occurring in sequence after the other, development operations are coordinated as parallel, simultaneous activities conducted by integrated task groups cooperating to cut development time and increase design and development efficiency. Feedback from all affected teams and at all levels flows back into the process to ensure optimum quality and correct errors before they have the chance to proliferate. Interactive systems are capable of reducing or even eliminating redundant operations related to design, data entry, manufacturing, shipping, billing and customer service functions. The medium for communicating this feedback is *reusable data* in the form of shared resources (Hannam 1996:113 ff.), e.g.:

parallel project development

- Product Development (PD) engineers use CAD programs to create draft designs. Ideally, they take advantage of customer-generated machine-readable data that can be fed directly into design systems without supplemental data entry.
- Design specialists, using the same data generated by PD, continue CAD activities to generate product drawings and specifications, while at the same time Manufacturing Engineering compiles Numerical Control (NC) programs for the one-off manufacture of tooling to produce detail parts.

reusable data applications

- The same basic CAD information is passed on to Materials Management for the generation of bills of materials and materials releases to suppliers for just-in-time (JIT) delivery of raw and semi-finished materials or out-sourced detail parts.
- Manufacturing Engineering coordinates the acquisition and installation of manufacturing equipment for production of the new design. Master process charts and routings are generated that document sequential material flow and the assignment of machine capacity throughout the process, again relying on existing design and logistical data.
- Production implements the manufacturing process as defined by upstream partners.
- Quality Assurance interacts at all levels to ensure defect and error-free production. The QA system, like other members of the team, utilizes design data to produce control plans, to design gauging systems, and to monitor the production process.
- Each of these steps involves the utilization of online resources and accompanying hardcopy and online production-support tools, which, under ideal conditions, will be used to assist in the creation of customer-friendly product-support documentation.

information "carriers"

SEE

Concept Systems 1.4.1:89

The coordinating mechanisms throughout these process stages consist of relational links accompanied by consistent data element concept identifiers which are essentially terminological designations and are linked to data element names, which function very much in the form of terms. Even part numbers act as terms in such systems. Concept identifiers, terms, and other designators are information "carriers" that transport data, but also assist in controlling and directing traffic in the system. Beyond the level of object names (designations), structural elements such as CAD parts families and bills of materials reflect both generic and partitive concept systems, while master process charts document and control work flow as represented in sequential concept systems (Weatherall 1988:78 ff.). QA and design methodologies such as Failure Mode and Effect Analysis (FMEA) integrate multiple terminological functions in that they are structured according to the careful definition of classified product characteristics. All these activities entail the need for systematic data element concepts that will function meaningfully throughout all levels of the system. Hence, the assignment and definition of data elements is a

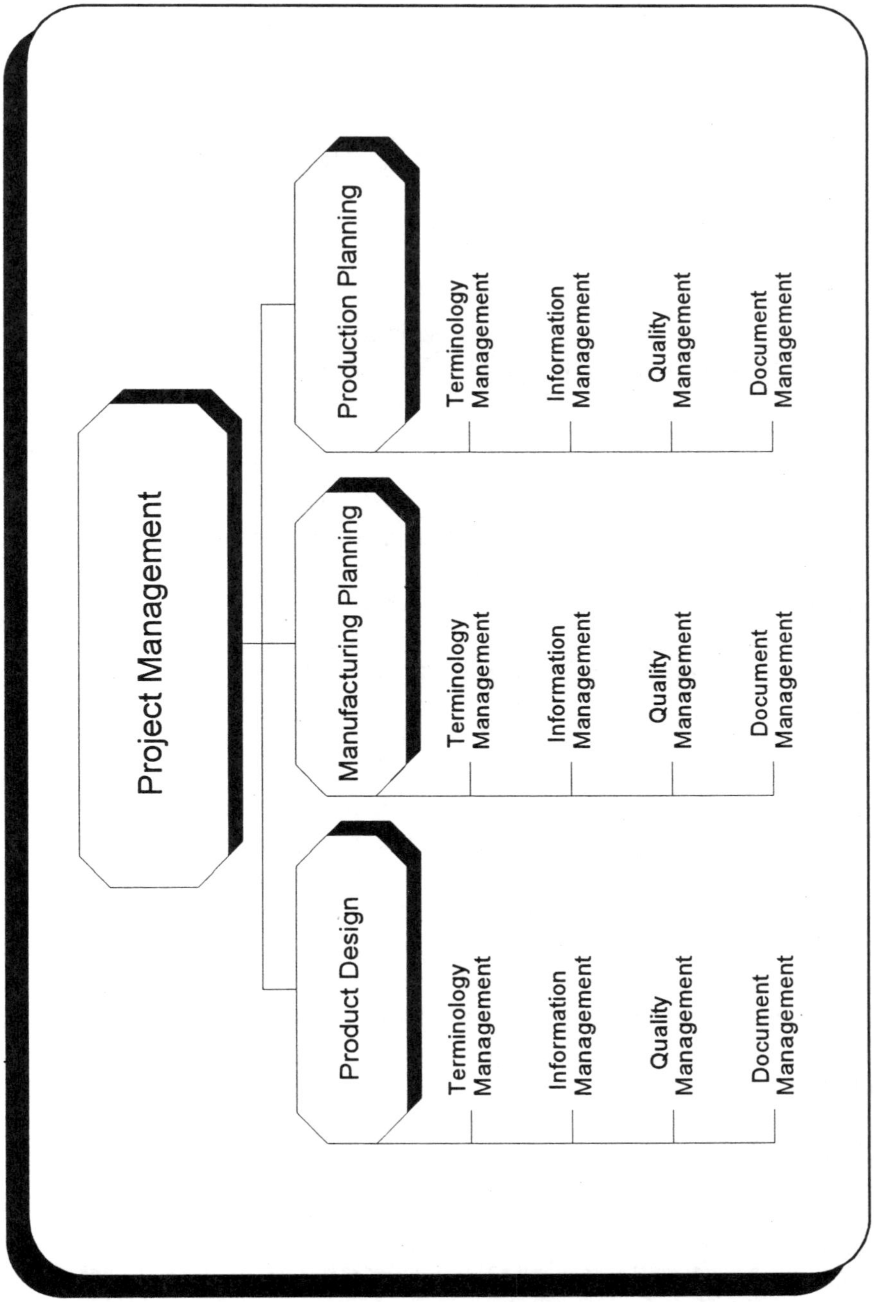

Figure 4: *Project Flow with Parallel, Simultaneous Project Management*

fundamentally terminological activity that must take place early in the development process, as illustrated in Figure 4.

3.5 Forging Physical Links in the Production Chain

Section 3.2 above described the physical linkage of hardware islands within enterprise structures. The resulting network is, of course, the first step toward integration; establishing the capability for all systems to *talk to one another* at the software level is the next stage.

As we have already seen in Section 3.1, the path to integration leads back to the master data files and data dictionaries. Not only are the files created for stand-alone systems likely to be quite different in structure. Similar data elements may well be defined and named differently. The only way to solve such inconsistencies in the long run is to implement data dictionaries and registries on an enterprise-wide or even industry-wide scale in such a way that all data elements that occur in the system can be mapped to units that are globally applicable and therefore convertible to all levels and applications. These problems were once further exacerbated by additional inconsistencies that arise when it becomes necessary to exchange data with different physical sites within the same organization or when enterprises encounter the need to exchange data with external partners, but today, with Web links and enterprise-wide intranets these problems of data transfer are more easily solvable. Again, catalogs of data categories, together with their definition and the specification of their critical attributes, must be created and mapping procedures developed so that data can be successfully ported from one environment to another. The creation of prenegotiated data element lists for use in specific work groups and agreement to employ specified industrial exchange standards further facilitate the implementation of open systems.

data dictionaries

catalogs of data categories

SEE *Internet Resources 8.5:845*

REFERENCES

Bray, Olin H. 1988. *Computer Integrated Manufacturing: The Data Management Strategy.* Boston: Digital Press.

Freedman, Alan. 1989. *The Computer Glossary: The Complete Illustrated Desk Reference.* Point Pleasant, Pennsylvania: American Management Association.

Galinski, Christian. 1989. *Terminology and Information Management.* Unpublished memo.

Gaylord, John. 1987. *Factory Information Systems: Design and Implementation for CIM*

Management and Control. New York and Basel: Marcel Dekker, Inc.

Graves, Ronald B. and Gillman, Daniel W. 1996. *Standards for Management of Statistical Metadata: A Framework for Collaboration.* Accessed 1998-07-08. http://www.scc.ca/jtc1/sc21/wg3/isis/isis.html .

Hannam, Roger. 1996. *Computer Integrated Manufacturing: from Concept to Realisation.* Harlow, England and Reading, Massachusetts: Addison Wesley.

ISO/IEC 11179-1. (In preparation) *Information technology — Specification and standardization of data elements — Part 1: Framework.* Geneva: International Organization for Standardization.

ISO/IEC 11179-2. (In preparation) *Information technology — Specification and standardization of data elements — Part 2: Classification of Data Elements.* Geneva: International Organization for Standardization.

ISO/IEC 11179-3:1994. *Information technology — Specification and standardization of data elements — Part 3: Basic attributes of data elements.* Geneva: International Organization for Standardization.

ISO/IEC 11179-4:1995. *Information technology — Specification and standardization of data elements — Part 4: Rules and guidelines for the formulation of data definitions.* Geneva: International Organization for Standardization.

ISO/IEC 11179-5:1995. *Information technology — Specification and standardization of data elements — Part 5: Naming and identification principles for data elements.* Geneva: International Organization for Standardization.

ISO/IEC 11179-6:1997. *Information technology — Specification and standardization of data elements — Part 6: Registration of data elements.* Geneva: International Organization for Standardization.

Scheer, August Wilhelm. 1994. *Business Process Engineering. Reference Models for Industrial Enterprises.* Berlin: Springer Verlag.

Weatherall, Alan. 1988. *Computer Integrated Manufacturing: From Fundamentals to Implementation.* London and Boston: Butterworths.

Wright, Sue Ellen. 1994. "Terminologie als Organisationsprinzip in CIM-Umgebungen." *Akten des Symposions: Terminologie als Produktivitätsfaktor in volkswirtschaftlicher und betriebswirtschaftlicher Sicht.* Bozan-Köln: Deutscher Terminologie-Tag.

7.2.2 Terminology-Based Knowledge Engineering in Enterprises

BERND DÜSTERBECK AND WILHELM HESSER

1 INTRODUCTION

The implementation and integration of computer-aided (CA) technologies on a company-wide level means that technical information management is becoming a factor whose importance should not be underestimated. Recent legislation concerning manufacturer liability and the quality assurance standards ISO 9000 - 9004 place special requirements on the field of technical documentation. Technical information must be collected and maintained in such a way that it can be easily found. This affects not only the strategic re-utilization of design-oriented know-how, but also the retrieval of information relevant to product liability and quality assurance. From the point of view of standardization, a company's object numbering system has until now been the main tool for retrieving technical know-how. (Objects in the system may consist of any item tracked in the system, from detail parts to subassemblies to individual pieces of equipment.) However, it is evident that conventional numbering techniques can no longer keep pace with the increasing complexity of modern products and the growing stream of technical information.

SEE

Quality Management 7.2.3:488

SEE

CIM 7.2.1:467

INFO

ISO 9000 INFB 23:868

2 INDEXING MODEL FOR THE DESCRIPTION OF TECHNICAL INFORMATION IN ENTERPRISES

indexing model

This chapter presents an indexing model designed to point the way towards a future system of coding and identification to accompany the complete process of product development. The indexing model provides a modular description of technical objects using terminological tools to support object specification. The model consists of a vocabulary pool, several terminology systems of varying structure, and also a system of product characteristics. The core of this model is the description of a product group within a loop of autarchic terminology systems based on the product life cycle. At present the following terminology systems are regarded as indispensable:

product group

- terminology system documenting product function
- terminology system documenting product structure
- terminology system documenting product assembly

terminology systems

The modular approach is fundamentally open and allows for implementing additional terminology systems. By introducing a generically structured terminology system, it is possible to describe the company's range of parts and assemblies independent of the product (Figure 1).

Company-specific vocabulary and the concepts it represents are documented in the vocabulary pool, which serves to integrate terminology systems and product characteristics.

vocabulary pool

The modular approach permits a strict separation of tasks. Each terminology system related to the product life cycle illustrates one unambiguously defined aspect of the product or product group. Hence the terminology system only requires limited structural complexity. The terminology systems interact to allow for encoding highly complex technical structures.

modular approach

3 The Indexing Model at the Company Level

Sections of the model presented here were tested as part of an industrial project. The aim of the project was to improve terminology management and at the same time develop a language-assisted classification for the spare parts division of a company within the steel industry. The following components of the indexing model have been tested and installed:

language-assisted classification

- vocabulary pool
- generic system of terminology
- system of product characteristics

With approximately 20,000 items in total, the spare parts range of the company remains relatively manageable and offers sufficient scope to test out the envisaged project method on a small scale.

Spares parts from the different manufacturers converge in the spare parts store. The names of the parts are adopted into the master parts catalog directly from the manufacturers' spare parts catalogs and lists. The terminology employed is therefore extremely heterogeneous. Control of spare parts terminology and the development of a spare parts classification facilitates the following procedures:

master parts catalog

- tracing of parts
- comparison of parts
- monitoring of the parts range

The system is designed to identify and eliminate duplicate and therefore redundant stock holdings. It should also be possible to employ the new spare parts classification in analyzing and controlling changes within the full parts range. From the technical point of view, the system should improve reaction times in the event a machine goes down by facilitating improved recognition and exploitation of opportunities for work-station substitution.

redundancies

Logical spare parts classification should make it possible to achieve a systematic reutilization of stock, thus reducing the workload and related warehousing and administrative costs. Improved data quality should speed up procurement processes and prevent false acquisitions.

improved data quality

3.1 System Implementation

So as to account for the conditions prevailing in the company, the system has been designed in accordance with the "bottom-up" procedure. The terms used to describe the spare parts were documented as part of a vocabulary analysis. A total of 5,000 individual names were recorded using PC-based database management software. The following file structure was used:

vocabulary analysis

- term
- synonyms
- description (i.e., written description) or a relation to standards or drawings)
- spare part identification numbers
- term identification number

In the case of this project, it was impossible to automate the documentation process because of the extremely unstructured file format of the master part catalog. An automated process will be tested during a follow-up project.

master parts catalog

classification elements

The vocabulary recorded during the first stage of the project formed the basis of the language-assisted spare parts classification. The terms and structures used as classification elements were obtained by a cluster analysis of the vocabulary recorded, i.e., directly from the parts range. The internal structure of the parts range has a direct influence on the structure of the spare parts classification.

spare parts classification

All classes of the spare parts classification are represented by class headings. The name of each class indicates directly which parts it contains. In addition to the class headings, the system employees two additional terminology-based classification functions, which are primarily aimed at guaranteeing a considerably higher ease of retrieval than conventional numbering systems. The synonym function incorporates the following into the class heading:

- synonyms (e.g., angle; bracket; support)
- stylistic conventions (e.g., hexnut; hexagon nut; 6-sided nut)
- inversions (e.g., high-pressure hose; hose, high-pressure)

polyhierarchical network

The materials classes are linked with one another in the form of a network reflecting polyhierarchical structures, which enable ambiguous terms and relationships to be detected and assigned to the correct class heading (e.g., key ⇒ typewriter key, Allen key, security key). Classification is based on technical similarity. The lowest level consists of classes that contain a manageable quantity of technically similar parts. If the classes are too large, a more detailed parts description is required, including additional information in the form of "lines" of subject characteristics.

3.2 Retrieval Structures

In a fully developed stage, the system will provide five methods of accessing technical spare parts information:

INFO

Classification INFB 9:334

1. **Access via the classification system**. The user selects the desired branch in the hierarchy and follows the route to the target class. This procedure triggers either a display of the key data in a line of subject characteristics or a direct display of the data from the master catalog.

classification index

2. **Access via the classification index.** The index contains all the class headings, their synonyms and terminology classes in alphabetical order. If users are not totally familiar with the class headings used in the classification, they can use synonyms and

polyhierarchical structures as alternative routes, which will automatically direct them to the correct class heading used in the system.

3. **Independent input of class heading**. In addition to selecting the desired class heading with the cursor or cursor bar, users may also type it in themselves. In this case, too, they will be automatically directed to the hierarchical structure of the spare parts classification. *class heading*
4. **Access via vocabulary pool**. The index contains all the controlled spare part terms and their synonyms. *vocabulary pool*
5. **Independent input of spare part terms**. With approximately 20,000 active parts, the spare parts range has a very broad distribution. Cleaning rags and screws are just as much a part of the range as transformers worth several hundred thousand USD. On the one hand, the parts range displays a high level of heterogeneity and, on the other hand, most groups of similar parts are relatively small. Hence the classification has a broad spread of classes (approximately 130 classes on the highest level) and generally a shallow structural depth (on average two or three levels). The vocabulary designated for naming the parts is to be linked to the classification at the lowest level. The spare parts catalog entries themselves are linked to the spare part vocabulary (Figure 2). *spare parts terms* *spare parts vocabulary*

This system design makes it possible to carry out an integrated program of terminology management from the catalog entry level through to the field of classification.

4 Results and Outlook

During the project period, it was possible to achieve considerable improvements in the spare parts management of the cooperating company. Spare parts stock was reduced by 25% from 20,000 to 15,000 items. Terminology analysis was especially effective in supporting the identification of redundant and unnecessary materials. Spare parts-related terminology employed in the system was reduced by 20% from 5,000 to 4,000 terms. In general, the installed information system enabled the company to realize a highly efficient utilization of the spare parts stock. *stock reduction*

As the pilot project has shown, the indexing model can be a productive tool for implementing terminology-based knowledge engineering throughout the company. A follow-up project is planned that will verify the methods tested by including a considerably higher amount of master parts catalog data together with the respective terminological data. In addition, it is planned to install and test the other sections of the indexing model.

indexing model

The experience gained in these activities is a useful contribution to the development and improvement of terminological knowledge engineering in enterprises. We are sure that the indexing model and the methods utilized will play a significant role in supporting the integration of information within the enterprise, which is one of the key tasks for industry on the threshold to the 21st century.

REFERENCES

Dahlberg, Ingetraut. 1974. *Grundlagen universaler Wissensordnung. Probleme und Möglichkeiten eines universalen Klassifikationssystems des Wissens*. Pullach b. München: Verlag Dokumentation.

Pahl, G, and Beitz, W., eds. 1986. *Konstruktionslehre. Handbuch für Studium und Praxis*. Berlin: Springer.

Scheer, August-Wilhelm. 1990. *CIM. Der computerunterstützte Industriebetrieb*. 4. ed. Berlin: Springer.

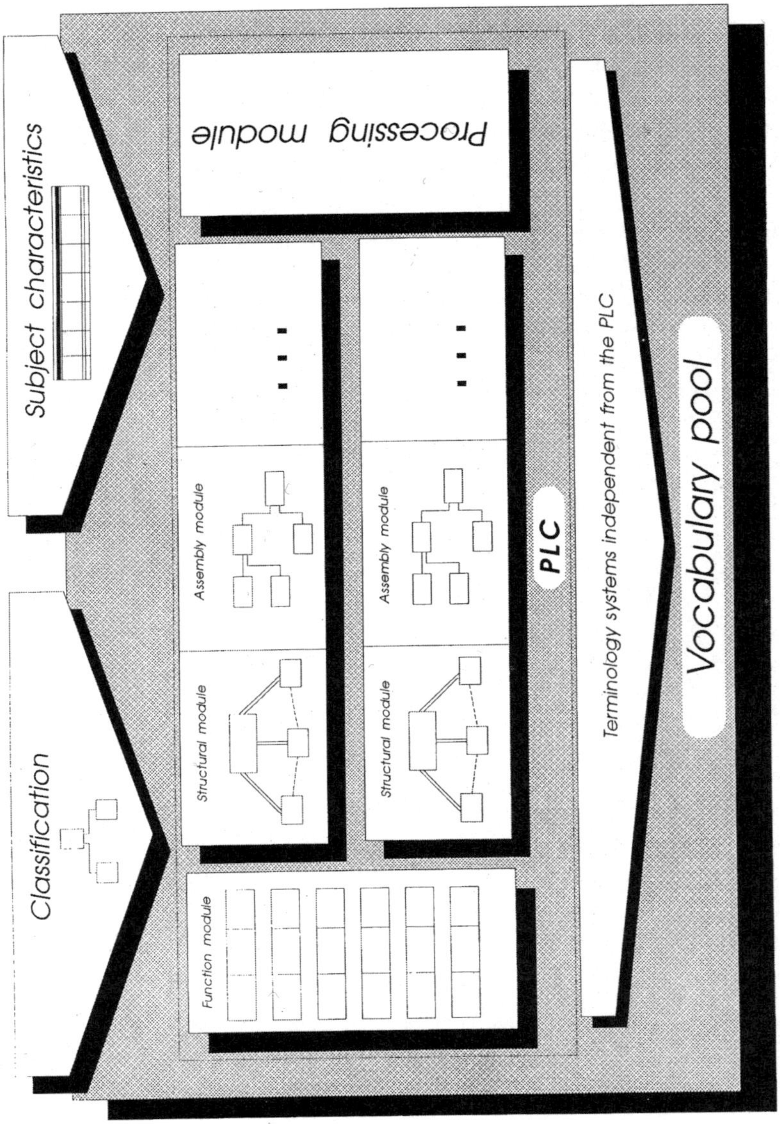

Figure 1: *Integrated Description of Technical Objects*

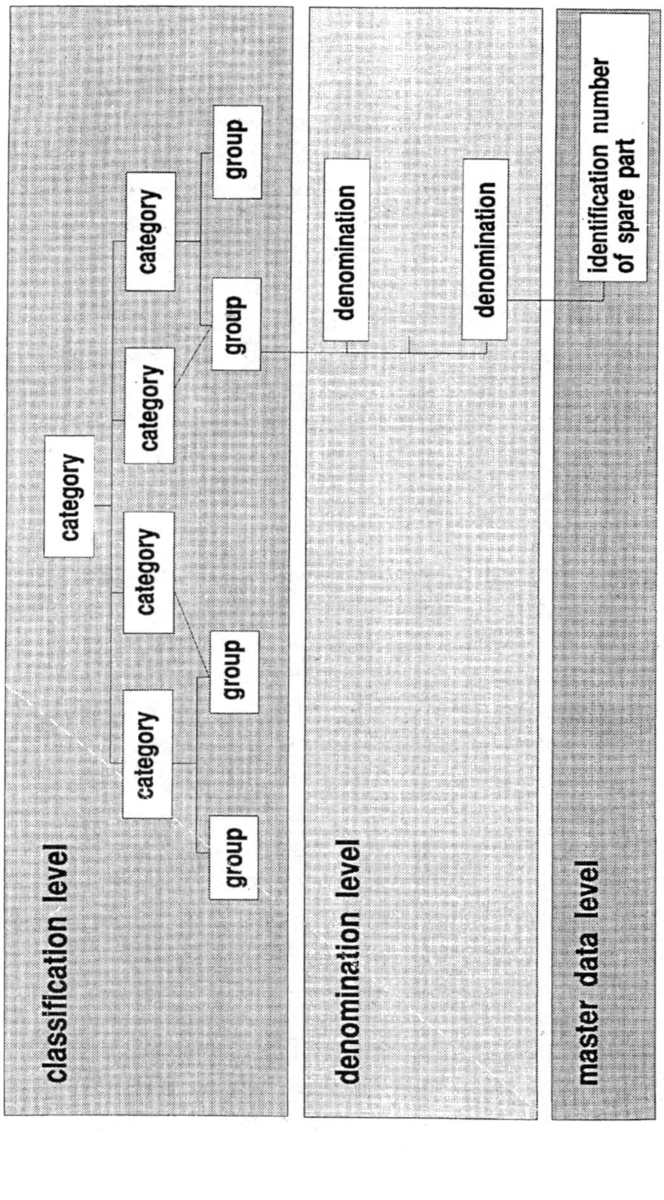

Figure 2: *Levels of Classification Activity*

7.2.3 Terminology and Total Quality Management

SUE ELLEN WRIGHT

1 TERMINOLOGY FOR QUALITY ASSURANCE

Quality Control

Quality Assurance

Total Quality Management

Since World War II, manufacturing industries have expanded their commitment to *Quality Control* (QC). In the last decade, increasing labor and material costs, enhanced consumer sophistication, and above all, global competition have dictated a shift toward *Quality Assurance* (QA). The current phrase is *Total Quality Management* (TQM), which is in essence an outgrowth of the move to Quality Assurance (QA). Other articles in this section treat the role of terminology in increasing industrial efficiency and document accuracy. While one can also speak of QA as applied to terminology itself, this discussion will focus on the role of terminology as a quality-determining production factor in document generation.

1.1 Quality Control

discrepant material

QC focuses on product inspection, either 100% or random sample inspection. If discrepant material (many QC experts are very reluctant to use *defect* or *defective*) is found during *end-item inspection*, the product run is sorted and any non-conforming material (another euphemism for *defective material*) is scrapped. The mission of this approach is to prevent at least all but a statistically acceptable number of defective parts from reaching the customer. Obviously, QC can involve considerable waste in worker-hours and material that have been invested in products that are later destroyed or reworked, and internal tensions are bound to crop up between departments due to conflicting commitments to meeting company quotas on the one hand and assuring quality on the other.

document production

The intention of this chapter is to compare document production procedures to those for industrial products. In order to simplify the train of thought, this discussion will treat translation as one type of document production, although many of the steps described in this context are relevant to other types of technical communication.

end-item editing

The translation procedure that is analogous to the end-item inspection cited above is the final editing process practiced by most reliable translation service bureaus and departments (Pilgrim 1991; Ganser 1989). Although efforts are made to ensure high quality through careful selection of translators, the primary quality focus is on final editing. Feedback to translators may improve future quality and is ideal, but it is not a universal practice.

1.2 Quality Assurance

prevention vs. detection

A commitment to *Quality Assurance* manifests itself in quite a different way. QA focuses on assuring product quality from the initial stages of design and development throughout the materials sourcing process to product manufacturing and shipment. End-item inspection remains a step in the process, but it is no longer the primary function of the system. Not only is quality monitored throughout the production cycle, but design features are also strictly analyzed at every stage of production to determine the probability of product failure due to error introduced at that point. Rigorous testing determines unforeseen causes for error, and equipment and methodologies are introduced to reduce this risk.

inspection philosophy

In contrast to the specialized Quality Control inspectors who are independently responsible for quality under a QC regime, QA enlists everyone who has any role in producing the product or in supporting its production. The goal is not just to prevent non-conforming material from being shipped—it is to guarantee that no discrepant product is manufactured in the first place.

QA procedures are designed to:

QA objectives

- Provide customer safety, for no company will willingly expose itself to potential liability claims resulting from product failure.
- Ensure product reliability, for customers will not repeatedly purchase products from any company whose products do not perform as advertised or as anticipated by the end user.
- Cut production costs, for no company can afford the inefficiency and financial losses inherent in scrapped materials and wasted worker-hours.

terminology and QA

The analogy has already been drawn between QC end-item inspection and the final editing pass performed when producing a quality translation product. This chapter will demonstrate that the determination, documentation, and effective application of terminology play a pivotal role in document production. This role is in many respects comparable to the step-by-step efforts introduced in industry at every stage of the production chain to prevent product failure or deviation from specified standards.

1.3 Total Quality Management

As noted above, TQM represents a direct outgrowth of the Quality Assurance movement. QA, especially within the framework of the ISO 9000 family, depends on complete commitment to quality principles at every level of an enterprise and throughout the complete life cycle of every product or service (including documents) that the enterprise produces. By the same token, terminology also reaches into every level of activity and thus constitutes a vital link in the total quality chain.

INFO

ISO 9000
INFB 23:868

2 Reliability as a Function of Repeatability and Reproducibility

Whether we talk about QC, QA, or TQM, product *reliability* is the ultimate objective. The *Compilation of ASTM Standard Definitions* defines *reliability* as "the probability of performing without failure a specified function under normal conditions for a specified period of time" (ASTM 1990). In manufacturing and specifically in gauge control, reliability can be expressed as a function of *repeatability* and *reproducibility*. A gauge or a process must yield the same results over a reasonable amount of time. It is critical that the same operator using the same gauge to measure the same part again will obtain the same reading (repeatability), and that a different operator using a different, but analogous piece of equipment will obtain the same reading when measuring a part manufactured to the same specifications (reproducibility).

reliability

Commonly called *Gauge R and R*, these principles can also be extended to entire processes. *System capability* can be defined in a highly precise, statistical fashion for systems that involve quantifiable values, but in a broader sense, *capability* describes a process that is capable of producing high quality results with near zero defects, a distinction that can be reasonably applied to document production as well.

system capability

text reliability

The analogy to translation or text production in general is, of course, a metaphorical one at best. The narrow tolerances demanded of a manufactured part are difficult to define for texts, although advances in the area of Controlled English and Controlled Language in general are creating environments involving a high degree of quantifiability. Despite these differences, repeatability and reproducibility are nevertheless critical factors in determining the reliability of texts. Here, too, we have to ask whether the same "operator" (i.e., translator) can replicate his or her results over time, or can a different operator produce the same degree of accuracy, either simultaneously or over a period of time? These are questions that plague both the individual translator doing repeat business with the same client and translation teams doing large volumes of work in team situations.

INFO

Controlled Language INFB 25:872

3 Significant Characteristics

inspection characteristics

Quality Assurance, like terminology, is concerned first and foremost with the *characteristics* that distinguish one object from other objects in its class and that must be tested in order to ensure proper performance. Of special concern are those properties of the product that can lead to customer injury (*critical characteristics)* or to seriously impaired product performance (*significant characteristics*). When manufacturing industrial objects, engineers devise a full range of tests and inspection procedures designed to check these characteristics. The goal is not just detecting discrepant material, but also preventing its incidence at any point in the production sequence.

language and content

In trying to maintain our analogy with the translation process, it is instructive to consider what the significant characteristics of a text actually are. Hohnhold implies that they are *language* and *content* (Hohnhold 1990:19). How do language and subject matter manifest themselves in a text? How can the translator (or the technical writer) recognize those elements of the text that are likely to create critical or significant problems if the text does not "function" according to the criteria demanded by the specifications set down for the text?

risk of failure

If we can assume that the translator or technical writer producing a technical text is a master of the general language in which the text is written, we can then conclude that the terminology that makes up the text comprises that aspect of the text that poses the greatest risk for failure.

4 Optional Methodologies

terms & texts

The premise that terminology reflects the critical and significant criteria inherent in the text implies two basic principles:

- Each term constitutes a significant (or potentially critical) characteristic that must be correctly rendered if the text is to function as intended.
- Each specialized term represents a potential departure from general language and hence a potential problem for the translators or writers that they must solve in order to assure capable performance as defined above.

probability of failure

The short text shown in Figure 1 is designed to illustrate the relation between terminology and text. The text itself describes features of a statistical process control chart. The underlined items reveal that the terms used in the text constitute over a third of the body of the text. It would, however, be presumptuous to postulate any kind of ostensibly objective statistical prediction of *probability of failure* based on this kind of value. Probability of error must account for the translator's prior knowledge of the subject matter, research sophistication, the availability of reliable resources (both human and printed), as well as the recurrence of terms in the text.

lexical approach

On the other hand, the *possibility* of error (and of associated product failure) is great indeed. Left to their own devices, our hypothetical translators may follow one of the following methodologies:

- Consult with a standard technical dictionary and try to calque any unfamiliar terms based on a general knowledge of the basic term elements involved (the lexical approach).
- Consult specialized monolingual glossaries or other source material (parallel and secondary texts) to find functional equivalents for the problematic terms that are commonly used among experts working in the target language (terminological approach).

terminological approach

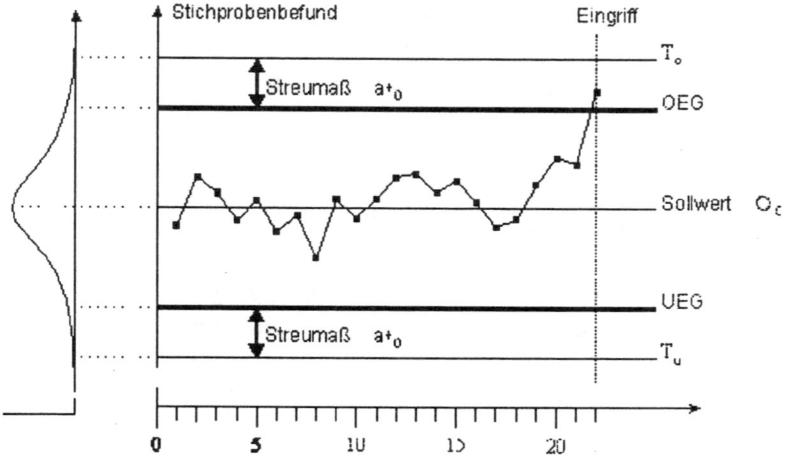

Berechnen der Eingriffsgrenzen

Bis zur Schulung des Bedienungspersonals werden die Eingriffsgrenzen nur von QS-Personal neu berechnet. Das Programm ist so auszulegen, daß eine Berechnung der Eingriffsgrenzen möglich ist, aber nur durch einen Berechtigungscode tatsächlich durchgeführt wird. Bei Regelkriterienverletzungen wird das Bedienungspersonal nicht die Möglichkeit eröffnet, neue Eingriffsgrenzen zu berechnen. Generell gelten auch für Stichproben als Eingriffsgrenzen die 2/3-Mittentoleranz. Bei einseitigen Toleranzen sind die Eingriffsgrenzen mit der QS abzusprechen.

Calculating the *Eingriffsgrenzen (?)*

[Until the operators have been trained, only QA personnel are allowed to recalculate the *Eingriffsgrenzen*. The program is to be designed so that it will be necessary to enter a password before calculating the *Eingriffsgrenzen*. The operating personnel should not have the option to recalculate the *Eingriffsgrenzen* if the control criteria are violated. The 2/3 average tolerance is generally also used as the *Eingriffsgrenzen* for random samples. In the case of one-sided tolerances, the *Eingriffsgrenzen* must be discussed and agreed upon with Quality Assurance.]

Graph call-outs: *Random Sample Findings; Intervention; Spread; Specified Value*

Figure 1: *Eingriffsgrenzen*

4.1 The Text as a Source of Information

methodology

Without trying to propose any predictive values, let us examine just one term from the possible terms identified as *significant characteristics* and project the methodology that a translator might adopt in attempting to produce the translation product.

statistical significance

INFO

Text Designations INFB 22:867

Before searching for information in any other resource, the first thing to do is to learn as much as possible from the text itself. Statistically speaking, we can see (give or take a few marginal terms that may not be considered to be technical) that roughly one third of the text is composed of terms or, put differently, that one third of the potential characteristics in the product are significant. The term *Eingriffsgrenze* occurs in the title and four additional times in the passage, representing in itself 7% of the total text. Furthermore, this is not only the most prominent term in the text, it is also one of the most difficult to decipher for an uninitiated reader. If not critical, it is at least clear that this is the most significant term in this particular text chunk. Given the frequency with which it occurs and its potential difficulty, it poses the greatest risk of error (failure) for the overall function of the text.

textual information

Reading the macro-text (not shown here) reveals that the general topic of the text is Statistical Process Control (SPC). Secondly, something (*Eingriffsgrenzen*) must be calculated, not just by anyone, but by an authorized person (someone with a *Berechtigungscode*—an *authorization code* or *password*). *Operators* are not initially allowed to do this. Thirdly, the values to be calculated are related to *random samples* and to *tolerances*.

text-imbedded information

In the graph accompanying the sample, the abbreviations *OEG* and *UEG* appear with *Eingriffsgrenzen*. *EG* for *Eingriffsgrenze* makes sense, but what are *O* and *U*? The text does not expand the full form, for the author assumes that the reader knows them. Examination of Figure 1, however, clarifies this issue: *OEG* appears at the top of the graph and *UEG* at the bottom, and both occur equidistant from T_o and T_u respectively, which the translator may know represent *obere (upper)* and *untere (lower) Toleranz*. This suggests that *O* and *U* refer to the *obere* and *untere Eingriffsgrenzen*, which appear as values on a graph and have a relation to the respective tolerances. These conclusions are textual, are supported by the graphics, and are for the moment limited to the conceptual level without reference to potential English equivalents.

Die *Eingriffsgrenzen* werden bei erstmaliger Fertigung einer Teilenummer auf einer Bandlinie manuell nach Angaben von QS-Personal eingegeben. Es werden von der QS festvorgegebene *Eingriffsgrenzen* eingegeben. Nach den ersten 10 Stichproben werden neue *Regelgrenzen* berechnet. Auf dem Bildschirm wandern die alten Eingriffsgrenzen auf der linken Seite mit jeder Stichprobe aus den Bildschirm und von der rechten Seite aus werden die neuen *Regelgrenzen* eingezeichnet.

[When any part number is run on an assembly line for the first time, the *Eingriffsgrenzen* are input according to instructions from QA personnel. QA inputs fixed default *Eingriffsgrenzen*. After the first 10 random samples, new *Regelgrenzen* are calculated. On the screen display, the old *Eingriffsgrenzen* move to the left side with each random sample and disappear from the screen, and the new *Regelgrenzen* are drawn in from the right side.]

Figure 2: *"Eingriffsgrenzen" and "Regelgrenzen" (Unresolved Terms)*

outside information sources

Further study of the text (Figure 2) reveals that other *Grenzen* occur in the document as well, specifically *Regelgrenzen*. The graph and the description in the text indicate a close relationship between the two, but the difference between them remains unclear. At this point, the text has yielded up about all the information available. The translator must now seek further knowledge from outside sources to solve the terminology problems.

4.2 The Lexical Problem-Solving Approach

lexical equivalents

The lexical approach to terminological problem-solving promises the translator ready-made equivalents for source-language terms. As succinctly stated by Zgusta (1984:147),

> ... the dictionary should offer not explanatory paraphrases or definitions, but real lexical units of the target language which, when inserted into the context, produce a smooth translation...

term components

Translators seeking their ready-to-use equivalents at the time this text was originally translated would have found that *Eingriffsgrenzen* did not exist in the major collections. Ernst, for instance, does not list the term, but does offer the following possibilities for the word *Eingriff*:

Eingriff *m* (Masch)/contact, gearing || Eingriffnahme *f* /intervention Eingreifen *n* (Getriebe) / catch[ing], gearing, engagement || (Zahnhöhe) (Uhr) depth || ~**e** *m pl* interference

The only subject fields addressed in the Ernst are *mechanical engineering* and *gearing*, so the cautious user will question these items. Among the compounds that follow, the only term that sounds remotely related might be *Eingriffslinie* (line of engagement), but the subject reference (*Zahnrad*) (gear) rules out this option. Kučera, however, provides a potentially interesting entry:

Eingriff *m* (z.B. seitens der Bedienungsperson)/intervention

The mutual presence of *Bedienungspersonal* (operating personnel) both in the subject reference and in our text is promising. To continue the process of componential analysis, examination of the second element in the compound yields the following options:

componential analysis

Ernst:
Grenze *f* (Staats- usw) / frontier, border || Rand *m* / margin || Beschränkung, Begrenzung *f* / limitation || (Bergb)/border of a claim, boundary || (ALGOL) (DV) bound || (Math)/limit, bound
Kučera:
Grenze *f* / boundary *n* || s. Rand || (which yields: edge, border, lip, margin)

In terms of mathematical possibility, there are (7 x 9) potential options offered by our dictionary study, unless we entertain the notion that the term elements just might reverse themselves (which they sometimes do), in which case the product is (2 x 7 x 9):

statistical probability

Eingriff	*Grenze*
contact	frontier
gearing	border
intervention	margin
mesh	limitation
engagement	boundary
depth	bound
interference	limit
	edge
	lip

If, as some empirical studies have indicated, dictionary users tend to go for the first or possibly the second item in the dictionary entry (Hartmann 1987:16), our hypothetical translator might be prone to opt for *contact frontier* or *gearing border*. We have, however, visualized this translator as a competent person who understands how specialized dictionaries are structured and has some knowledge of the subject matter. A rank novice might opt for an implausible solution, but seasoned professionals will avoid incongruent selections, linking dictionary cues to their own experience using terminology. Thus mathematical probability is only marginally related to the translator's probable choice because, although the process is not precisely defined, it is certainly not random.

implausible solutions

Experience shows that *contact, gearing, mesh,* and *depth* are too specifically related to gear mechanisms to qualify in this context. Although *engagement* and *interference* are also often identified with gearing, they also have broad applications in many other areas. This narrows options to *intervention, engagement*, and *interference*. Reversion to the initial *Bedienungspersonal* link from Kučera makes *intervention* look like a very good option.

componential analysis

The alternatives for *Grenze* are harder to sort out. The astute translator will discard *frontier* and *lip* as implausible options in this context. *Bound*, aside from occasional usage such as *out of bounds* and *without bound*, is rarely used as a noun in modern English. Unfortunately, the other potential choices have merit. Let us then endow our translator with the good sense (and the luck and the time) to take a broader look at the way *Grenze* performs when combined with other term elements. Unfortunately, hard-copy dictionaries do not allow us to search for imbedded or terminal elements as such, but good lexicographers do frequently include such information in their entries, and the CD-ROM versions of dictionaries like the Ernst allow for truncated searches.

permuted lookup

In Ernst's full page of *Grenz-* and *-grenze* compounds, we find "GMEQ (Stichprobe) / average outgoing quality limit, AOQL." This is not our term, but it does document that *Grenze* means *limit* in conjunction with quality sampling. Further scanning of *Grenz-* compounds reveals a collection of instances having to do with mathematics, inspection, and quality control in which *Grenz-* represents a *cut-off*, a *boundary*, or a *margin*, but let us again give our translator the benefit of a doubt and allow him or her to decide that the *Eingriffsgrenze* is an *intervention limit*. If the requester of the translation is anyone other than an SPC expert and conducts a quality control edit of the text, this solution may very well pass

plausibility standards

through to the customer because it matches critical plausibility standards built into good editing: based on the graph, we can assume that if the values reach this limit, the operator must *intervene* in the process to prevent violation of the tolerance limits.

4.3 The Terminological Problem-Solving Approach

Despite the plausibility of the solution reached above, there are those who believe that terminology management begins where the dictionary leaves off. Both Hohnhold and Dubuc stress the importance of comparative textual analysis (Hohnhold 1990: 64ff.; Dubuc 1985: 65). Hohnhold lists a wide range of text types that are used for textual documentation, ranging from monolingual subject-specific books, technical journals, research and development reports, product documentation, and specialized technical standards. Today we should add specialized web resources. Monolingual lexical resources round out Hohnhold's list.

comparative textual analysis

If we turn to Hofmann's *Meßwesen* (1980:134), a monolingual source of German Quality Assurance terminology, we find that *Eingriffsgrenzen* refers us to *Kontrollgrenzen*:

> **Kontrollgrenzen**—*Regelgrenzen, 3σ-Grenzen, Eingriffsgrenzen.* Für eine bestimmte Kontrollkarte charakteristische Grenze eines Bereichs, in den aus den Stichproben berechnete Kennwerte mit einer bestimmten statistischen Sicherheit fallen, wenn der Prozeß in statistischer Kontrolle ist; Linie auf der Kontrollkarte, die die zulässigen Abweichungen, des zu regelnden statistischen Parameters in den Stichproben oder Proben begrenzt.
>
> [... For any given control chart, a characteristic {boundary? margin? edge? limit?} of a range for which there exists a specific probability that the characteristic values calculated for random samples will fall if the process is in statistical control; line on a control chart that {borders, limits; forms a boundary/margin/edge} of the permissible deviations for the statistical parameters being controlled in the random samples or tests.]

This entry equates *Kontrollgrenzen, Regelgrenzen,* and *Eingriffsgrenzen* as synonyms. Thus the interplay between the two terms in Figure 2 is an example of purposeful stylistic variation by the German writer, which can confuse an English translator trained to use consistent terminology throughout a technical text.

synonyms

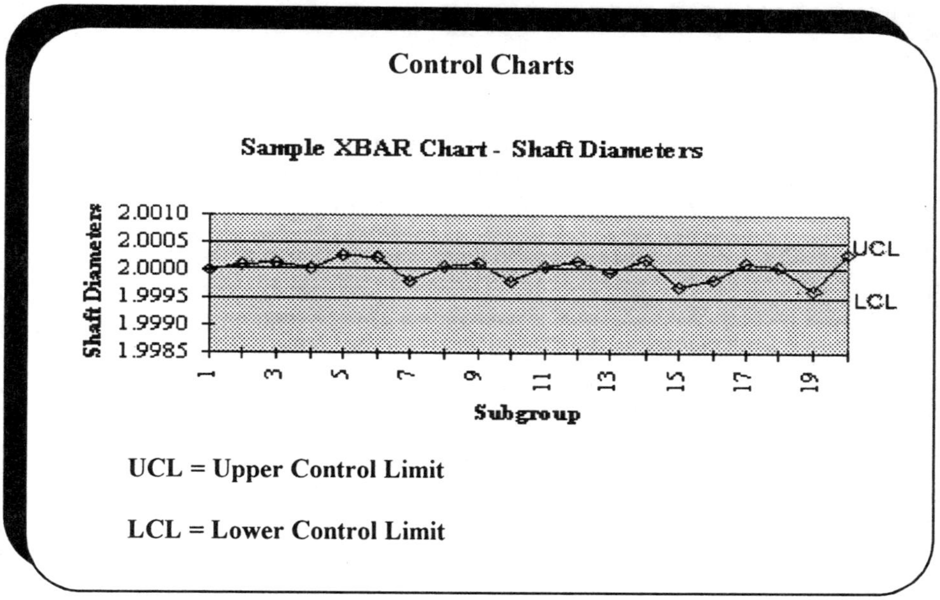

Figure 3: *English Control Chart Showing Control Limits*

control limit

Having identified the relationship between the terms in German, the next step is to locate the appropriate English equivalent. An English "hidden glossary" in an automotive industry publication reveals a definition for *control limit* (Ford 1983), accompanied by the chart shown in Figure 3, where the *UCL* and *LCL*, expanded here as *upper* and *lower control limit* parallel the *OEG* and *UEG* call-outs found in the German text. Further exploration in Juran (1988:24) and VDA materials (1986:66) substantiate these findings. The prominence of *Eingriffsgrenzen* in VDA materials simply underscores a difference in usage between engineers working in the former West and East German automotive industries.

Web search

The scenario described above recounts a terminological search that was conducted before the ready availability of resources on the World Wide Web. Although access to the hardcopy materials described would be extremely useful and would serve to further corroborate the findings of any search, a Web search for *Eingriffsgrenze* immediately turns up the University of Hannover's excellent website for the Mechanical Engineering Department's course in statistical process control (Hannover 1999). Finding the

Subject Field: QA

Lang: de **Term:** Eingriffsgrenze
Synonym[1]: Regelgrenze **Synonym[2]:** Warngrenze
Synonym[3]: Kontrolgrenze **Synonym[4]:** 3σ-Grenze

Definition: Linie auf dem Arbeitsstreifen der Kontrollkarte, die die zulässige Abweichung des zu kontrollierenden Parameters (der Charakteristik) gegen den zu erwartenden Wert abgrenzt. Diese Grenze ist ein Kriterium für die Qualität des Erzeugnisses sowie die Genauigkeit und Stabilität des technologischen Prozesses auf Grund der Lage der Merkmalswerte bezüglich des Verlaufs der Linie. **Source:** HOFMANN 1980

Context: Ist ein Prozeß beherrscht, so müssen sich alle Datenpunkte, annähernd einer Normalverteilung, innerhalb der zugehörigen Eingriffsgrenzen bewegen und ungefähr die gleiche Anzahl Punkte oberhalb, also auch unterhalb des zugehörigen Gesamtmittelwertes fallen. **Source:** VDA 1986

Superordinate Concept: Kontrollkarte
Subordinate Concepts: obere Eingriffsgrenze, untere Eingriffsgrenze

Lang: en **Term:** control limit

Definition: A line (or lines) on a control chart used as a basis for judging the significance of the variation from subgroup to subgroup. Variation beyond a control limit is evidence that special causes are affecting the process. Control limits are calculated from process data and are not to be confused with engineering specifications. **Source:** FORD 1983

Context: A central line is drawn on each [control chart] at the grand average $\bar{\bar{X}}$ and at the average range $\bar{R}$. Control limits are established at the average ± 3 standard deviations. These are the upper control limit (UCL) and the lower control limit (LCL). **Source:** JURAN 1988

Superordinate Concept: control chart
Subordinate Concepts: upper control limit, lower control limit

Figure 4: *Terminological Entry*

appropriate English equivalents is not quite as easy since the researcher does not know the precise term to search. Searches on "SPC" and "control chart" actually turn up useful information, but the solution is not as directly apparent as it was in the hardcopy materials used above. Once a suspicion arises that "control limit" might be a solution, however, specific searches produce excellent materials, notably Harvill 1997 and Busitech 1999.

4 Term Reusability

retrieval

If length of discussion is any indication, it took no less time to arrive at a plausible (but nonetheless incorrect) solution using multilingual dictionaries than it did to find the correct solution using monolingual parallel texts. Figure 4 consolidates the information collected during the research phase into a terminological entry. It equates *Eingriffsgrenze* with *control limit* and lists all the synonyms found. Definitions and contexts were revealed during the search, and the inclusion of superordinate and subordinate concepts positions the concept in its subject field. This entry, if recorded in a terminology management system, provides for ready retrieval of the information so that the same translator can ensure using the same solution in all ensuing translations, and other translators working in a team can also access the same information for their work, thus ensuring repeatability and reproducibility. This scenario has been designed to demonstrate how terminology management practices provide quality assurance support for multilingual document generation.

5 Quality Assurance for Terminology Management

INFO

Quality Checklist INFB 24:869

The focus of this chapter has been on documenting terminological information for the purpose of assuring quality in the document production chain by validating the equivalents used for translation and multilingual technical writing. Another important aspect of quality control and assurance involves the validation of the data entry process itself. For this purpose, quality manuals and checklists can be created as guidelines for auditing terminology management methods and practice.

References

ASTM Committee on Terminology. 1990. *Compilation of ASTM Standard Definitions*. 8th Ed. Philadelphia: ASTM.

Busitech. 1999. Quality Glossary. http://www.busitech.com/glossary.htm. Accessed 1999-11-14.

Dubuc, Robert. 1985. *Manuel pratique de terminologie*. 2e édition, revue et augmentée. Quebec: Linguatech.

Ernst, Richard. 1989. *Dictionary of Engineering and Technology, Vol. 1: German-English, 5th ed.* Wiesbaden: Brandstetter.

Ford Motor Company. 1983. *Continuing Process Control and Process Capability Improvement: A Guide to the Use of Control Charts for Improving Quality and Productivity for Company, Supplier and Dealer Activities.* Detroit: Ford Motor Company.

Ganser, Doris. 1989. "Exploitation by Translation Agents." *ATA Chronicle*, XIX, 3, 21-23.

Hannover: Universität Hannover, Fachbereich Maschinenbau. 1999. *Statistische Verfahren des Qualitätsmanagements.* http://130.75.138.31/vorlesung/qs1/KAP05/k05b.htm . Accessed 1999-11-14.

Hartmann, Reinhard. 1989. "Lexicography, Translation and the So-called Language Barrier." Mary Snell-Hornby, Esther Pöhl, and Benjamin Bennani, eds., *Translation and Lexicography*. Papers read at the EURALEX Colloquium held at Innsbruck 2-5 1987. A Special *Paintbrush* Monograph. Amsterdam: John Benjamins, 9-20.

Harvill, Jane L.1997. *Lectures for Statistics 414/514; Statistical Quality Control.* http://129.252.51.247/sealand/qualityg.htm ; accessed 1999-11-14.

Hofmann, Dietrich; Meinhard, Regina; and Reineck, Horst. 1980. *Meßwesen, Prüftechnik, Qualitätssicherung: Begriffe und Definitionen.* Berlin: VEB Verlag Technik.

Hohnhold, Ingo. 1990. *Übersetzungsorientierte Terminologiearbeit: Eine Grundlegung für Praktiker.* Stuttgart: Intra.

Juran, J.M., and Gryna, F. M. 1988. *Juran's Quality Control Handbook.* Fourth ed. New York: McGraw Hill, 1988.

Kučera, A. 1982. *The Compact Dictionary of Exact Science and Technology, Vol. II: German-English*. Wiesbaden: Brandstetter.

Pilgrim, Sarah H. 1991. "Letter to the Editor." *ATA Chronicle*, XX, 6.

VDA. 1986. *VDA Qualitätskontrolle in der Automobilindustrie: Sicherung der Qualität vor Serieneinsatz*, 2. grundlegend überarbeitete Ausgabe. Frankfurt: Verband der Automobilindustrie e. V. (VDA).

Wright, Sue Ellen. 1991. "The Role of Terminology in Translation-oriented Quality Assurance." *Horizons: Proceedings of the 32nd Annual Conference of the American Translators Association.* Medford, NJ: Learned Information, Inc, 227-246.

Zgusta, Ladislav. 1984. "Translational Equivalence in the Bilingual Dictionary." Hartmann R.R.K., ed.. *LEXeter '83 Proceedings: Papers from the International Conference on Lexicography, Exeter, September 1983* (Lexicographica. Series Maior 1), Tübingen: Niemeyer, 147-154.

7.2.4 Project-Integrated Terminology Management for Technical Writing and Translation

GERTRUD GRAUBART CHAMPE

1 INTRODUCTION

The following case study documents the evolution of terminological working methods at the University of Iowa Translation Laboratory. The Laboratory engages in practice, teaching, and research in translation, terminology, and translation pedagogy. The core of the Laboratory's activities is a full-spectrum translation service that concentrates on best practice rather than volume and serves as a test bed for procedures in translation project management.

The methods used in terminology management in a translation environment vary considerably, depending on the purpose for which the translation is being done. When the work done by pragmatic (non-literary) translators is classified by text type, four classes and four main approaches can be distinguished:

SEE
Technical Translation 2.1.1:147

SEE
Term Formation 1.1.1:25

INFO
Text Designations INFB 22:867

- *Documentation of industrial equipment and processes:* This activity requires emphasis on detailed source language (SL) terminology, an understanding of the process involved, and collaboration with the client, who is usually an exporter. It is often necessary to give careful consideration to the need to coin new terms because different languages require different approaches to the formation of neologisms. Maximal collaboration, preferably in the form of draft review with the end user, is desirable because a terminology probably already exists in that environment and must be respected if the translation is to be considered of high quality.
- *Use of parallel texts:* The translation of expository texts such as scholarly and general articles, political reports, or monographs requires extensive use of parallel texts in the target language (TL). Parallel texts can be texts that deal with the same or similar subject matter, in which case they can provide

excellent sources of technical terminology and of collocational usage. They may also take the form of texts that illustrate the same or similar text type in the TL (i.e., product literature, users manuals, etc.), in which case they provide stylistic cues. Besides preventing error, this kind of research ensures that the translator does not reduce to paraphrase concepts that the end reader is accustomed to encountering as well-defined terms.

- *Technology transfer:* When preparing texts for use in developing areas, terminologists must be aware of the area's language planning practices and to act in accordance with them, both in terms of linguistic standards and in the area of creating neologisms. This is particularly true in locales that are still developing technical terminlogies in their language(s) following periods of colonial rule, which may even entail the avoidance of internationalisms.

SEE *Multilingualism 3.1:245*

- *Evolving technologies:* When documents are based on research results, especially when that research dates from an earlier time, particularly a time of innovation or transition in the relevant domain, it is important to use parallel texts in a number of languages and disciplines so as to identify emerging terms accurately. These considerations also apply for currently emerging technologies, where several competing strains of development may exist, each with its own peculiar terminology. Such lack of standardization and harmonization frequently characterizes the development of individual projects within enterprises, such that it becomes necessary to document the terminology used over the course of short periods of development time or among professionals working in different functions within the same work group or enterprise.

emerging technologies

The present chapter will concentrate on the first category, specifically, the translation from English to some other language of documentation for machinery and equipment intended for export. Terminology management for such projects involves a range of operations requiring collaboration between translators and clients. Despite the time constraints involved in such projects, effective action develops slowly and there is a widespread need for education on both sides. Ample evidence exists that it is not only the client who must be educated. Frequent queries to

client/ translator education

various terminology listserves demonstrate that many translators are not isolating and defining concepts before searching for target language equivalents. Agencies, bureaus, and localization services often tend to supply their freelance translators with bilingual word lists rather than *terminological* support, and manufacturers vary greatly in their willingness to discuss SL terminology. Many of them are reluctant to approach the foreign customer for consultation regarding TL terminology.

just-in-time terminology

The development of a procedure for just-in-time delivery of reliable terminology has become a principal focus of activities in the Laboratory. Since outside cooperation is critical but unpredictable, this procedure aims at maximal self-sufficiency. The present report chiefly concerns the considerations involved in preparing translations of long, highly technical documents, such as the manuals related to a complex piece of machinery. It also touches briefly on the process of developing terminological information for third-party translation groups.

"blind" translation

The realization that system, order, and initiative in the management of terminology are essential for technical translation became a concrete reality for the working group as a result of experience. Early in the Laboratory's existence, the technical writer of a neighboring manufacturer submitted a short list of machine labels for translation. The writer, inexperienced in technical work, was reluctant to let us inspect the machine to which the labels would be affixed because he did not understand what was to be gained by an exercise that would undoubtedly cost him time and money. The group, on the other hand, showed its ignorance by agreeing to translate the labels blind. Two years later, the writer submitted a manual for translation for a machine similar to the first one. It was then that it became clear that a *brake*, in the context of this machine, is not a mechanism for stopping or slowing something down, but a mechanism for bending sheet metal. This sobering revelation was instrumental in enforcing the need to develop sound terminological practices.

2 Source Language Terminology

The first task in preparing to translate a text is to work toward absolute clarity in understanding the SL document. This can be a task of significant size and scope, since most clients have not developed a system of terminology management within their own companies. Resulting ambiguities, redundancies, and proprietary uses of language can be a major source of translator error. The work of clarification must be performed within the manufacturer's time requirements and without excessive imposition on the time of the manufacturer's staff. Clearly, this effort is more likely to run smoothly if the client-translator relationship is a long-standing one. Failing this, it is important to have quality assurance procedures in place and to articulate this process effectively to new clients. The following protocol has evolved for large industrial projects:

terminological clarification

2.1 Technical Orientation

The translator and terminologist are introduced to the equipment being documented by a member of the manufacturer's publication department. This step is obligatory; if it is necessary to collaborate with a translator from another city, his or her expenses are paid for visiting the factory.

2.2 Labels and Screens

By request of most manufacturers, the first material to be translated is anything that stays permanently with the machine, such as labels or any computer screens related to control automation or error messages because these items are needed early in the production process. Although some translators would prefer to handle long descriptive materials first, this order of events can actually be favorable for the terminologist and the translator. The research required for a successful translation of these items lays the foundation of a term collection for the project and sets the theme for a bird's eye view of the machine and its functions.

integrated text

Later, once the full text of the manual is in hand, an electronic search can locate all mentions of each label in the text of the manual. On the basis of this information, the work order, and communication with the client's publications department, a terminological record is prepared for each label, containing the following data categories:

- *Part number*
- *Text of label*
  If the label is for a switch or selector, all possible positions must be listed.
- *Location* of the item on the machine
- *Object identified*: a switch, indicator light, etc.
- *Function* of the entity labeled, e.g., mention of such information as whether actuating the switch a second time has a different effect from actuating it once
  For languages in which ON and OFF are not as broadly applicable as in English, it will be important for a translator to know what is being turned on—water, gas, compressed air, electricity, or a machine function.
- *Contexts*
  Attestation of the term in parallel texts, such as a competitor's literature or more general literature, such as industry journals or standards-related materials
- *Source* of parallel texts
- *Equivalents* of the term in other languages
  This information is usually provided at a later stage.
- *Source* of the foreign language equivalent, e.g., name of an informant or citation of a text
  Particular note is made if the translation equivalent is generally usable or, as occasionally happens, applicable only to one client.

SEE *Data Categories 8.1.2:552*

Careful reading of the manual usually provides information for the data categories in the terminological entry. Information can also be obtained from the manufacturer's technical writing department. Questions to this department should be submitted in writing and responses suitably stored.

Screens for the control system and for error messages are processed in analogous fashion, with special attention given to defining function, which is necessary because the language of these items is terse and often arbitrary. The train of thought in the SL leading to certain nomenclature may be quite foreign to the TL and what is more, many screen phrases tend to be proprietary and impossible to translate literally. The translator must be provided with an unambiguous explanation.

2.3 Text of the Manual

Terminology documentation in the SL continues based on the text of the manual. At this point, the terminologist already has some understanding of the machine's physical layout and function. Terminological entries are created for items in the text. Entries are required even for items that have already been mentioned in entries for labels or screen messages, with attention given to their use in collocations. In addition to all the information already mentioned, it can often be advantageous to scan in line drawings, which provide positive identification of a part or an assembly in the machine.

SEE
Term Entries 8.1.2.572

After terms have been extracted from the text of the manual, there are many ways to obtain information about them. It is increasingly possible to find these terms in English on the World Wide Web, using a combination of search engines and directory queries. Information retrieved by such a search can provide explanations, definitions, descriptions of function and, what is particularly important when looking for or coining TL equivalents, synonyms in the SL. However, the final, and most authoritative source is always the cooperative client.

SEE
Term Extraction 8.4.1:725
Internet Resources 8.5:845

In addition to being examined for terms, the text must be searched for potential sources of translator error. These fall into several categories and can never be eradicated entirely, but once they are understood, they no longer constitute traps for the translator:

stylistic precision

- *Style*: It is very likely that the translator, almost always a non-native speaker of English, will have some degree of difficulty reading the documentation for a piece of equipment. This results largely from a difference between the translator's training and that of the technical writer. The technical writer has been trained to achieve precision in ways such as extensive naming, intentional redundancy, and the use of very specific industry jargon—devices meant to avoid a false move on the expert user's part. The translator, on the other hand, has a different perception of language, having been trained to achieve precision by means of consistency and economy of means. For this reason, it often falls to the terminologist to clarify the writer's words for the translator.

SEE
Technical Writing 7.1:445

- *Misleading term*: In the string "Back off the 3/4 inch squeeze plate travel adjusting capscrew by one-half turn", good sense suggests, and inquiry confirms, that this fastener is no different from any other capscrew. However, without a query, the translator will be hoping to find a very particular word in the target language, and the word will not be there. The next step for the terminologist is to determine whether there is more than one capscrew involved in the process, in which case the delimiters used in the string are important to understanding, or whether a simple "capscrew" will do. With respect to screws and fasteners, however, the form and standard designation for this screw in the TL will be the determining factor in finding the appropriate equivalent. The modifiers chosen from the English perspective may be quite irrelevant or misleading. Perhaps the single most frequent cause of error by an otherwise competent translator is the inability to recognize which word in an English string is the head word and which words are modifiers. Making this clear is the task of the terminologist.

INFO

Head word INFB 6:330

- *One term referring to more than one concept (polysemy)*: This situation may result from an oversight in editing the source text or from a similarity of function for two very different objects, where the source language may focus on a conceptual similarity such as appearance or shape and the target language focuses on very different features, such as function. Whereas the overlap in terms may go unnoticed by the technical writer, it can pose a potential disaster for the translator. The writer and the terminologist must confer and, wherever possible, line drawings of the two items should be scanned into the termbase. The conversation with the technical writer is much easier when the terminologist has been able to glean a hint about the conceptual differences involved from some third source such as a termbase, a Web page, or a print source. It is then possible to offer the writer some alternatives for the two ambiguous terms and receive confirmation. This way of creating order was successful in a manual where the word *actuator* was used to label both something small and simple, a part of a spool valve, and a much larger, more complex object, which turned out to be a pneumatic transducer used to position a control valve. A dictionary, offering six or

sources of ambiguity

eight unmarked possibilities, was of no help here. As soon as the terminologist was able to supply the writer some options gleaned from a termbase, he was able to conclude that the second instance of *actuator* could easily be replaced with the less ambiguous term *transducer*.

- *Several terms used to refer to the same concept (synonymy)*: It will sometimes happen that as a translator works on a long text, there is a growing suspicion that several different words are being used to refer to the same concept. The chain of synonyms in one publication can be as long as four or more. This multiplication of terms results from the fact that often, a document such as an operations or maintenance manual has developed to its present state over many stages, along with the machine. Each of the many authors involved has left his or her mark on the text. And if one considers that sometimes the writer is an engineer, sometimes a repair technician, and sometimes a line foreman, it is not surprising that different passages demonstrate different registers and use different words to name the same thing. *univocality*

standardization

The ideal solution to problems of synonymy and polysemy in corporate terminologies is to convince the client that the standardization of English terms can be as helpful to the company as a whole as it is to terminologists and translators. Long-term terminology research involving the publications department working together with company engineers can ideally result in a coherent company-wide vocabulary to be used at all levels of the enterprise. Unclear or ambiguous proprietary terms can be replaced by more general words, polysemic and synonymous terms can be identified and harmonized, and definitions should be established for all the terms on the list. Where several registers exist within the company, e.g., terms used by Product Development, Manufacturing Engineering, Sales, and Service departments, the resulting synonyms can be identified, and, if possible, one term can be chosen as the one preferred for use in the manual. On the broader scale, just documenting the existence of conflicting terms improves overall communication within the company, even if no single term can be found that will meet everyone's needs—a highly descriptive technical term may not satisfy marketing's need for a quick, catchy buzzword, for instance. Independently of this kind of team project, we maintain an inventory of collocations from various general and preferred usage, "loosen/remove the screw". The ultimate manifestation of an effort to rationalize in-house terminology usage is to implement Controlled English, which of course requires an enterprise-wide commitment to total reorganization of the document production chain.

INFO
Controlled Language
INFB 25:872

3 Target Language Terminology

3.1 Completing Terminological Entries

On the basis of the understanding derived from documenting the SL terminology as described above, the terminologist, perhaps working with the translator, must discover whether the relevant concepts are known in the TL, and if so, how they are expressed. To a great extent, the expression which is finally chosen as the best TL equivalent will be arrived at by a number of exploratory steps. This process is especially problematic if the terminologist and the translator are living in the country of the SL, where TL material is not always readily available. It cannot be repeated too often that the results of every inquiry should be documented and entered in the termbase as soon as possible. In this way, the disappointment and time lost at each wrong turn can be avoided the next time, and each discovery will be readily available for reuse when needed.

terminology validation

There are two greatest difficulties encountered in establishing TL terminology: one is the variation that exists between the ways that terms are coined in the two or more languages involved, and the other is the scarcity of documentation as to how people *do* say things, as opposed to how a dictionary may suggest they *should* say them. Of course, one problem with dictionary definitions and equivalents is that they pertain to generic lexical meanings that may be quite irrelevant or only peripherally related to terminological applications in a given field. The terminologist must consult authentic materials that were produced in the TL with no thought to translation or, what is sometimes even better, an informant who is both a native speaker of the TL and an expert in the content field of the translation.

target-language validators

The best person for the job may, for instance, be employed in the company buying and using the equipment in question, although care must be taken to ensure that the TL informant is linguistically and technically competent to render a valid opinion. It is important that in-country informants and validators feel that this activity is a valuable part of their role in the company and that they do not resent it as an unreasonable imposition on their time. It is also important to recognize that they may feel that it is incumbent on them to report errors in the text or terminology not because there are truly errors there, but rather in order to enhance their own prestige.

neologisms

Because the technologies treated in the source language texts have evolved in an English environment, there may well be terms that do not yet exist in the target language. Hence the terminologist must often be prepared to work with the translator to find an acceptable paraphrase, or to coin a new term, where there is an obvious need. The appropriateness of neologisms and paraphrases must be verified, if possible, in existing literature.

3.2 Terminology Resources

A rich variety of resources exists, and the practiced terminologist knows how to evaluate and apply them.

bilingual dictionaries

- *Reference works*: In some respects, specialized bilingual dictionaries ought to be the tool of last resort for serious terminologists because they can be misleading unless they have been specifically prepared for the narrow field in question. Nevertheless, even non-specific texts can still be consulted first because a terminologist who is experienced in the content area at hand will usually know quite easily whether the translation equivalent is useful. Furthermore, even when lexical information is incomplete or not really pertinent, it can nevertheless provide clues for further research.

SEE

Web Resources 8.5:845

- *Term banks:* Official terminological resources are available in CD-ROM form and on the Web. Some offer the advantage that they have been created on principles of sound terminological methodology and provide adequate documentation to demonstrate whether one has found the right equivalent. Of course, not every project involves languages that offer these kinds of terminological resources.

- *Proprietary termbases*: The software is available which will permit translators and terminologists to make excellent reference tools for themselves, on the basis of their own earlier work. Especially for those who choose to restrict their work to a small number of specialty areas, building such a database is an important component of professional development.

termbase creation

- *The client's own bilingual word list*: These lists often seem inadequate to a terminologist because they frequently lack such useful supplemental information as definitions and contextual references, but the equivalents may be obligatory, especially in cases where the client is a large, multinational manufacturer, or has some other reason for standardizing terminology. It is an inescapable reality that in some circumstances, standardization is more valuable to the client than absolute authenticity.

bilingual equivalents

- *Texts*: Although the use of parallel texts is more important for some types of projects than others, it is useful for any project where the resources exist.
    - In some subject fields, searchable corpora exist, and software is available to align occurrences of the term in question. Such corpora, used in the TL, provide information about the exactness of term equivalence and proper collocations.
    - Individual publications in the subject field are valuable, but occasionally difficult to find, especially technical publications in languages of limited distribution or where the technology in question is undergoing rapid development.
    - A growing number of texts can be found on the Internet, but more often in English than in other languages.
    - Gray literature, i.e., texts without ISBN numbers—catalogues, user instructions that have not been translated, earlier documentation of the end user, or of the end-user's competition—are all useful.

SEE
Collocations 8.4.4:788

- *Human resources*: Communication with experts is essential since many special language terms are taken for granted by practitioners in a field and not overtly defined in support literature. Communication among translation colleagues is invaluable. Terminology is not a commodity that improves from being kept secret. It is of great mutual value for both

SEE
Artificial Intelligence 8.4.4:809

terminologists and translators to exchange work with their colleagues in other countries, either during networking sessions at conferences or through listserves. Of course some information is proprietary to a client, and the terminologist will make note of that, but the majority of work can advantageously be shared. Colleagues can answer individual queries for each other, send each other reference material, and share database entries. As mentioned before, useful contacts can also be made, with the client's permission and help, with the client's customer.

SEE *Terminology Interchange 8.1.5:613*

Additional, very interesting contacts can be made by those who specialize in providing just-in-time terminological support to translators. In some countries, independent terminologists working on a freelance, ad hoc or systematic basis provide research support on a contract basis. In some countries, particularly those with a tradition of language planning, government language agencies will suggest candidate terms. More and more translation and localization firms offer this service as a supplemental support function for their long-term clients. The Translation Laboratory accepts this kind of commission, usually for translators in other countries who need help understanding American technical writing. Even if this work proves difficult, with last minute queries and little or no context, this kind of research can be extremely useful in the long run if a reciprocal relationship can be worked out with colleagues in other locales.

- *Quality control*: The most important quality control mechanism for a translation is a draft review by a native speaker competent in the subject field. Some clients make the time to have their customers review texts, particularly documentation, in a no-fault atmosphere. Care should be exercised here, however, because customers do not always provide competent review, and reviewers sometimes feel constrained to find fault in order to prove their own worth. If the ideal situation for client review does not exist, two other methods can be introduced. One is to find an editor locally or to work with a network of freelance editors. The other is to engage the translator in a conversation based on questions prepared by the terminologist. These are on the order of "Did you use the same word here and here?" "Does the word you used for

SEE *Quality Assurance 7.2.3.488*

water here mean the same thing as drinking water?" "Did you translate this word as a noun or as a verb?" "How did you decide which word to use for this term?" The usual result of such a session, lasting from 20 to 30 minutes, is that the translator makes a change here and there, improving the usefulness and accuracy of the work.

4 THE ECONOMICS OF TERMINOLOGY MANAGEMENT

The work described in these pages is detailed and laborious, and many clients find it difficult to justify the added expense and effort involved in creating extensive terminology resources. Part of the terminologist's job is to educate the client to understand that up-front investment in terminology documentation pays off in reduced variable costs over time, but once done, and correctly documented, the resulting resources continues to serve for future projects.

INFO

Economics
INFB 26:873

7.2.5 Software Terminology and Localization

LICIA CORBOLANTE AND ULRIKE IRMLER

1 OVERVIEW

non-English software

Until the early 1980s, computer literacy was still the restricted domain of specialists. Software was aimed at users with good computer knowledge and adequate knowledge of English, which was the accepted language in which programs were written and in which they worked. In recent years, however, computers with user-friendly, graphic-oriented software, have become widely available to users who might not know English or most likely prefer to work in their native languages on a regular basis. Current legislation in many regions, such as the EU or parts of North America, requires that all on-the-job tools be made available in a worker's native tongue.

INFO

Localization INFB 27:877

With the rapid expansion of this lucrative, competitive market, it has become necessary to produce versions of the same program in many different languages. This process is called localization and requires an enormous amount of terminology management effort, often in many languages simultaneously, and within very tight time frames. Furthermore, computer programs today are no longer developed just in English, but written in many different languages, so localization now involves the adaptation of computer software and documentation from and to a wide variety of languages. Localization concerns not only very user-friendly applications, with basic, informal terminology, but also highly specialized programs and operating systems. As a result, software terminology is extremely varied and reflects the multitude of domains it is derived from. The continuous development of computer technology also results in the introduction of new concepts that require naming and consequent equivalents in the target languages, which means that terminologists and translators must coin new terms more frequently than ordinarily common in other, more stable domains.

The speed of development and the relatively short time that software terminology has been established as a field means that, apart from a few exceptions, it has not been standardized but varies greatly between different companies and institutions. Software terminologists, localization specialists, and software translators are therefore faced with a demanding and yet exciting task that requires the ability to research terminology in the most disparate fields thoroughly but in limited time, to satisfy technical requirements, and to solve language- and culture-specific problems. This article will highlight the most common software terminology problems in a localization context. The focus will be on the problems terminologists and translators of computer applications and documentation face in their daily work and will suggest some of the most common solutions based on the practical experience with software companies and localization agencies. Additional issues, not strictly terminological as such, but which might need to be resolved at the initial stage of software terminology management, will also be covered. All examples will assume software originally created for the American market and consequently English as a source language (SL).

terminology development

INFO

Text Designations INFB 24:867

2 Software Terminology and Localization

In the extremely competitive software industry, new applications appear on the market on a daily basis. Computers, in fact, are already a fact of every aspect of life: applications targeted at the home market that do not require any computer literacy coexist with sophisticated programs catering to specific professional, technical and scientific needs. However, the borderline between low-end and specialized software is becoming more and more blurred, and the actual terminology found in a software product might often be drawn from different domains with different degrees of specialization. Computer jargon is used along with acronyms and neologisms, and general terms often acquire new meanings in the computer context. Software terminology is difficult to define and classify, it spans different domains, and it is continually expanding.

SEE

Term Formation 1.2:25

Technical Translation 2.1.1:147

When a software application is transferred from the SL into different target languages (TLs), it might also require a considerable degree of localization. Resolving the language-specific, SL technical, cultural, and marketing factors is all part of the localization process and involves not only translation but also a

adaptation

considerable amount of adaptation, and in some cases reworking original material so that the end-product will also suit the needs of users in the target culture. The localization of software into Japanese, e.g., might require a considerable degree of re-design of the user interface (UI) to accommodate the features necessary for the Japanese language, such as the size of text boxes and the number of dialog boxes. Localization is not a straightforward process, although a variety of tools has greatly reduced technical difficulties and sped up translation, there still remain numerous language-specific issues. A software terminologist must therefore not only be able to find acceptable and adequate equivalents in the TL, but might also be asked to solve language-related problems at the glossary compilation stage.

3 The Concept of Implied User

3.1 Locale-Related Expectations

cultural aspects

Language-specific problems in software terminology are not necessarily only morphological, grammatical, or syntactical, but they can also have strong cultural aspects. When confronted with culture-specific material, it is essential to try to define who the implied users of the localized product will be, i.e., the hypothetical ideal users of a program, and reflect on their expectations and background so as to configure tailor-made solutions. Departing from the SL material and implementing replacements might be a necessary step in order to achieve an equivalent effect in the localized product. In other words, the users of the localized product should respond to the product in the same manner as the user of the original product. For instance, users in some parts of the world, require software and related materials to be in a neutral tone with straightforward examples (yet without lapsing into an excessively formal approach). Americans, on the one hand, seem to prefer a friendly, colloquial tone and elaborate examples, even in software. A Japanese audience, on the other hand, may need precise examples and a more visually appealing presentation. Literal translation and very strict adherence to a design geared toward the US market are oftentimes not the best solution for all languages. Although adaptation problems caused by cultural factors are more frequent when dealing with sets of examples, scenarios, and ready-made models and templates, they can also occur with single terms. Defining the implied user is therefore an important step toward achieving better localization and facilitating terminology management.

3.2 The Implied User's Level of Competence

level of competence

Once the expectations of the end users with respect to their response to the product have been assessed, it is essential to define their level of competence and encyclopedic knowledge, which will determine the choice of product-specific terminology. Evaluating user competence is particularly important with popular packages that are targeted at large numbers of users, whereas the specific knowledge needed for more sophisticated programs is more easily defined. The user's level of competence, in fact, might be the determining factor in deciding the translation to adopt when different alternatives are available, e.g., when choosing the translation for the word processing term *font size* into Russian between **КЕГЕЛЬ,** a specialized term used by experts, and **РАЗМЕР,** a very general word, meaning *size*, that might be user-friendly for non-experts, but perhaps not specific enough for specialists.

encyclopedic knowledge

Encyclopedic knowledge (information and experiences shared by implied users, such as sports metaphors or references to regional historical figures that are part of local common knowledge) is another criterion that can assist in finding appropriate terminology. A mouse pointer shaped like a *no entry* traffic sign, for instance, indicating that an action is not allowed, might be called *Ghostbuster cursor* in the U.S., a very effective name because any American user would be familiar with the *Ghostbusters* film and its logo, but a literal translation would probably not be acceptable in a TL where the film was not generally familiar. In such cases, a more neutral translation such as *action not allowed*, or with a different figurative meaning, e.g., the name of a traffic sign, would be more suitable because its reference would be familiar to a wider range of users throughout the world.

Figure 1: *Ghostbusters Logo*

merged audiences

With the growing popularity of computer applications and the growing competence of users in this area, the border between expert and non-expert users is becoming more and more blurred. In order to be appealing to a general market, software products need to be both easy to use for the beginner and sophisticated enough to appeal to the more advanced user. Technical concepts that remain untranslated (as programming concepts often do) in a software application for programmers are often translated in software for a more general audience. This creates conflicts and terminology inconsistencies when the two target groups are merged, as is more and more the case, since many applications now contain programming functionality.

4 Typical Features of Software Terminology

computer-specific terminology

product-specific terminology

As has been mentioned above, software terminology is not restricted to a well-defined domain, and is therefore not easily classifiable. A first distinction can be made between computer-specific terminology that applies exclusively to computers (e.g., hardware and operating systems), and product-specific terms (e.g., referring to applications such as word processors and spreadsheets), which can also be found in non-computer contexts. Such a distinction is relevant because the terminology found in hardware and operating systems is often not suitable for localization. Acronyms and abbreviations are widely employed and the original English term might already be well established in the computer literature of the TL, so that translations would be unacceptable and might lead to compatibility problems between different programs. Not infrequently, common English abbreviations are widely lexicalized in the TL, although the English expanded form may be unfamiliar or may have been replaced with a translated full form in the TL, e.g., *FTP* and *protocolo de transferencia de archivos* (es). In contrast, applications usually draw specialized terms from fields whose terminology is likely to be already standardized in the TL and translation is usually possible. Certain practices that were commonly employed in programming, such as the communication between programs via certain terms in the software, have now been discontinued, precisely because they cause translation problems. This so-called hard coding of language in the software is regarded as bad practice from an internationalization standpoint.

structural diversity

The most obvious difficulties when dealing with software terminology are caused by the diversity of structures between the SL and TLs. In ordinary written texts, in fact, grammatical and morphological differences hardly represent a translation problem, but they can become a tricky obstacle in software. At this point, however, rather than attempting more detailed definitions of software terminology, it might be more useful and relevant to outline issues that characterize it. As it will be shown by the following examples, solutions to language-related localization problems can often be provided only on a case-by-case basis. However, awareness of the difficulties involved and of areas that need particular attention can contribute to faster and more effective terminology management.

general language terms

- ***General terms*** such as *icon, desktop*, and *file* are very commonly used in software, where they can acquire a more specific meaning. They are relatively simple to translate—in fact most TLs would have suitable equivalents—and any additional connotation given to general terms in a software context is usually easily assimilated by the end user. However, in some TLs it might not be acceptable to choose the same ordinary equivalent because of its lack of prestige (especially if computer applications enjoy a very high status) or because of extra connotations that a term might have in the TL. A literal translation of *mouse*, e.g., would not sound prestigious enough in some languages, or it might carry ambiguous connotations in others. In some instances, a whole metaphoric system of terminology, such as the English term *tree* for an expandable navigation structure and its related terms *branch*, *leaf node*, *grove*, and *forest* could not be reproduced in German, since the translation initially chosen for *tree* was *Struktur*. Translation of related terms in a terminology system should always be taken into account when choosing a corresponding term in the target language. A core terminology of general terms has already been largely established and standard translations exist in many languages, but with the rapid development in the computer industry new terms borrowed from general language are continually being introduced.

- ***Scientific or technical terms*** (e.g., drawn from statistics, project management, publishing etc.) can be easily reproduced in the TL, generally because they are already well documented in the TL.

scientific and technical terms

In most scientific taxonomies, in fact, naming of new elements is achieved through Greek and Latin word elements, which are universal in most Western European languages. TL equivalents are usually adaptations of the original words to the appropriate spelling conventions, but care must be taken when translating to languages that do not share this tradition of scientific naming, or where Latin or Greek roots are avoided in the name of language purity. With regard to specialized, non-computer-specific technical terminology, the only real difficulty would be accessing accepted standard terminology efficiently and rapidly. Research time might be reduced if reference material used by the software developers is known: an existing TL translation of such material and detailed bibliographies would greatly speed up preparation work. In this way, it would also be possible to detect at an early stage features in the software that might only be applicable to the SL market or that might require extra localization work and adaptation.

short forms

• ***Acronyms and abbreviations*** are very common, especially in hardware and operating systems contexts, and often represent key concepts. In most cases, it would be advisable not to translate acronyms and abbreviations if they have technical significance. Terms such as RAM, CD-ROM are already well established internationally. Moreover, it is important to be aware of possible technical issues that might render translation unfeasible: DLL, for instance, is not only the acronym of Dynamic Link Library, but also corresponds to a file extension which must be shared by all programs, irrespective of their language. Although some acronyms might not be translated, it is an important terminological task to define the rules according to which they will be explained to the end user, for example by providing their description both in the SL and in the TL at their first occurrence in the documentation, e.g., OLE (Object Linking and Embedding [en], *objecten koppelen en insluiten* (nl), and by standardizing such explanations. In some rare cases, serendipitous orthographic similarities can be exploited in order to retain familiar acronyms, for instance, *Failure-Mode-Effect-Analysis* (*FMEA*) in English becomes *Fehler-Mode-Effekt-Analyse* in German, even though *failure* and *Fehler* are not at all complete equivalents in other contexts.

Abbreviations pose fewer difficulties. They can be translated, but the resulting TL terms might be less transparent than the original SL abbreviations. Sometimes abbreviations and concise combinations of words can also acquire a unique meaning (e.g., *bit*, originally meaning *binary digit*), which is hardly reproducible in the TL, and should be treated as neologisms (see below).

acronyms abbreviations

• ***Term formation***—Different languages have different traditions for new term formation. The strategy applied for a new form in the SL may not work well at all in the TL. For instance, when working from English, new compounds such as *bubble memory,* neologisms such as *byte* or *pixel* (picture element) or other new forms, such as *dithering* (a term from computer graphics that in ordinary English only conveys the idea of hesitation) are quite common in software programs and probably pose the majority of translation problems. In fact they are usually difficult to research and in some cases it is almost impossible to establish their source. When the new form describes a special feature of an application, early access to a beta version of the program is essential in order to attempt a definition of the new feature and possibly rename it according to TL rules. In this process, it might be useful to obtain the alternative names that were discarded at development stage in favor of the final term. The decision whether to create a corresponding TL term or adopting the new SL form will also depend on the degree of specialization of the term itself and on the preferences for translation or borrowing. There is also a large degree of variation between TLs in the tolerance towards English terminology integration into the TL system. In some languages terminology is integrated as is, i.e. without adapting the term to the spelling convention of the foreign language system, but in other languages, such as Japanese, the term is phonetically recreated, such as the term, ｶﾝｾﾙ(ka-n-se-ru), taken from the English *cancel*.

SEE *Term Formation 1.2.1:25*

term adoption

• ***Marketing factors*** can , in fact, greatly influence software terminology. Some keywords (e.g., *cut & paste*, *drag & drop*, *double click*) describe special features of a program and are aimed at increasing the appeal of a product. They should be translated in a way which makes them keywords in the TL as well. They often carry figurative or metaphorical meaning, so that it is important to provide an adequate TL term that will have the same impact on the TL user. Keywords require extra research—in fact they often contain important marketing elements in that advertising campaigns are often centered around such features because they

keywords

uniquely characterize a product, and there might be legal implications if they are copyrighted or trademarked. If a key term is copyrighted in the SL country, it might make economic sense not to translate it into a TL and extend the original copyright to other countries rather than using a non-copyrighted translated term or going to the trouble and expense to acquire a copyright for the translated term.

cultural implications

When keywords have cultural implications because of their imagery and metaphorical meaning in the SL, it might be impossible to use the equivalent term in the TL, so that it becomes essential to define the symbolical value and figurative meaning of the original term to determine whether a semantic equivalent would be acceptable in the TL, or whether an adaptation should be made. A good example is the term *wizard* (a feature that enables semi-automatic execution of otherwise complicated tasks), which in English evokes ideas of magicians and also recalls the term *computer wizard.* In some languages, however, it would be impossible to use a literal translation because it would not be suitable to the expectations of the TL users; possible alternatives would be *assistant, guide* or also *guided creation,* where the connotative meaning of the original term is neutralized and a less figurative term is chosen, which, however, still describes the characteristics of the feature. These kinds of solutions, however, have the potential for opening up new problems. If, for instance, the original use of *wizard* is replaced with *assistant* in a TL, and the English programming environment later introduces another functionality that is then called an *assistant,* the terminologist is confronted with the problem of having to come up with yet another new term in the TL.

adaptation

• ***Register and style***—The combination of formal and informal style, which can often be found in U.S. software, might not be acceptable in the TL. It might therefore be necessary to establish the correct register level expected by the TL user and adapt SL terms accordingly, by making them more neutral if necessary. In some languages, e.g., *drag & drop* might sound too informal or even too formal, and might have to be substituted by a more neutral expression like *move selection*. Similarly, adjectives such as *smart* might be inappropriate if translated literally and could be replaced by, for instance, *logical*.

market expectations

• ***Specialized vs. nonspecialized terminology***—Some applications have a mixture of specialized and nonspecialized terminology, e.g., *kerning* and *dropcap* in a word processing context, which might be acceptable in English but not in all other languages. The decision whether to maintain the same combination also in the TL, to upgrade nonspecialized terms and substitute them with a more specific TL translation, or only to adopt more general terms will largely depend on the definition of the implied user, i.e., the expectations of the market for that particular product.

toolface

• ***Visual elements***—When translating software terms, it is important to remember that more and more commands often have a visual equivalent, i.e., icons or buttons that become part of a self-explanatory toolface. However, some TL terms might not share the same semantic components as the respective icon: an icon with an open door can well represent the command *Exit* in English, but would not be suitable in those TLs where such a command is translated as *terminate*. In this case, the choice will be between localizing the icon/button by changing its toolface, or adopting a different translation that is semantically closer to the SL term. Marketing and economic considerations usually determine which is the most feasible option. In general it is good practice to avoid word or visual puns in the SL. For instance an icon representing different users on the same administrative level as *under one hat* (*unter einen Hut bringen* [de] by depicting people who are standing under a large hat is a poor example of a globalized visual representation, since it relies on an idiomatic expression specific to German or possibly English.

offensive visuals

Visuals are often the source of localization problems, since there is also the potential of offence and they are often produced by non-natives to the TL. It is important to have all visuals reviewed for the target market, since even seemingly harmless clip art can be inappropriate. The depiction of cartoon figures with 4 fingers instead of 5 is not appropriate for the Japanese market, for instance, since it is reminiscent of the practice of severing a finger by members of the Yakuza crime syndicate. Especially geopolitical content, such as maps and flags, should be carefully reviewed before being included into a software release in any market to avoid unpleasant surprises.

• ***Synonyms***—In any given situation, the SL may possess several near synonyms that can be used to finely differentiate between similar concepts (e.g., English *flip* and *rotate*). The words *connect, link, join* and *log in*, e.g., can all express the idea of establishing some type of connection. In some TLs there might not be four different terms to express the different meanings of the concept. Possible solutions are to adopt the same term for different contexts, to differentiate by adding adjectives or explanations, or to introduce or adapt a loan word (e.g., the Finnish verb *linkittää* as a loan-word to translate English *link*). *near-synonyms*

As a basic rule, if there are TL terms which share the same etymology as the SL term and are not overtly inappropriate loan translations, it might be advisable to choose such terms even if stylistically there might be better options. This is especially true if you need to save a good term to use for a different, but closely related concept. For example, *collegare* would be a good Italian translation for *connect*, but it might be advisable to choose *connettere* instead so as to be able to translate *link* as *collegare*. *etymology*

• **Word length**—In English, for instance, words can be very short and can easily fit in many elements of software (buttons, icons, menus). It is advisable to check the maximum word length allowed by a program interface before establishing TL terminology. Resizing and adaptation of software elements are in fact possible, so that even long TL words might be used, but there are cases in which word length becomes a determining factor in the choice of terminology, e.g., the word *size* on a small cursor can hardly be replaced by the Italian standard translation *ridimensiona*. Abbreviations are usually acceptable, but could create comprehension problems or could be visually less appealing. It might also be very difficult to reproduce the effectiveness of some SL short words (e.g., file names and file extensions like *readme.now*), because in some TLs abbreviations would not be very meaningful. If the abbreviation of a term in the TL is not acceptable, the translator or terminologist may find it necessary to contact developers to urge them to implement the change in the design of the software itself. In general it is good practice to allocate extra space for any linguistic entity that needs to be translated and to provide flexibility in box or graphics sizing to accommodate expansion factors. *box resizing* *expansion factor*

• ***Conciseness***, which is typical of the English language cannot always be easily reproduced (a German sentence, e.g., can be up to 60% longer than its English equivalent while containing the same number of words). In a TL, a combination of words might be necessary to reproduce the meaning of a single SL term. However, if technical restrictions only allow one single word also in the TL, alternative solutions are essential. The English verb *toggle* ("switching back and forth between two states"), for instance, lacks a direct translation in many TLs, where more than one word might be needed to convey the same concept. With a graphical alternative solution, however, two words could be combined, such as *activate / deactivate,* and no meaning is lost.

In cases where the comprehension would have to be sacrificed for the sake of conciseness, it may be advisable to redesign the dialog box to accommodate longer strings and allow for a linguistically appropriate rendering of the SL string. Ideally software designers should take localization issues into account when designing the SL software, since the cost for re-designing the software in several TLs far outweighs the cost of a original global interface design. The minimum requirement for the treatment of dialog boxes is that no text or interface element be coded so that it cannot be changed. This point underscores a fundamental principle of software adaptation: the more problems that can be solved at the internationalization stage, the less effort will be necessary during localization. Another way to state this guideline is that internationalization needs to be done only once, whereas localization must done and redone for every locale and every product release. Consequently, the more problems that can be solved through proper internationalization, the greater the efficiency and cost savings during localization, and the easier it will be to ensure successful simultaneous product release.

agreement

• ***Grammatical issues*** or differences between SL and TL cannot always be overlooked in the localization of software. Many software messages, for instance, are similar but have different referents. For instance, "New *%s* will be saved", where *%s* is replaced at runtime by different terms, such as *name, folder, file,* which vary in gender in many languages and thus would lead to grammatical errors in the TL. In English these strategies cause no problem at all and constitute an economical solution in which only one string and a set of changing elements are necessary. Here the verbal form, the article and the adjective are unaffected by the number and gender of "%s", whereas in other languages these grammatical features cannot be ignored. Similar problems can arise if the SL software is designed to parse for plurals and forms them automatically by adding "*s*". A well globalized software application is designed to take these localization issues into account, e.g., by increasing the number of strings and/or of sets of words in the localized version; or by choosing only referents that share number and gender. At the very least the occurrences of awkward and ungrammatical solutions should be flagged, if no alternative is possible and pointed out to the user in printed and online documentation. Figure 2 shows a typical dialog box illustrating this problem.

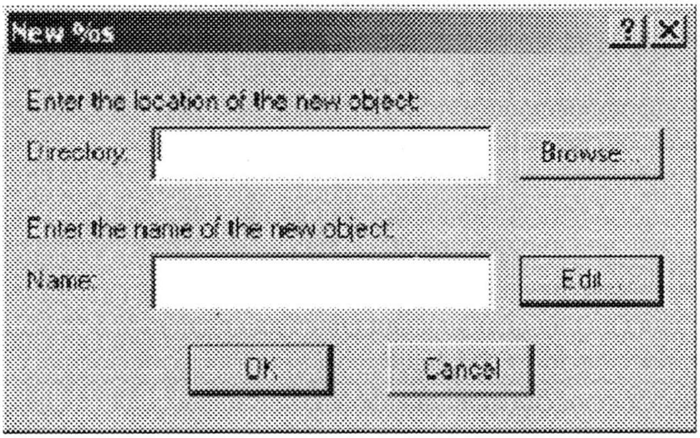

Figure 2: *Dialog box for "New %s"*

concatenated strings

Another maddening example involving so-called concatenated strings featured a situation where a string at the top of the dialog was compiled into text at run time: **The <ordinal number><name of day>.** Here the article *the* would remain stable, while the other two elements would be introduced as selected from a combination of drop-down boxes, to yield, for instance, *the third Sunday*. For several languages this was a nightmare, for example in Italian, where possible options include *il primo lunedì*, *la prima domenica*, and *l'ultima domenica*. The changing articles and word endings drove localizers to find a creative solution involving the rearrangement of the dialog items.

ambiguities

• ***Word order***, the formation of compounds, the number of words used to express a single concept, and spelling conventions all vary from language to language. If it is impossible to modify the application code so as to accommodate for differences in the TL structure, alternative solutions will have to be found. English allows for lengthy compound formation, e.g., "Web-based Enterprise Management Common Information Model Object Manager Tester", to give an extreme example. In many other languages the ambiguities inherent in such a term must be resolved in order to express the relationship between the different elements of the compound. For instance, the translation of *small computer system interface* requires a minimum of technical knowledge to determine that *small* can only refer to *computer* and not to *interface*. A term such as *last member query interval* could have multiple interpretations and requires communication between the localizer and the US developer in order to resolve the ambiguity. *Standard Toolbar Color* could be interpreted as the color of the standard toolbar, the standard color (used) for the toolbar, the color of standard toolbars, or standard color for toolbars, but only the final interpretation is correct.

case conventions

• ***Capitalization*** of nouns within a sentence is acceptable in English, so that nouns with initial capitals from the same word list can appear in any position within a sentence (e.g., "*Document1* is being printed" and "Now saving *Document1*"), whereas in other languages such usage would be totally inappropriate. Capitalization issues should not be underestimated at the terminology management level. In fact entries in a glossary might be capitalized irrespective of their actual use in the software. In some TLs, for instance, uppercase initials in the names of months and days of the week are errors. Although they do not hinder comprehension, they are a typical example of inaccurate localization.

character sets

• ***Extended characters***—Although most software now supports extended characters (accents and diacritic signs), there were instances in the past in which character support limitations of the interface influenced terminology. The Italian translation for the field *company* in a database, for example, might turn out to be *azienda* rather than the more adequate *società* if extended characters were to be avoided. There has been a major shift in awareness regarding character sets in the last few years and many companies are using Unicode now, which allows the storage of characters for many more TLs, including most Asian languages. Although newer approaches solve these problems, localizers should be aware of these issues because they may recur in situations where the decision is made to localize older programs with legacy code elements.

- ***Untranslatable elements***—There are also instances of items that should never be translated, such as commands in some programming languages and some operating systems. In such cases, terminology management involves establishing standard explanations of the commands whose meaning would be transparent only to English speakers.

commands

5 Working with Terminology

In the localization industry, terminology recycling by means of translation memory (TM) tools has become a common practice. For these purposes it is important to maintain termbases with consistent terminology that can easily be reused. This benefits the individual translator working on small projects as well as the large corporation with the need for large-scale terminology recycling. It also contributes to the standardization of terminology in a field where standardization is still in its beginnings.

translation memory

5.1 The Importance of Detailed Glossaries

Software companies with localization departments have developed tools that build glossaries of SL terms automatically, extracting the relevant information from the software and arranging it in a standard format that allows easy access and the possibility to use the glossary online and to integrate automatic translation tools. Some of these tools are also now available commercially.

SEE *Term Extraction 8.4.1:723*

In small projects involving software localization, however, glossary preparation work might still be manual and translators might be asked to build their own glossaries, which can be a very time-consuming task if no specific tools are available. In such cases, it is important to plan the glossary structure carefully and decide which type of information should be included. The most obvious type of glossary would be an alphabetical list of the terms used in the software interface and in the documentation of a specific

manual preparation

project. These project-specific glossaries can then be merged to create a general glossary comprising different products, which would constitute a good working basis for successive projects.

Lists of words or tables in word processors are the most basic type of glossaries. They are very easy to create, but they tend to be slow when they are sorted alphabetically or searches are made, and are in fact useful only for very small projects. Spreadsheet and generic database packages are probably the best solution when proper terminology database systems are not available. They can generally handle thousands of entries which can be easily sorted, and which can be classified according to different criteria. As noted in Chapter 8.1.3 of this book, however, generic database or spreadsheet programs are rarely truly effective in providing the long-term functionality needs of translators and localizers.

SEE

Entry Structures 8.1.3:572

A software glossary, however small, should not be simply a list of words with their translations in the TL, but should always include additional information about the context and the use of terms. Tools that extract all text strings from software and build up a glossary can automatically add a large amount of additional information. If such tools are not available and a glossary needs to be created manually, it is possible to compile a collection of the terms used in the software by following a test script, i.e., a description of all the procedures and steps needed to display all menus and dialogue boxes of program. Alternatively, terms might be extracted using a software localization software tool, in which case a list of all interface items and error messages could be built, but the necessary context might be lacking and several duplicates might be encountered. It is also very important not to forget the *situational* context of terms and strings, for instance in interface boxes and on buttons, because potentially ambiguous terms can mean different things depending on the functions they represent. For this reason many localizers prefer to actually run the application parallel to localizing the product so that they see on-screen precisely what is happening in any given case.

SEE

Contexts 1.3.3:80

Text Production 8.4.3:772

5.2 Managing Additional Information

SEE

Data Categories 8.1.2:552

Whichever method is used to create a glossary, it is essential to include additional information, which can obviously be classified in several ways according to one's needs. What follows is a suggestion for possible fields, all of which are self-explanatory, that could be added to a glossary:

- ***String Categories*** enable the classification of terms according to their function in the software, thus providing an essential additional contextual reference. Possible categories that can be used are *menu name, command, dialogue box title, check box name, function name, general terms* (terms used in the documentation, not included in the interface, but specific to software and hardware), etc.
- ***Word Class***—i.e., parts of speech or grammatical markings (noun, verb, adjective, etc.).
- ***Environment or Platform***—e.g., DOS, Windows, OS/2, Macintosh, etc.
- ***Product Name***—The type of program can provide another kind of contextual information: *subscript*, for instance, might refer both to character formatting and to matrices, and might require different translations according to the type of product.
- ***Product Version***.
- ***Comment*** or ***Term Note***—It might be necessary to provide additional explanations to correctly identify terms that can appear in different contexts and might require different translations. *Left* and *right,* e.g., are often used in check box labels and sound like very obvious terms, with only one translation possible, and yet in Italian they might be *a sinistra* and *a destra* if they refer to alignment, but *sinistro* and *destro* if they refer to a border. Sometimes it is essential to note the complement to a toggle element, e.g., the German particle *auf* can mean *open* if the toggle is *auf/zu*, but it can mean *up* if the complement is *ab*.
- ***Time and date stamp***.
- ***Updates***.
- ***Access keys*** (see below).

INFO

Part of Speech INFB 7:331

5.3 Software Glossaries

Many current localization tools combine two different functions: on the one hand, they are translation tools that allow for viewing menus items and for viewing dialog boxes and resizing them. On the other hand, they also can serve as the environment for terminology management of the product glossary organized to reflect

common hierarchies

the exact hierarchy of the various commands and options used in the program interface. Each string has its own unique ID number, so that the strings for the translated version of the software can be easily identified and the TL software build can be generated. With the growing localization market, the selection of different localization tools has also become wider, and there are tools available for a variety of budgets and purposes.

5.4 Additional Terminological Tasks

access keys

Software terminology management might involve dealing with additional non-linguistic tasks that require some knowledge of localization procedures. A good example of these new tasks is represented by access keys, i.e., key combinations that allow access to menus or commands from the keyboard, such as a combination of the ALT key plus an underlined character in a menu, command, or option name. Although there are tools that can automatically assign access keys and check the occurrence of duplicates, access keys are still mainly defined manually at the glossary creation stage. In deciding on access keys, it is important to be aware of technical restrictions. It is inadvisable, e.g., to choose letters with diacritic marks or letters with descenders (e.g., *g* and *p*) because the underlined character might not be sufficiently visible. In Asian languages, access keys are added in Latin characters after the term itself, since Asian characters cannot be created on a keyboard with a single stroke. Consistent use of access keys within the product should be ensured, and access key conventions should be respected: choice of access keys is determined by specific criteria, e.g., in order of importance, using the first letter of the most important word of the command or dialogue box label; the first letter of the first word; a distinctive consonant (e.g., *x* in *Exit*); or any other unique letter.

language-specific concerns

Identifying national language standards, specifying sorting orders, and defining shortcut keys are other tasks of this kind. Managing software terminology, therefore, is not simply a lexicographic task, but involves functional aspects as well.

6 Consistency and Standardization

SEE
Terminology Interchange 8.1.5:613

Consistency between related products and between different versions of the same product is extremely important. Already existing glossaries should always be the starting point for terminology research on a new project. Even when glossaries are not in a termbase but are contained in a spreadsheet or in a database file, it is relatively easy to write a macro that extracts terms from an existing glossary and to export them into a new project-specific glossary. In this way, a large part of terminology management will be carried out automatically, and it will be possible for terminologists to concentrate their work on new features and terms. The most important benefit of methodical terminology management that takes advantage of modern terminology tools and processes is terminology consistency within and across projects, and, as integrated tools environments become more and more common in localization projects, between different tool types, e.g., among termbases designed for human translators, terminology resources created for machine translation, and a variety of other localization tools. Evolving interchange modalities will enhance this kind of capability in the future.

consistency check

A basic consistency check, however, can be carried out even with very limited resources. Glossaries, in fact, can be sorted in different ways to identify different problem areas, e.g., according to TL rather than SL to detect translation inconsistencies; according to category and class, to check for stylistic consistency; and according to access keys (see above), to check for duplicates.

SEE
Standardization 2.2.1:197

Consistency naturally leads to standardization, and standardization, in fact, greatly facilitates communication and comprehension because it creates familiarity with software concepts. The user will immediately recognize commands and options common to different products, and learning times will be reduced. Standardization should be possible, not only within a product and between products of the same family (e.g., word-processing elements in various programs), but also between programs belonging to the same environment; and, ideally, between environments and platforms.

As was pointed out earlier on, however, in the competitive software market it is important to characterize a product as much as possible, and the right terminology, because of its uniqueness, of its specificity, can greatly enhance the marketing of new features.

As a result, the same item might be given different names in different products. Additionally, products with common features might be developed at the same time and be subject to different naming conventions. Generally in such cases, when localizing inconsistent terms into a TL, marketing considerations and the concept of the implied user will assist in deciding if it is more appropriate to maintain the differences of the SL versions, or to standardize and choose one single term to facilitate the end user.

In fact, the idea that a user should be able to move from one product to another with the minimum amount of learning time has made large software companies aware of the importance of standardizing terminology, and has lead to the decision within Microsoft to use standardized localized terminology for the main interface elements of its products. This effort was originally supported with the publication of a comprehensive guide for terminology translations, originally in 14 languages for the Windows Interface (Microsoft 1993). Today the Microsoft English glossaries are available online, and the multilingual glossaries can be downloaded from the Microsoft FTP site in more than two dozen bilingual modules (Microsoft current).

SEE *Internet Resources 8.5:845*

With the introduction of software that facilitates terminology recycling and computer-aided translation, the need for standardized software terminology is not only an issue of user-friendly, consistent translations, but also one of saving costs in localization. Computer-aided translation tools are used in all large-scale translation operations in different industries and are essential in maintaining consistency from one version of the software to another.

REFERENCES

Esselink, Bert. 1999. *A Practical Guide to Software Localization*. Amsterdam and Philadelphia: John Benjamins Publishing Company.

Microsoft. 1993. *The GUI Guide. International Terminology for the Windows Interface*. Redmond, Washington: Microsoft Press.

Microsoft. *Microsoft Glossary*. http://support.microsoft.com/support/glossary/default.asp (Monolingual English terms and definitions). Accessed 1999-08-01.

Microsoft. *Microsoft Glossaries*. FTP://ftp.microsoft.com/developr/msdn/newup/glossary (Downloadable in bilingual modules for 20+ language pairs). Accessed 1999-08-01.

Section 8

Computer Applications for Terminology

8.1 Database Management Methods

8.1.1 Criteria for Evaluating Terminology Database Management Programs

KLAUS-DIRK SCHMITZ

1 INTRODUCTION

software tools

Today there is a wide range of dedicated software tools for supporting translation, documentation, and technical writing as well for accessing subject-specific information. These programs include computer-aided translation systems, information and documentation systems, electronic dictionaries, and terminological databases on a variety of media, such as CD-ROM or DVD, and terminology management programs, all of which house data on the user's own network, as well as various forms of terminological data accessible on the World Wide Web. This chapter is confined to an examination of terminology management systems (TMSs), although some of the aspects discussed also apply to evaluations of other types of programs. The concentration here on dedicated TMSs reflects the frequently confirmed experience that users who try to use standard spreadsheet, database, or word-processing programs to manage terminological data almost inevitably run into problems involving compromised data integrity due to inadequate data modeling features, in addition to difficulties manipulating large volumes of data as resources grow over time.

SEE *Internet Resources 8.5:845*

terminology management program

Terminology management programs are defined as software products (programs and program packages) that are designed for the management of terminological data. They enable the user to collect, store, manipulate, and retrieve terminology.

user groups

Besides the traditional user groups for terminology management programs, such as technical translators, interpreters, terminologists, standardization specialists, and language planners, we now find other user groups like technical writers, subject-field experts, documentation specialists such as compilers of thesauri and information specialists, and knowledge engineers.

Different users in these groups employ terminology management programs in various organizational environments: they may work alone (e.g., as freelance translators), or be employed in an individual company or department (e.g., in a translation bureau or language service), or in some division of a large enterprise or organization with homogeneous information processing facilities.

SEE *Technical Translation 2.1.1:147*

evaluation criteria

This chapter will deal with criteria used in evaluating terminology management programs. Some of the criteria are very general and applicable to all program types, while others are very specific and geared to the main characteristics of terminology management programs. The envisaged user group(s) and work environment(s) will determine the extent to which any one criterion is important for a particular evaluation.

2 Technical Aspects

hardware and software environments

The technical aspects of evaluating terminology management programs involve the hardware and software environment required for using a certain program. The extent to which these technical features are crucial in selecting a terminology management program depends very much on the existing or planned electronic data-processing equipment in a certain work environment (see Section 5 below).

platforms and operating systems

Most terminology management programs offered on the market run on IBM-compatible personal computers. Only a few require an Apple Macintosh computer or a UNIX-based minicomputer. Today virtually all serious PC-based programs run under MS Windows™ '95/'98/'00, etc., or NT™. From the software point of view, the type and version of the operating system under which a terminology management program runs are important for its evaluation, as is the type of processor used.

quasi-multitasking

In most practical applications, the terminology management program will be used together with word-processing software, translation memory, or perhaps with special localization software designed to allow for text processing without disturbing programming code or desktop publishing and SGML-style tagging. Since the original personal computer was designed as a single-task machine that could not run more than one program at the same time, mechanisms for quasi-multitasking were originally needed.

terminate-and stay-resident programs

Most older terminology management programs were developed as terminate-and-stay-resident (TSR) programs. After loading a TSR program into RAM, the word processor was started. If the user needed to look up a term in the terminology file while working with the word processor, a "hot key" activated the TSR program. Other terminology management programs made use of a memory management/multitasking tool such as DESQVIEW™ for interaction with the word processor.

memory management tools

Today the graphics user interface (GUI) offered by MS Windows™ allows different application programs to be run quasi-simultaneously in different windows. Such functions as copy and paste are standard features of the Windows environment, and the trend is to integrate TMS functions directly into the word-processor toolface.

multi-user environments

If the terminology management program is to be used in a multi-user environment, the software should allow both the program and a common terminology database to be stored and accessed by all users on a network file server. Besides taking software aspects into account in evaluating terminology management programs, it is also important to consider some of the technical hardware details required of the computer system.

clock speed

The microprocessor's clock frequency, as expressed in MHz, is usually an indicator of an application's processing speed. Since programs for managing terminology contain more commands for accessing data on the hard disk than time-consuming commands for calculations, the access time rating for the external storage device is much more important than the processor's clock frequency. Of course, operating speed has grown much more affordable in recent years, with standard chip speeds running from 200 to 400 MHz and beyond. Consequently, although operating speed remains an important consideration, it is not the serious stumbling block that it once was.

program search speed

The user's operating speed, however, is perhaps less critical than the actual speed of the application itself. Translators and technical writers require instant gratification. Standard searches requiring more than a fraction of a second are intolerable. What is more, this level of performance must be maintained even after the termbase has reached very large proportions (thousands or hundreds of thousands of entries). If a program offers free-text searches or utilizes fuzzy logic to retrieve information, the amount of time it takes to search the database can be increased considerably if the program has to search each entry. Programs that can periodically build auxiliary indexes increase the speed of operations that would otherwise require time-consuming string searches.

main memory capacity

The capacity of main memory (RAM), expressed in kilobytes (KB) or megabytes (MB), must accommodate the terminology management program. When earlier TMSs were used in memory-resident mode, additional programs (e.g., the word processor) also had to fit into RAM as well. For TSR terminology software, 1 MB RAM was satisfactory, whereas programs running under early versions of MS Windows or under other memory management tools such as DESQVIEW preferred a main memory capacity of 4 MB or more. Again, with new computers being shipped with ever-increasing standard RAM capacity, these considerations are no longer so critical unless a user is locked into outdated, low-capacity equipment. Any computer that meets a user's needs for photo and other graphics manipulation programs is likely to be more than adequate for most TMSs.

hard disk capacity

Hard disk capacity is needed for storing the application programs and the data files. For terminology management, data file size depends on the system-specific entry structure and the corresponding data categories, the expected number of terminological entries, and the data storage architecture of the terminology management program. To evaluate a specific program, it is necessary to create a small set of typical terminological entries as they occur in practice and to multiply the reserved bytes in the resulting data files by the number of entries expected in the next few years. Given the formidable storage capacity of most modern harddrives (i.e., multiple gigabytes), all but the most massive data collections will fit nicely on the average drive. Other considerations, such as color graphics display and mouse function control have become standard to Windows applications and no longer merit special discussion here. The actual implementation of graphics images in individual terminological entries, however, is not a universal feature of all programs.

3 TERMINOLOGICAL ASPECTS

task suitability

The most important criterion in evaluating terminology management programs is the suitability of the software for the intended terminology task. Since there are various kinds of terminology management, great variety can be found in the software products available, which may be based on different system philosophies.

Bilingual or language-pair-based terminology management programs are able to meet the requirements for term-oriented or lexicographic work, but these programs mostly offer only a limited number of terminological data categories and a very simple entry structure. Multilingual terminology management programs come closer to a concept-oriented approach and are much more applicable to multi-user environments. Some of these systems support the idea of term autonomy, where every synonymous term can be fully documented with data categories such as grammatical information, contextual examples or project code (8.1.3: 572; Schmitz 1998).

INFO *Lexicography vs. Terminology INFB 3:328*

The third type of terminology management programs with respect to system philosophy includes all systems with a freely-defined entry structure. These terminology management programs allow users to define their own data categories and their own entry structure so that the software can be exactly adapted to the user's terminological needs and can grow as future requirements change.

freely defined entry structure

To evaluate the utility of a particular terminology management program in a specific environment, it is necessary to examine in detail whether all the necessary terminological and administrative data categories are supported by the system, whether the system's entry structure meets user requirements, and whether the system philosophy is in line with the philosophy behind a user's own work. ISO 12620:1999 provides descriptions (i.e., definitions) and harmonized names for data categories used in terminological working environments.

SEE *Data Categories 8.1.2:552*

It is also important to ensure that the maximum volume of the data occurring in practical terminology management can all be stored in the system's data categories and that multiple data of the same type (e.g., more than one contextual example for one term) can be handled by the system (repeatability of data categories).

system capability and flexibility

For terminology management programs with a predefined set of data categories, the user-friendliness of the system can be supported by suppressing or ignoring any data categories not needed in certain applications or for certain users (see Section 4 below). Some terminology management programs on the market include additional modules for creating and managing classification

INFO *Classification INFB 9:33*

systems, concepts systems, or thesauri. In evaluating a program, users have to decide whether they need these features now or whether they may want to use them in the future.

data integrity

An additional important criterion for evaluation is the system's support for data integrity. The best moment for doing this is when the data are entered, updated, or deleted by the user or when data are imported automatically. In some systmes, however, special validation routines are implemented that could be periodically initiated by the user. Validation procedures should check whether types and forms of data elements are allowed for a certain data category, i.e. whether the structure of a date conforms to the prescribed format or if the permissible instances for the grammatical gender comply with termbase or exchange format specifications.

validation

Such validators can also be configured to detect the presence of double entries (doublettes). If the user tries to enter a term that already exists in the database, the system should give a warning and allow the user to decide whether the doublette is desirable (i.e., a true homograph), or whether to edit an existing entry (i.e., the concept is already documented in the system). It is more complicated for validation procedures to control the integrity of references (links) between entries, i.e., for concept relations, antonyms, or to shared resources such as bibliographical entries, because to do this, the validation utility must examine all the entries in the database or—to make matters even more difficult—to verify entries in other databases.

4 User Interface Aspects

user documentation

interface language

In assessing user interfaces for terminology management programs, the general criteria for evaluating other kinds of user programs also apply. The operating manual, help function, and online tutorial should support the user in learning to operate the terminology management program or in dealing with errors and troubleshooting. Since terminology management programs are often used in a multilingual environment, the languages in which the manual, help function, and tutorial are available and the language of the user interface (commands and messages), as well as their quality, completeness, and readability, should be included in the evaluation procedure.

user interaction

The user can interact with the terminology management program via commands, function keys, traditional menus, or pull-down/pop-up menus by using keyboard or mouse commands. The type of interaction should be convenient for the user and conform with the type of interaction used for the word processor or other programs. This objective of common functionality and even seamless interaction has been greatly improved in current Windows environments and especially with fully integrated tools packages.

screen displays

The information displayed on-screen (system operation and terminological entry) should be easy to read and clearly arranged. Frames, colors and screen display attributes can help here. Fixed templates for data categories and the terminological entry can support readability if the information does not have to be distributed among many screens. In addition to the terminological entry, the dictionaries selected or the subject fields, source and target language and other options (e.g., insert/overwrite mode) should be displayed on-screen. In displaying a specific entry, some terminology management programs inform the user of the preceding or following alphabetical entry/entries (so-called *zoom windows*.)

input and retrieval

A very important aspect of the user interface and the retrieval efficiency of a terminology management program is how a look-up term is input. Does the user have to type in the complete term or only the first letters? Can one use wildcards for truncation or a character pattern if the term is not completely known? Can the user look up parts of a multiword term or of a string compound? Does the software support a free-text search in other data categories than in the term category (e.g., in the definition or context field)? Does the program allow for the automatic generation of multiple keys to the same entry? Does it provide for the compilation of fuzzy-logic retrieval indexes? Is it possible to restrict the look-up procedure by using filters (e.g., to search according to subject fields, specific users, dates, etc.)? As noted above, speed is a critical factor with respect to such specialized searches.

system response

Closely connected with the search facility is the question of how the system's response to an unfound term or to a wildcard search is evaluated. Some terminology management programs only display a message *term not found* on the screen, while others automatically show the term that exhibits the closest match to the search term based on alphabetical proximity, truncation, or fuzzy matching. After a wildcard search, a hitlist of matching terms should be presented from which the user can select the desired term. Generation of Keyword In Context (KWIC) indexes can be highly useful in collections with many compound or multiword terms. Some programs offer the option of automatically generating a new term entry based on an unfound term or generate history lists of unfound terms that can be used as guidance for creating new term entries.

INFO
Termbase Terminology INFB 28:878

information transfer

Since in practice most terminology management programs are used along with a word processor or some other text processing system, the various ways of selecting and transferring information must be considered. In selecting information, the desired data element can be chosen by naming the corresponding data category, by moving the cursor symbol or bar to the element, or by marking with the mouse. As for the type and the quantity of transferable information, some terminology management programs only allow one of the target language terms to be selected, while others permit other data elements or the whole entry to be chosen. Of course, it is also possible to use the standard Windows clipboard, but this can be more time-consuming than the automated transfer routines described above. Clipboard transfer can also inadvertently import embedded codes from source text (e.g., from HTML pages) to the TMS, such as when a soft-carriage return manifests itself as a string of extraneous hard spaces in the target terminological entry.

SEE
Term Extraction 8.4.1:723

Besides the transfer of information from terminology management program to word processor, it is also useful to reverse the process, i.e., to transfer the look-up term or contextual information from the word processor to the terminology management program. This feature has become increasingly important as we move to environments where the text to be translated is more frequently available in machine-readable form. The next obvious step is to be able to generate new starter entries based on automatic term recognition and extraction.

5 ORGANIZATIONAL ASPECTS

From the organizational point of view, users have to assess whether a given TMS fits into their work environment. As a general precondition, all hardware and software requirements of the program, as discussed in Section 2 above, should be compatible with the existing or planned hardware and software infrastructure. In some cases this may, for instance, entail implementing a "second choice" TMS system because of client preference or the compelling need to use the system with its companion TM.

hardware-software infrastructure

The maximum number of languages and the maximum number of dictionaries or glossaries that can be handled by a terminology management program should be greater than or at least equal to the actual number of languages and dictionaries that are to be used in practical terminology management. The maximum number of entries that can be stored in a dictionary is not such an important criterion for evaluation as it once was. Modern terminology management programs are only limited in the number of entries they can handle by the capacity of the hard disk, or the maximum number of entries they can accommodate is so high that there are no restrictions for most practical work. Furthermore, the operating speed of well-designed systems should not diminish appreciably with increased database size, not even for very large collections.

hard disk capacity

All working languages to be used by the terminology management program have to be supported by standardized alphabetical sorting and by the correct character set. Since program developers cannot be expected to foresee all the languages that might be employed, some terminology management programs allow users to define their own alphabetical sorting arrangements. For non-Latin character sets such as Greek or Cyrillic, additional program modules or predefined sets are available. To cope with languages like Arabic, Japanese or Chinese, modified operating systems or other special utilities may be required, even in the Windows environment. Although many new programs are moving rapidly towards the implementation of Unicode (ISO 10646-1:1993), it will take a number of years before the universal application of the standard can be taken for granted. Some languages, such as Urdu and Thai, still pose unresolved problems related to directional tracking and other special character set and layout features.

alphabetical sorting

SEE *Multilingualism Vol. 1, 3.1:252 ff.*

If the terminology management program is to be used in a multiuser environment, the software should support a network application (see Section 2). This includes correct handling of situations where different users might simultaneously compete to write to the same entry. From the organizational point of view, the program should allow the system administrator to define different access rights, different subsets of data categories, and different screen

SEE *User Profiles 8.1.4:600*

displays or filters for different user groups, such as foreign correspondence clerks, translators or terminologists.

If interchange of terminological data is planned with other departments or colleagues (using the same terminology management program or other data processing systems), terminology management programs will have to allow the import and export of data. This interchange facility should permit the user or the system administrator to define the extent, structure, and layout of the data or should import or export data corresponding to an interchange standard such as ISO 12200:1999 (MARTIF).

SEE *Terminology Interchange* *8.1.5:613*

integrated environments

In current high-performance translation and localization environments there is a growing trend to require full integration of terminology, translation memory, and machine translation systems. In such situations, ease of interactivity is essential among the various major components as well as other localization tools (i.e., alignment tools, gateway software solutions, or special localization programs designed to enable translation of interface or help information without jeopardizing the integrity of actual program code). Interchange solutions also play a role here if it is desirable to share terminological information among different applications.

hardcopy publishing

The interchange capability of a terminology management program can also be used for exporting terminological data to a word processing or a desktop publishing program (with the necessary layout tags) for printing a dictionary or glossary. If users decide to replace their (old) terminology management program with a new one, the interchange mechanism saves them from the potentially onerous job of rekeying all the data by hand.

precompiled dictionaries

Precompiled terminology files or dictionaries are available for some terminology management programs and can only be accessed by the appropriate program. The availability of such dictionaries may be a very important criterion for choosing a particular program, but it must be borne in mind that this existing terminology is in most cases only a first step in terminology management and does not spare users the effort of creating their own entries.

6 ECONOMIC ASPECTS

auxiliary tools

The economic aspects of program evaluation pertain primarily to the purchase and operating cost of a terminology management program. When customers decide to buy a certain TMS, they have to consider whether their present hardware (microprocessor type and hard disk capacity) and application programs (e.g., the word processor) can still be used. In comparing prices for different terminology management programs, it is necessary to assess whether important auxiliary tools, e.g., import-export routines and character set or classification modules, are already contained in the software package and included in the price, or whether they must be purchased separately.

For multi-user and network environments, a multiple software license for a given number of users can be cheaper and more efficient than buying several single copies of the program. In some cases, auxiliary programs or additional dictionaries have to be bought only once. It is also possible in specific organizational environments for some users to be given only a cheaper read-only version of the terminology management program, whereas other users need to work with the full version.

customer support

Apart from acquisition costs, additional expenditure may be necessary for installation, service, online product support and training courses. The reputation of the software developer and the local dealer is an important evaluation criterion, but this information is not usually easy to assess. The number of installed terminology management programs and the experience of other users may help indicate the quality of the software and the future support that can be anticipated from the developer. A history of periodic upgrades and a clear indication that the existing program is undergoing constant research and development in the anticipation of a future upgrade are important factors in judging a product's long-term viability. No one wants to "get stuck" with a product that is no longer adequately supported by the original publisher.

7 EVALUATION SCENARIOS

This discussion has described all the important criteria used to evaluate terminology management programs, although this account cannot claim to be exhaustive. For certain programs and in specific application environments, other aspects may have to be considered as well, such as in-house restrictions on program use (i.e., rules that dictate the universal use of a common, nondedicated database management program for all applications in an organization) or the difficulties involved in providing important program modules to a broad freelance subcontractor base outside the parent company. New circumstances can arise that require rapid reorientation on the part of clients and software developers, such as was the case in the mid 1990s when the rapidly evolving need for Web-based terminology management services sent everyone scurrying for solutions..

These criteria for evaluating terminology management programs can be used to conduct a market analysis and survey of existing terminology management programs, e.g., as a basis for selecting and purchasing a certain software that fits a user's specific needs and work environment. But the discussion could also be used—and I hope it will—in the design and development of new systems and in efforts to improve existing systems.

There is a fairly extensive literature available for planners to consult as they try to make their decisions. Freigang et al. (1991), as well as Mayer (1990) and Schmitz (1991 and 1993) were among the first to work out standard criteria for evaluating terminological databases. Blanchon (1991, 1993, 1994) has been instrumental in preparing detailed evaluations of virtually every system that has come on the market in recent years. GTW (1994 and 1996) and TermNet (1993) publications have placed special emphasis on these concerns as well. Articles by Schmitz (1994a and b, 1995, and 1996 a and b) provide a more recent view of related issues.

REFERENCES

Blanchon, Elisabeth. 1991. "Choisir un logiciel de terminologie." *La banque de mots,* Numéro spécial 4, 5-96.

_______. 1993. "Terminological Database Software: The Changing Market." Schmitz, Klaus-Dirk [ed.]. *Supplement Proceedings of the Third International Congress on Terminology and Knowledge Engineering,* Cologne, 25-27 August 1993, 10.

_______. 1994. "Inventaire de logiciels de terminotique." *RINT*. Available at: http://www.uhb.fr/~lemeur_a/inv94.htm. Accessed 1998-07-13.

Freigang, Karl-Heinz; Mayer, Felix; Schmitz, Klaus-Dirk. 1991. *Micro- and Minicomputer-based Terminology Databases in Europe.* Vienna: TermNet (*TermNet Report 1*).

GTW. 1994. *Criteria for the Evaluation of Terminology Management Software.* Felix Mayer, ed. Bozen: GTW Report.

GTW. 1996. *Guidelines for the Design and Implementation of Terminology Data Banks.* Christian Galinski and Klaus-Dirk Schmitz, eds. Saarbrücken: GTW Report.

ISO 10646-1:1993. *Information technology — Universal Multiple-Octet Coded Character Set (UCS) — Part 1: Architecture and Basic Multilingual Plane.* Geneva: International Organization for Standardization.

ISO 12200:1999. *Computer Applications in Terminology—Machine-readable Terminology Interchange Format (MARTIF)—Negotiated Interchange.* Geneva: International Organization for Standardization.

ISO 12620:1998. *Computer Applications in Terminology—Data Categories.* Geneva: International Organization for Standardization.

Mayer, Felix. 1990. "Terminologieverwaltungssysteme für Übersetzer." *Lebende Sprachen,* 3, 106-114.

Schmitz, Klaus-Dirk. 1991. "Rechnergestützte Terminologieverwaltung am Übersetzer-arbeitsplatz." *Akten des Symposions "Terminologie als Qualitätsfaktor."* Köln: Deutscher Terminologie-Tag, 89-106.

_______. 1993. "Interaktionsarten bei Terminologieverwaltungssystemen." *Sprachdaten-verarbeitung für Übersetzer und Dolmetscher. Akten des Symposiums zum Abschluß des Saarbrücker Modellversuchs.* Ingeborg Fischer et al., eds. Hildesheim: Olms-Verlag, 109-132.

_______. 1994a. "Verarbeitung fachsprachlichen Wissens in Terminologieverwaltungs-systemen." *Fachkommunikation: Kongreßbeitrage zur 24. Jahrestagung der Gesellschaft für Angewandte Linguistik (GAL).* Bernd Spillner, ed. Frankfurt: Peter-Lang-Verlag, 45-55.

_______. 1994b. "Überlegungen zum Einsatz und zur Evaluierung von Terminologie-verwaltungssystemen." *Lebende Sprachen,* 4/94: 145-149.

_______. 1995. "Evaluating Terminology Management Programs." *TAMA '94. Proceedings of the Third TermNet Symposium, "Terminology in Advanced Microcomputer Applications."* Vienna: TermNet, 311-315.

_______. 1996a. "Tools for Managing Terminology." *The ELRA Newsletter.* 1/3 (October 1996): 11-12.

_______. 1996b. "Terminology Management Systems." *The Translator's Handbook.* Rachel Owens, ed. London: Aslib, 221-246.

_______. 1998. "Über wichtige Aspekte bei der Einrichtung einer rechnergestützten Terminologieverwaltung." *Proceedings of the 11th European Symposium on Languages for Special Purposes.* Lita Lundquist, Heribert Picht, and Jacques Qvistgaard, Eds. Copenhagen '97. Copenhagen: Handelshojskolen i Kobenhavn, 391-398.

TermNet. 1993. *Proceedings of the 2nd TermNet Symposium "Terminology in Advanced Microcomputer Applications — TAMA '92"*, Avignon, 5-6 June 1992, *TermNet,* Vienna.

The Unicode Consortium. 1997. *The Unicode Standard: Version 2.0.* Reading, MA: Addison-Wesley Developers Press.

8.1.2 Data Categories for Terminology Management

SUE ELLEN WRIGHT

1 TERMBASES

As described in the previous chapter, today terminological data are most frequently managed in dedicated terminology management systems, which are commonly called TMSs. When an individual or organization creates a computer-based collection of terms, this resource is usually called a terminology database or a termbase for short. Physically speaking, termbases can consist of a single data file or two or more physical data files that comprise the virtual database file. Generally, when multiple files exist, users are unaware that the system is using more than one file at any given time.

INFO

Termbase Terminology INFB 28:878

Termbase files are made up of concept-oriented terminological entries. The difference between terminological and lexicographical entries has already been treated in this book. In some systems, virtual term entries can be made up of one or more physical records. These systems treat subsets of the term entry (the languages segments) as separate, but linked, records in the system, whereas other systems use a single record for each entry, but divide it up into subordinate segments. In systems like those described by Sager in this volume (8.4.2:761), terminological information can reside in various kinds of tables and files to be collected as needed to build user-defined output records. Users are frequently unaware of this kind of internal, system-specific segmentation.

term entries

INFO

Lexicography vs. Terminology INFB 3:328

2 DATA ELEMENTS AND DATA CATEGORIES

Data files, entries, and records can be viewed as packages or containers into which data are packed (Wright 1991). Individual units of data included in databases are most often called data elements. They manifest themselves in data records and entries as data items or instances. (The exact terms used by different schools of thought vary, and efforts to unify usage are futile.) Data

data elements

elements are the smallest indivisible units of information in a data entry and generally reside in individual data fields. Data elements (fields) can be packed inside data segments, which in turn form components of the data entry.

data categories

Terminologists distinguish between the instantiation of the individual element (i.e., the instances where it is used) and the general category to which elements belong. For instance, a term entered in a database will generally reside in a data field with the data element name *term*. Many terminologists consider any single instance of this data field to be a data element, while all instances of this particular data element comprise the data category identified by the name *term*. Data categories represent a class of terminological elements and can be viewed from a functional standpoint as *instructions for interpreting a given data field* (ISO 1087-2:1999).

2.1 Sets of Data Categories

selecting data categories

Generally each database defines a set of data categories to be used in that database. Some free-form indexed text databases allow for the addition of data categories on an ad hoc basis, but this article will not deal with this practice. Many of the articles included in this book treat specific data categories, e.g., terms, definitions, contexts, and data categories designed to represent the position of concepts in concept systems. Efforts have been made in the past to define a single ideal set of data categories that everyone would use or a short list of data categories from which everyone would choose, but experience has revealed that termbases are designed to meet very different needs in different working environments, which gives rise to variant data structures (GTW 1996).

data category list

ISO Technical Committee 37, *Terminology (principles and coordination)*, Sub-Committee 3 for Computer Applications, Working Group 1 has collected a broad selection of the terminological data categories used in many different termbases, a list that forms the basis for ISO 12620:1999, *Computer Applications in Terminology — Data Categories*. This collection is based on many years of preliminary documentation (Felber 1980 and 1982; Galinski et al. 1982 and passim; Galinski 1983 and 1984; Budin et al. 1994, Budin and Wright 1994, and Wright 1993 and 1995). Until its finalization as a standard, new data categories continued to be added to the list, and more will undoubtedly be proposed in the future, which implies that the list will always be incomplete at any given time, although it already contains most of the data categories that are likely to occur in termbases. It is especially important to realize, however, that any given termbase will almost certainly use a relatively small subset of the full list. It is the purpose of this standard to provide a global resource for the selection of data categories, but not to specify which categories should be used in any given environment. This choice remains the responsibility of individual data modelers based on their evaluation of user needs within their respective environments. It should also be noted that the specification of data category names in ISO 12620 applies to names used in interchange formats. Although the standard can be used to select names for new database designs, the work group recognizes that individual existing applications already have their own naming conventions and there is no recommendation that changes should be made. In this regard, the data category names in the standard can function as data element concept identifiers that can be used for mapping between applications, provided that the categories in questions meet the criteria set down in the data category descriptions included in the standard.

2.2 Data Category Specifications

data dictionaries

Data categories possess a number of features that significantly affect their accuracy and effectiveness with respect to data maintenance and manipulation. In order to ensure the integrity of data within databases, data categories must be clearly specified and individuals entering data must guarantee that the data they enter conform to the data category descriptions specified in data category lists or data dictionaries. ISO/IEC JTC 1 / SC 32 has written a series of International Standards governing the documentation of data elements in data dictionaries (ISO 11179), although the list given in ISO 12620 was developed prior to and independent of these specifications. In particular, the names specified in ISO 12620 reflect a consensus basid on real names used in a wide variety of termbases and do not conform to the naming principles set down in ISO 11179-5:1995. The following features represent minimum criteria for the information that should be documented in such data dictionaries (Schmitz 1998).

2.2.1 *Data Category Name*

monosemous names

Terminologists have long followed the practice whereby each data category has a single name that should ideally be used consistently throughout a given TMS. They generally take care to ensure that integrated subsystems do not introduce new names for the same or closely similar content, which can lead to ambiguity in interpreting the data, a situation that is similar to polysemy in general language environments. As noted, ISO 12620 does not attempt to standardize what data category names will be used in different termbases. The purpose of the standard is to designate the data category names that should be used during data interchange, which means that if data category names do not comply with this standard, they must be mapped to standardized forms before exporting them for interchange (Wright 1996, 1998). The alternative to maintaining monosemous names, should there be demand for different naming conventions within an enterprise or work group, is to maintain a table of correspondence designed to map variable names to the same data category concept identifier. This capability may be particularly important for situations where different applications (TMS, TM, MT, etc.) would share the same terminological data in integrated environments (see Newton 1999).

2.2.2 *Data Category Description*

definitions

Data category specifications contain descriptions as one of their principal attributes, although some database management experts prefer to talk about data element definitions. Terminologists call them descriptions instead of definitions in order to distinguish them from definitions in terminology collections. A data category specification may stipulate that a particular category "contains xyz...", but it is also important to define or describe xyz in order to ensure that users clearly understand what the content of the category really is. For instance, a data dictionary specification for data elements used in TMSs could state for the data category *synonym*: "Contains a synonym," but this is not necessarily useful if the user does not know what is meant here by a synonym. Hence the description in ISO 12620 reads instead, "Any term that represents the same or a very similar concept as the main entry term in a term entry." Descriptions (or definitions) can, however, reside in a parallel terminology standard where they can be referenced, provided that they are readily accessible, e.g., on the Word Wide Web.

Data category descriptions are also important in order to avoid ambiguity. For instance, in terminology databases, the categories *example* and *context* are sometimes confused with one another. A dictionary of data categories to be used as a basis for interchanging data must clarify which possible interpretation is required in order for interchange partners to arrive at a *normalized* form that they can both or all interpret correctly. Thus the descriptions play an important role in the mapping process cited above. Such descriptions should ideally meet the requirements for terminological definitions, i.e., by indicating the relationship of the concept represented by the data category within its concept system and by distinguishing each data category from closely related categories.

SEE *Terminology Interchange 8.1.5:613*

2.2.3 *Data Type and Permissible Instances*

dates

Data dictionaries specify more than just the content of the data entered into a specific field. They can also prescribe the data type (e.g.,. text, dates, numbers, etc.). For dates in particular, ISO 12620 recommends the use of the standard ISO date format (e.g., 1998-08-04), but does not require this form because so many existing databases use other formats. Given the growing interest in data interchange and system integration, however, it is highly advisable for managers of current databases and designers of future databases to reexamine non-conforming procedures in order to ensure consistent reliable database performance in the future.

standard formats

Data type designations can also be associated with references to standardized formats for representing information, such as standards for country, language, and script symbols, or international identification numbers used as bibliographic references. In materials databases, for instance, one of the most common forms of representation is to require the use of SI units. In this regard, some databases may also specify field length for such standardized elements, although the trend in terminology database design is to avoid restricting field length for such terminological elements as *term, definition, context, etc.*

pick lists

Restriction of element content to a specified data type and standard constitutes the specification of the permissible instances that can occur in that data category. Another way of specifying permissible instances is to create a picklist from which users select the appropriate content for any given case. The advantage of using predefined pick lists where appropriate is to prevent the introduction of different forms for representing the same meaning or the inclusion of undocumented content. For instance, a system could define the English parts of speech used in the system as: n, v, adj, adv, thus ruling out the option of *noun, verb, adjective,* and *adverb*, or for using *substantive* instead of *noun*.

2.2.4 *Granularity*

grain size

Defining data categories is also important because different termbases treat similar data categories in different ways. For instance, Figure 1 illustrates an example of a data category for *address*. Figure 2 shows the same information, but with information units divided into finer categories. To ensure that the correct data are input into any of the fields, or that users can accurately interpret these data, data entry must conform to data element descriptions. This difference between methods for subdividing information is called granularity in analogy to the notion of *grain size* in materials science and optics, where a higher degree of granularity corresponds to smaller grain size and greater precision.

retrieval & manipulation

Increased granularity facilitates retrieval and manipulation. For instance, the entry in Figure 2 can be sorted by last name, town, state, postal (zip) code, or country. (Theoretically, it could also be sorted by the given name, but this option is probably rare.) In a database set up by a local municipality (or by a catalog company plotting customer demographics), it might, however, be desirable to sort by street name and number, in which case the data category *street address* could be subdivided into *street number* and *street name* to achieve higher granularity. A typical example of variation in granularity in a terminological data category involves the decision whether to include all grammatical information in a *grammar* data element or to specify *part of speech* and *gender* separately.

Data element	**Content**
Name	John Doe
Address	267 Prospect St. Kent, Ohio 44240 USA

Figure 1: *Minimum Granularity*

Data element	Content
Last Name	Doe
First Name	John
Street Address	267 Prospect St.
City	Kent
State	Ohio
Zip Code	44240
Country	USA

Figure 2: *Increased Granularity*

data reusability

Each database system will define the granularity of data elements in keeping with perceived data management needs. System designers must think through these needs carefully when modeling data in order to ensure that the structures they implement at the outset will indeed meet future requirements for data manipulation and maintenance. It is important to bear in mind that in interchange environments, differences in granularity tend to limit data reusability. Allowing for broader categories facilitates interchange, but reduces the power and freedom to manipulate data. One could write a conversion routine that would combine elements from Figure 2 to fit into those in Figure 1, but there is no automatic way to split the elements shown in Figure 1 so that they could be assigned to the more granular model shown in Figure 2. Thus, when planning an interchange formalism or creating a data element dictionary, it will be desirable to indicate the degree of granularity required for automatic interchange.

2.2.5 *The Elemental Nature of Data Categories*

data integrity

Some instances of granularity, such as the *address* example cited above, reflect the needs and philosophy of the system designers. Other choices, however, are not so optional. For instance, some termbase models include bibliographic source information in the same field with a text segment or place the page numbers together with the source identifier that references a separate bibliographic entry as a shared resource. This practice violates another principle of data management, the *elemental* nature of data categories, which requires that only one kind of data and only one instance of that particular data category be included in a given field. Failure to observe this maxim also compromises the essential *integrity* of the database and prevents efficient manipulation of the data. Consequently, designers should avoid "shopping basket" style data categories such as undifferentiated *notes* or *comments* that can result in many different kinds of information being stored in the same category (Wright 1993), a danger implied by Le Néal's *note* field in her discussion of the Teleglobe Canada dictionaries (8.2.1:653)..

SEE *Entry Structures 8.1.3:572*

2.2.6 *Redundancy and Shared Resources*

data maintenance

A cardinal rule of database management states that redundancy should be avoided. For instance, it is possible to include complete bibliographic references for each text element included in a terminological entry (definitions, contexts, etc.), but if information needs to be corrected or updated, it becomes necessary to find each citation to the same source and make the desired change. Furthermore, repetition of the same information unnecessarily occupies storage capacity and requires additional data maintenance. Consequently it is more economical from the standpoint of database management to enter the bibliographic reference once in a single bibliographic entry or table and to link all the individual entries that cite that resource to the bibliographic entry. This approach also contributes to efficiency with regard to increased granularity because it is much more feasible to enter and maintain articulated data in a single bibliographic entry that will be used as a shared resource than it is to do so in each individual term entry. The exception to this rule is to include URLs (universal resource locators) directly in each entry, both because they are frequently unique to a few entries and there for not shared and because it is very convenient to be able to use them as live links to go directly to the original source on the World Wide Web.

hyperlinks

Bibliographic entries are not the only kind of shared resource included in termbases files. Links in terminological entries can point to responsibility entries, graphics, video and audio files, tables and charts, diagrams of concept systems, lists, materials management systems, thesauri, and other types of related resources. It is also more practical to maintain these kinds of "foreign" data files as shared resources because they usually must be stored as separate objects somewhere in the information environment, which may include links to the WWW. They are then referenced from individual term entries using relational or hypertext links.

3 ISO 12620:1999, *Computer Applications — Data Categories*

data category selection

As noted above, ISO 12620 is designed as a global list that accounts for all types of data categories that are used in various kinds of databases in different terminology management environments. The original list has been augmented with recommendations from terminology database managers working in translation agencies and language services, standardization and language planning environments, term banks, academic institutions, and individual companies. This collection covers both prescriptive and descriptive environments, and allows for interaction with classification, thesaurus, and concept system management. The basic purpose of the standard is to support individual system designers in selecting the subset of categories that will best support their needs regardless of the environments in which they are working (Wright 1998).

The standard is divided into three major groups reflecting the three major types of information contained in termbases:

- Terms and term-related data
- Descriptive information pertaining to the concepts treated in term entries
- Data that provide information relevant to database administration

meta data categories

While terms stand alone and have no sub-categories, term-related, descriptive, and administrative designations are sometimes referred to as *meta-data categories* because they represent high-level superordinate categories that are then subdivided into a number of other categories, together with their permissible instances.

CLS Framework

These groups are then divided into subgroups as shown in Table 1. The following overview outlines the specific data categories contained in the individual subgroups. The definitions for individual data categories and additional information on using them can be found at http://www.ttt.org and are not reproduced here because of lack of space.

3.1 Terms

Any terminological unit can be stored in the *term* category, a single-word or multiword term, a phrase or collocation, or possibly even a longer text segment for which it is desirable to document an equivalent. In terms of the standard, the *term* category is freely repeatable as needed, which means that a data element can recur within the same data entry, although repeatability may well be limited by the constraints built into a given database environment. The term is also potentially free to combine with other data categories, which means that the principle of term autonomy can be supported within the framework of the standard, but there is nothing in the standard that imposes this principle (see 8.1.3, Section 2.4:583 for further discussion of term autonomy).

term
autonomy

repeatability

3.2 Term-related Information

By the same token, *term* is the only data category that can have as its content a term representing the concept treated in a term entry. (One exception is when a term is used as a pointer to link, e.g., an abbreviated or short form to the expanded form of the term. Subsidiary information such as whether the term is a synonym, a preferred term, or a deprecated term, whether it occupies a certain register, or whether it is an abbreviation or a full form is all treated as attributive information associated with (i.e., combinable with) the term. This term-related information is treated as the first of the meta-data categories and is divided into subgroups such as:

term types

- Indication that a term is a symbol, formula, equation, or a materials management category (sku, part number)
- Indication that a term is a full form or some type of abbreviation
- Indication that a terminological unit represents a certain type of phraseological unit
- Grammatical information concerning the term itself (i.e., the word or words involved)
- Regional, temporal, register-related and proprietary restrictions
- Etymology, pronunciation, syllabification, hyphenation, and morphological information
- Specific term-related administrative information associated with the processing of terms in standards and language-planning environments

Table 1: *ISO 12620 Data Category Groups*

Term and term-related data categories

Subgroup 1 consists of the data category *term* and contains a term or other information treated as if it were a term (e.g., phraseological units and standard text).

Subgroup 2 specifies data categories for term-related information.

Subgroup 3 specifies data categories for information relating to equivalence between or among terms assigned to the same or very similar concepts.

Descriptive data categories:

Subgroup 4 specifies data categories for the classification of concepts into subject fields and subfields, along with other classification-related information.

Subgroup 5 specifies data categories for concept-related description, i.e., different kinds of definitions, explanations and contextual material provided to define or otherwise determine the subject field and concept to which a term is assigned.

Subgroup 6 specifies data categories for indicating relations between pairs of concepts.

Subgroup 7 specifies data categories used to express the position of concepts within concept systems.

Subgroup 8 the data category *note*. This category stands alone because it can be associated with any one of the other categories and therefore cannot be subordinated to any other specific subgroup.

Administrative data categories:

Subgroup 9 specifies data categories for documentary languages and thesauri.

Subgroup 10 specifies data categories for all other strictly administrative information.

Supplemental ISO 12200 Data Category Groups

Subgroup 11 specifies special codes used in the MARTIF standard.

Subgroup 12 specifies bibliographic data categories.

Many of these items (and of the other items listed below) can be configured as picklists containing permissible instances, whereby users can select the relevant specific content of the data category, for instance:

pick lists

grammatical number

singular
plural
dual
mass noun

3.2 Equivalence

Conceptual equivalence between or among terms assigned to the same or very similar concepts is implied by the very presence of a term in a given terminological entry, i.e., all terms in an entry can be assumed to be at least quasi-synonyms or quasi-equivalents of the other terms included in the entry unless otherwise indicated. If additional equivalence-related information is required, the categories in this subgroup can be used to indicate the degree of equivalence that exists between or among the various items, as well as indicating the existence of false friends (*faux amis*), directionality factors, and reliability codes. Perhaps the most useful category provided here is the *transfer comment*, which provides for a free-form note in which users can record important information on target language usage, directionality, or specific reliability-related concerns.

degree of equivalence

transfer comment

3.4 Subject Field Classification

The second meta-data category, called simply *descrip* in ISO 12200, provides for descriptive information used to define and describe the concepts treated in terminology entries. The first set of categories under this header designates subject fields (domains) and subfields, thus indicating the discipline to which a concept belongs and providing users with an orientation for relating the concept to other neighboring concepts. Categories in this section allow the user to indicate multiple levels of classification or to identify concepts according to differing criteria. For instance, the *diaphragm spring clutch* shown in Figure 1 of Chapter 1.4.1 (90) of this *Handbook* can be classified as belonging to the subject field

INFO

Classification INFB 9:334

automotive engineering, and to the subfield *drive train components*. It is also possible to provide links to classification systems or to indicate the notation assigned a concept in such a system.

3.5 Concept-related Description

In addition to categories such as definition, context, and example, this subgroup includes items for nontextual illustrations, which are usually referenced via hypertext links to external data files, as well as special items such as *unit* (as in SI units, for instance) and *range*, which are important for terminology-related materials databases. There is some difference of opinion in the terminology community as to whether *contexts* should be classified as term-related information or descriptive information. Several authors in this volume argue very persuasively that contexts effectively anchor the term conceptually within the flow of discourse and thus determine its current (as opposed to its potential) information content. Nevertheless, this issue does not necessarily have to be resolved in order to ensure effective data interchange because the assignment of the data category to any one data model can be implemented during data interchange (Wright 1999).

SEE

Term Extraction *8.4.1*:725

Terminology Compilation *8.4.2:761*

Text Production *8.4.3:772*

3.6 Relationships between Pairs of Concepts

This subgroup provides categories that designate the kinds of relations that can exist between individual pairs of concepts, e.g., generic, partitive, sequential, temporal, spatial, and associative relations. For instance, the relation between a diaphragm spring clutch and clutches in general is a generic one: it is a kind of clutch, whereas the relation between the clutch and the automobile is a partitive or meronymic one because it is a part of the automobile. The list of types of relations is actually an open one. Indeed, in the future it would be highly advisable for ISO TC 37 to explore the addition of a more comprehensive list of concept relations, e.g., cause and effect, producer-product, etc. It may also be useful to include categories used in creating ontological systems.

INFO

Conceptual Relations *INFB 10:337*

3.7 Position of Concepts within Concept Systems

SEE
Concept Systems 1.4.1:89

This subgroup contains the items used to differentiate levels in hierarchical concept systems, specifically *broader, superordinate, subordinate,* and *coordinate concept*. Although ISO 12620 separates positions in concept systems from types of relations, in actuality they frequently function together: A sailboat is a subordinate concept with respect to the superordinate concept boat in a generic concept system, whereas a clutch is a subordinate concept to the superordinate concept drive train in a partitive concept system. In both Geneter and MARTIF implementations of the data categories, the data categories described here in 3.6 and 3.7 are used together.

3. 8 Note

The data category *note* stands alone because it can be associated with any one of the other categories and therefore cannot be subordinated to any other specific subgroup.

3.9 Documentary Languages

administrative data categories

SEE
Thesaurus Management 6.1:381

Data categories used in documentary languages and thesauri (e.g., *thesaurus descriptor, top term, broader term, narrower term*, etc.) are listed here and classified as administrative data categories. They are grouped together under the *admin* meta-data category. Care has been taken to distinguish these items from both *classification systems* and *concept system* designators. The three types of resources have evolved out of different traditions, although they are all designed for knowledge representation. Despite their many obvious similarities, the elements in 3.6 and 3.7 were felt to be primarily descriptive and concept oriented, as opposed to the administrative role that thesaurus-style systems play in organizing, storing, and retrieving information. The current evolution of inferential (smart) ontologies in the field of artificial intelligence (AI) adds yet another trend that incorporates aspects of existing models with new insights to create highly powerful systems designed not only to represent knowledge structures and to retrieve information, but also to manipulate it and to make decisions as well.

3.10 Other Administrative Information

Section 10 of ISO 12620 specifies data categories that are related to the administration of termbases. These factors include such aspects as *terminology management transactions* (organization, modification, approval, etc.), *terminology management functions* (kinds of dates and responsibility codes), various *subset codes* (customer, project, product, security, etc.),and *authorization codes*. *Term qualifiers* identify items as *entailed terms, sort keys,* and *search terms*. The two- and three-letter language "symbols" provided by ISO 639 are included here. They are so called because in ISO, many letter-based abbreviations and units are designated as symbols, much in the same way that mathematical and chemical values are also considered by be symbols (e.g., Σ, π, H_2O). A data category designed to be used for marking text as "foreign text" when text segments in a Language B are embedded in surrounding text in a Language A.

INFO

Entailed Terms INFB 14:342

Section 10 also provides for categories that specify *collating sequences*, which pertains to alphabetical or other forms of ordering, and is particularly important when working with alphabets or languages where the collating sequence varies from standard Western European sequences. The designation of *entry type* can be highly useful in databases that incorporate several kinds of entries (terminological, lexicographical, bibliographic, responsibility, etc.). The *element working status* categories can be used to mark the current status of records, entries, and data fields (starter, working, consolidated, archive, etc.). Various types of *identifiers* can be associated with files, concepts, entries, and records.

linking categories

Some data categories in this section typically involve links between term entries. Although these items can be viewed as relational functions rather than as true data categories, they are nevertheless perceived or used as data categories with special linking features in some databases and as informational categories that require human action in others. They include *cross-references, see* and *see-also* links, *inverted* and *permuted forms* of complex terms, *homograph* and *antonym* references. In this context, it is important to remember that homographs that have different meanings will appear in separate term entries in concept-oriented termbases, and that the full documentation of an antonym will also appear in its own entry because this is by definition a different concept from that treated in the entry where a term is cited as an antonym.

Source identifiers are distinguished according to whether the source (bibliographic) information involved is included inside an entry or whether it is treated as a shared resource that is merely cited in the terminological entry. Of course, as noted above, the latter is the more desirable methodology to use in an efficient, well-modeled termbase. Special categories are provided for marking *namespace identifiers* such as *universal resource locators (URLs)* and *formal public identifiers (FPIs)* (which are used to identify documents on the World Wide Web, as opposed to webpage locations—the concept of an FPI is that if a document were moved from one URL to another, one could still find it because the FPI would remain the same). Obliquely related to the source categories are the *originating entity* categories, which are specific data categories used in language planning environments and provide markup for individuals, institutions, and databases that originally introduce terms or propose terms or concepts to be included in prescriptive terminological collections.

source markup categories

3.11 The Representation of Data Categories for Data Interchange

ISO 12200 specifies SGML markup for each of the data categories and permissible instances listed in ISO 12620. These items are cataloged in charts in Annex A of the standard. In addition to the ten sections described above, ISO 12200 also includes two additional sections that provide supplemental markup for use in the dynamic interchange environment. Section 11 lists the special tags that are used in the MARTIF standard for grouping, referencing, and retrieving other data elements.

MARTIF tags

3.12 Links

Section 12 outlines the kinds of links that can occur in MARTIF and provides examples illustrating how they can be used together with the data categories to achieve hypertext connections within interchange documents.

3.13 Bibliographic Data Categories

A number of data categories must be combinable with source documentation in termbases. Indeed, the provenance of the infor-

documentation

mation included in terminological entries constitutes the strongest evidence that there is for the quality of the equivalents and definitions contained in these entries. In some databases, terms themselves are linked to source information, and such items as *definitions*, *contexts*, and *notes* generally involve the citation of source information.

ISO 12083

As noted above, some databases provide source information or source codes directly inside the data fields being documented, which is an undesirable practice because it is impossible to automatically separate the different kinds of information involved. Other databases provide separate fields for bibliographic information, but do include complete citations for each documented item in each term entry. This practice results in high data storage requirements and difficulties in maintaining accurate information over time. Ideally, bibliographic information is stored in a bibliographic file or file segment that can serve as a shared resource for the entire termbase or even for several other integrated resources as well. An informative annex of ISO 12083 provides a list of SGML tags for common data elements that can be used in marking up bibliographic references in standard text. Informative Annex B of both ISO 12620 and 12200 mirrors these categories by citing a subset that is specifically applicable to the documentation of terminological information.

4 APPLYING ISO 12620 DATA CATEGORIES

SEE *Entry Structures 8.1.3:572*

The wealth of categories listed in the standard is evidence of the great variety in the approaches and needs that termbase creators address when creating their resources. This kind of diversity can have negative affects if it prevents the reuse of data via data import and export or if it makes it difficult for even human users to interpret data (see Budin and Wright 1994). The current standard represents the culmination of a process that has lasted several decades and that is likely to continue into the future as developers respond to the ever-changing needs of terminologists working in new environments. The intent of the standard is neither to simplify nor to complicate terminological entries, nor does it presume to legislate what data categories must be called in individual termbases. Its purpose is to present a catalog from which data modelers can select those data elements they want to use in their termbases.

Differences in data category names are not overly problematic because names can always be converted during interchange, but failure to take the definitions of individual elements into consideration or to ensure the integrity of the data assigned to those categories will not only cause problems with exchanging data. Bad data management practices can also be the cause of increased data management cost or even the need to re-key data in the future. The selection and organization of data categories is dependent on criteria that will be discussed in the following chapter on entry structures.

REFERENCES

Budin, Gerhard; Melby, Alan K.; Shreve, Gregory M.; and Wright, Sue Ellen. 1994. Chapter 13: "Terminological Databases." *TEI P3, Guidelines for the Encoding and Interchange of Machine Readable Texts.* C. Michael Sperberg-McQueen and Lou Burnard, eds., Chicago & Oxford: Text Encoding Initiative, 371-390.

Budin, Gerhard and Wright, Sue Ellen. 1994. "Data Elements in Terminological Entries: An Empirical Study." *Terminology*, I/1, 41-59.

Felber, Helmut. 1984. *Terminology Manual*. Paris: UNESCO and Infoterm.

Galinski, Christian. 1982 and passim. *Infoterm Guidelines on Terminography*. Unpublished manuscript. Vienna: Infoterm.

GTW. 1996. *Guidelines for the Design and Implementation of Terminology Data Banks.* Schmitz, Klaus-Dirk, and Galinski, Christian, eds. Saarbrücken: Association for Terminology and Knowledge Transfer.

ISO 639:1988. *Code for the representation of names of languages*. Geneva: International Organization for Standardization.

ISO 639-2:1998. *Code for the representation of names of languages—Part 2: Alpha-3 code*. Geneva: International Organization for Standardization.

ISO 1087-2:2000. *Terminology work — Vocabulary — Part 2: Computational aids in terminology*. Geneva: International Organization for Standardization.

ISO 8601:1988. *Data elements and interchange formats — Information interchange — Representation of dates and times.* Geneva: International Organization for Standardization.

ISO/IEC 11179-3:1994. *Information technology — Specification and standardization of data elements — Part 3: Basic attributes of data elements*. Geneva: International Organization for Standardization.

ISO/IEC 11179-4:1995. *Information technology — Specification and standardization of data elements — Part 4: Rules and guidelines for the formulation of data definitions.* Geneva: International Organization for Standardization.

ISO/IEC 11179-6:1997. *Information technology — Specification and standardization of data elements — Part 6: Registration of data elements.* Geneva: International Organization for Standardization.

ISO 12083:1994. *Information and documentation – Electronic manuscript preparation and markup.* Geneva: International Organization for Standardization.

ISO 12200:1999. *Computer Applications in Terminology — Machine-Readable Terminology Interchange Format (MARTIF) — Negotiated Interchange.* Geneva: International Organization for Standardization.

ISO 12620:1998. *Computer Applications in Terminology — Data Categories.* Geneva: International Organization for Standardization.

ISO/TC 37/SC 3 N 316:1999. "Computer Applications in Terminology — Generic Model (GENETER) for SGML-based Representation of Terminological Data." ISO working document.

ISO/TC 37 /SC 3 N 318: ISO 12200:1999. "Computer applications in terminology — Machine-readable terminology interchange format (MARTIF) — Negotiated Interchange — Amendment 1: (to include a normative annex H): Extended MARTIF. ISO working document.

Melby, Alan, and Wright, Sue Ellen. *CLS Framework.* http://www.ttt.org/clsframe/index.html. Accessed 2000-07-07.

Newton, Judith. 1999. *Data Element Naming Standards.* National Institute for Standards Technology and the Global Information Locator Service (GILS). http://www.net/naming.html. Accessed 1999-12-21.

Schmitz, Klaus-Dirk. 1998. "Über wichtige Aspekte bei der Einrichtung einer rechnergestützten Terminologieverwaltung." *Proceedings of the 11th European Symposium on Languages for Special Purposes.* Lita Lundquist, Heribert Picht, and Jacques Qvistgaard, Eds. Copenhagen '97. Copenhagen: Handelshojskolen i Kobenhavn, 383-390.

Wright, Sue Ellen. 1991. "The MicroMATER Tagset: Proposed Data Categories for the Exchange of Terminological Data between Terminological Database Systems," *Proceedings of the International Conference on Knowledge Organization, Terminology and Information Access Management*, Bratislava, Czechoslovakia: NISKO (The International Society for Knowledge Organization, Czechoslovak Chapter), 147-159.

_______. 1993. "Term Entry Formatting Conventions," *Keystones of Communication: Proceedings of the 34th Annual Conference of the American Translators Association,* Edith Losa, ed. Medford, N.J.: Learned Information, 179-196.

_______. 1995. "Creating a Data Element Dictionary for Computer-Aided Terminology Work," *Harmonizing Terminology for Translation and Information Retrieval: Theory and Practice, ASTM STP 1223,* Sue Ellen Wright and Richard A. Strehlow, Eds., American Society for Testing and Materials, Philadelphia, 169-188.

_______. 1996. "Mapping Local Data Categories to Categories Defined in ISO 12620," *TKE '96: Terminology and Knowledge Engineering.* Christian Galinski and Klaus-Dirk Schmitz, eds. Frankfurt/M.: Indeks Verlag, 259-271.

_______. 1998. "Data Analysis for Data Modeling and Mapping Data Categories based on ISO 12200 and ISO 12620", *Proceedings of the 4th TermNet Symposium Terminology in*

Advanced Microcomputer Applications, Tools for Multilingual Communication (TAMA '98), January 15-16. Vienna: TermNet, 273-302.

______. 1999. "The Representation of Terminological Elements in Termbases: Divergent Definitions of Data Category Scope and Content – Contexts." In: Budin, Gerhard, ed. *Proceedings of the IITF Cfonference in the Framework of ProCom '98 – Professional Communication and Knowledge Transfer*, Marking the Centenary of Eugen Wüster's Birth, Vienna, 24-26 August, 1998, Volume III. Vienna: TermNet.

8.1.3 Terminology Management Entry Structures

SUE ELLEN WRIGHT

1 ENTRY STRUCTURES AND TERMINOLOGY MANAGEMENT MEDIA

1.1 Paper Media

manual files

For decades, translators who recorded their own terminology entered their information in word lists or on file cards (*fiches*), mostly manually, but sometimes with typewriters. In many cases, this meant that after a number of years, a translator would have collected thousands of file cards. Generally, only the creators of these files could find their way around their collections, rendering them useless to anyone else.

cards and fiches

Most paper-based approaches were less than rigorous. Terminologists creating texts and translations on typewriters could neither perform terminology management tasks interactively using text processing nor automatically copy textual material, so the information recorded on a card was frequently minimal: terms, equivalents, subject fields, plus possibly dates or contextual references were likely to comprise the extent of these early entries. The critical feature of such entries is that the paper format, the physical limitations of the card itself, and the significant effort involved in creating the card writing by hand or using a typewriter dictated the choice and length of the fields to be included and the presentation of information on the card.

card layouts

Despite many haphazard efforts, a number of early models were developed to give terminologists guidelines for efficient, well organized terminology management (Wüster 1968:228, ff.; Dubuc 1978:43 ff.; Felber 1984:189 ff.). Wüster's discussion, which actually harkens back to 1952, when the *Machine Tool* project was initiated, features a complete description, including separate cards for each language in an entry, plus additional cards for illustrations. Layouts designed in Canada and at Infoterm in Vienna share the following features. (See Figure 1. Table 1 provides guidelines for interpreting the fields and abbreviations used on the card.)

207						113	
en	HE	VA	AB	LA	SY	**Planning and implementation of terminology projects**	

	illustration	
NOT		19.... ../ / /
DEF DSR EXP ESR CTX NOT	Illustrations are often used to clarify a definition or to make it more precise. Illustrations may be in the form of photographs, drawings, graphs or diagrams. They complement the definition. Exl: A picture of a triangle may be interpreted as a geometrical figure, a triangle, or a scalene triangle. Nevertheless, the use of such an illustration may often be helpful.	19.... ../ / /
BR G/P		S ISO 704-87
SP G/P		1987 08 10 /gb / /
SB G/P		
SB G/P		208

Figure 1: *Infoterm Paper Fiche (actual in-process terminological record on a file card, 1987-08-10)*

(**HE** = **HE**adword, **VA** = **VA**riant, **AB** = **AB**breviation, **LA** = **LA**mella, **SY** = **SY**nonym; **NOT** = note, **DEF** = definition, **DSR** = definition source, **EXP** = explanation, **ESR** = explanation source, **CTX** = context, **BR** = broader, **SP** = superordinate, **SB** = subordinate, **G/P** = generic/partitive, **S** = standard)

card layouts

- Although synonyms and equivalents may be cited, complete documentation tends to be limited to a single term on any given card.
- Full documentation of additional information associated with a concept requires additional cards. This means that each card constitutes a physical *record*, which taken together with other records for that concept can be construed as the entire concept-oriented terminological *entry*.
- The number of data categories treated on a record is limited.
- Maximum space is devoted to terms themselves and concept-descriptive textual information (primarily contexts and definitions, in many cases truncated to achieve brevity).
- Abbreviated forms are used for data category names.
- Codes are used for bibliographical sources, language-related information, and any pick-list-type information that can be prespecified for use in the work.

three letter codes

SEE *Data Categories 8.1.2:552*

The specification of standardized abbreviations for data categories to be used on cards (and later in early computerized environments) resulted in several lists of mnemonic short forms, notably (in addition to Dubuc and Felber) those associated with the Nordic Terminology Record Format and the Infoterm/Micro-MATER data categories list. The latter carried over into the early stages of the Text Encoding Initiative (TEI) (Hjulstad 1991; Wright and Melby 1993: 214-216). Eventually, however, as the list grew in the direction of the current set documented in ISO 12620:1999, maintaining three letter category codes designed for mnemonic recognition grew unwieldy. Furthermore, the expansion of computing power and the move away from paper environments and paper mentality in designing on-screen interfaces decreased the demand for such codes.

1.2 Computer-Based Terminology Management

computerized workstations

In contrast to the scenario described above, almost all translators today use computers, primarily for word processing, which saves considerable time in creating and editing their translations. The computer environment offers three options to translators for managing their terminology (Mayer 1990, 1993; Schmitz 1990, 1996).

1.2.1 *Word-Processing Programs*

word-processing files

Modern word-processing programs allow the user to open multiple screen windows and to work with several texts at the same time, translating in one window and recording terminological data in another. Merge and sort capability can be used to generate alphabetical lists and even to manage data-field-like text segments. Tests carried out within the framework of the Test Model for Linguistic Data Processing conducted by the University of the Saarland (Fischer et al., 1993), however, revealed the limitations of this method. To begin with, the user must possess fairly extensive data-processing and word-processing knowhow and must exert considerable effort to create meaningful files efficiently. Furthermore, the amount and scope of the data that can be effectively manipulated in this way are limited, and retrieval of each entry becomes increasingly slower as the quantity of information contained in the file increases. Such files have been known to become totally unusable when they reach a certain size, necessitating importation into more powerful data management systems.

columnar formats

The typical format used in word-processing systems, however, is a columnar treatment reminiscent of printed bi- or multilingual glossaries. This methodology reflects the retention of a paper-oriented presentational view of terminological information. Proponents of this approach often stress the ease with which columnar material can be converted to comma or tab-delimited notation in order to import or exchange data, but since filling in slots in the columns is left up to the time and discretion of data entry personnel, the tendency to leave out tabs for empty fields frequently leads to logical errors in the notation that can render whole resources unusable without considerable human editing.

1.2.2 *Off-the-shelf Database Systems*

general purpose systems

Under certain conditions, general purpose database and spreadsheet systems can be used for managing terminological data. Various analyses have demonstrated, however, that a great deal of effort is also required in this case to configure the database to meet terminology management needs (see Mommenthal 1990, Groß 1990, and Schmidt 1988). Interaction between the word-processing system and the database can be awkward, even in Windows™

environments, and the use of database management models with short, fixed field lengths poses disadvantages.

spreadsheet programs

It is not necessarily impossible to use a standard database program to configure intelligent terminological structures, but one frequent shortcoming of such systems has been the tendency to try to accommodate terminological data in a single data table. This approach generally limits the flexibility of data structures and prevents the introduction of multiple data categories within a given entry or provisions for extended field lengths. The same criticisms that can be leveled against database management programs apply to spreadsheets, with the additional problems that spreadsheet data can be awkward to view and that many systems limit string length. The main advantage to off-the-shelf solutions is that users are usually familiar with the programs, or company information management policy dictates that all databases, regardless of their purpose, will use the same software. It should be noted, however, that the Reltef™ model cited in Figure 11 provides a structure within which the information storage and retrieval advantages of standard database solutions can be positively exploited (Hardman 1997).

1.8 *Dedicated Terminology Management Systems*

structural formats

SEE

TMS Evaluation 8.1.1:539

In the broad sense, TMSs are database management systems that have been developed or configured specifically for managing terminological data, usually at translator workstations, although they may also be used by terminologists or by technical writers and standardizers, for instance. They also possess features that make them specifically appropriate for these applications. Chapter 8.1.1 has already considered the criteria for selecting a TMS. This article will focus on the evolution of the structural formats used in terminological records and entries.

2 Views on Terminology Management Systems

WYSIWYG paper

Any discussion of structures in TMS applications must define which specific view of the TMS we are talking about. In a paper fiche, ***What You See*** is indeed ***What You Get***, although the human user is frequently left to infer some of the essential information needed to interpret the record. For instance, Figure 1 is marked to indicate that part of the textual material on the card is an *expla-*

nation and part is a *context*, but users must distinguish what part of the text on the card is what. They must also observe that the *context* does double duty as both an *example* and a *note*.

multiple views

In TMSs, however, there may be several different presentations of the same data. There can be a data entry screen, one or more data display screens (See Figures 2 and 3, *MTX/Termex*™ or 4 and 5, MultiTerm for Windows™ display and editing screens[1]), and yet another internal organization of the data that can be configured for multiple views of the data. In fact, a look at the evolution of TMS practice reveals development from paper-like approaches to more flexible, customizable screen layouts.

TMS features

Furthermore, it is possible to view TMSs in terms of different features and modeling criteria. For instance, one system can be classified as similar to a second system in terms of basic modeling technique but together with another when it comes to presentational style. Finally, TMSs should avoid the kind of requirement for human inference described in the previous paragraph and adopt explicit markup procedures. The following treats a variety of views on termbases and their entry formats.

2.1 Fundamental Structures

fixed vs. definable structures

In looking at fundamental structures, two approaches prevail: systems with fixed and those with freely definable entry formats (Schmitz 1998b; Mayer 1998). In systems with fixed structures, the data categories are pre-defined and the user cannot change the format of the terminological entry or the assigned length of the data fields. Traditionally, database programs with fixed data entry formats have been SQL-based relational databases. These systems generally provide the basic categories to meet the needs of translation-oriented terminology management, but they may prove frustrating if users need to add any of the less-common data categories, to repeat categories within entries, or to accommodate terms or text segments whose length exceeds specified limits.

freely defined structures

Systems with freely definable data entry formats are usually based on indexed textbases. Users decide how the terminological entry will look by selecting data categories, bearing in mind that individual TMSs enforce certain conventions on data category

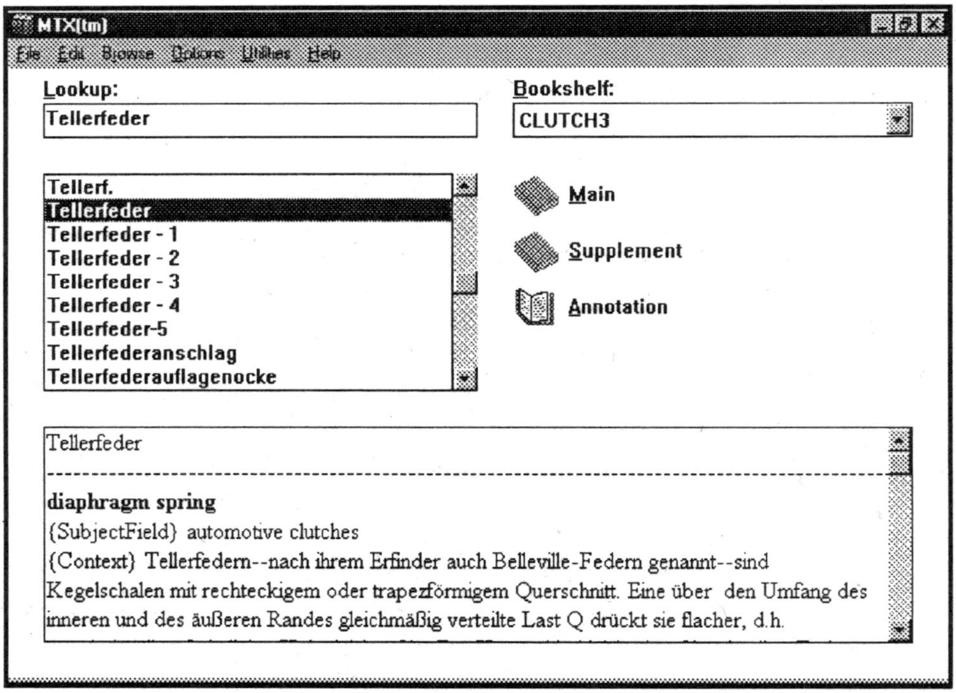

Figure 2: *MTX/Termex Data Display Screen*

assignment and behavior. These conventions may apply to the entire database, or they may apply only to the entry itself, which means, at least theoretically, that each entry could actually contain different data categories. Of course, reconfiguring the data structure for each entry would be highly impractical. Even most freely-configurable systems allow the user to use pre-defined input formats or template structures that ensure a relatively uniform data structure throughout the database file. Some do allow the introduction of additional, optional data categories, even when using a predefined template, which means that seldom-used, repeatable data categories can be introduced or reintroduced as needed.

2.2 Concept Orientation

one entry/ one concept

Sometimes beginning terminologists, whether working on paper or on a computer, start with a lexicograhical approach by documenting terms and their multiple meanings. Especially when working in

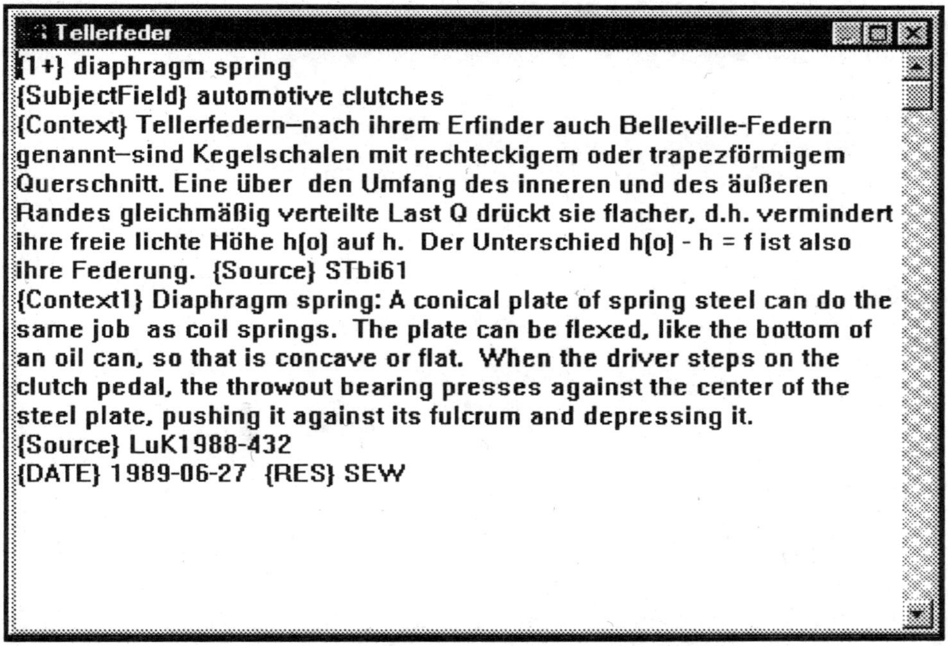
Tellerfeder
{1+} diaphragm spring
{SubjectField} automotive clutches
{Context} Tellerfedern–nach ihrem Erfinder auch Belleville-Federn genannt–sind Kegelschalen mit rechteckigem oder trapezförmigem Querschnitt. Eine über den Umfang des inneren und des äußeren Randes gleichmäßig verteilte Last Q drückt sie flacher, d.h. vermindert ihre freie lichte Höhe h[o] auf h. Der Unterschied h[o] - h = f ist also ihre Federung. {Source} STbi61
{Context1} Diaphragm spring: A conical plate of spring steel can do the same job as coil springs. The plate can be flexed, like the bottom of an oil can, so that is concave or flat. When the driver steps on the clutch pedal, the throwout bearing presses against the center of the steel plate, pushing it against its fulcrum and depressing it.
{Source} LuK1988-432
{DATE} 1989-06-27 {RES} SEW

Figure 3: *One Possible MTX/Termex Data Entry Screen*

multilingual environments, the shortcomings of this approach appear as soon as one needs to document multiple equivalents for multiple meanings of a term. The standard terminological approach of focusing on concepts and documenting all the terms associated with those concepts proves much more manageable in almost all environments.

INFO

Lexicography vs. Terminology INFB 3:328

virtual entries

Some TMSs make it necessary to treat detailed information on synonyms or equivalents in separate physical term *records*, a feature that reveals in part the influence of "paper mentality" on the evolution of early computer solutions, but that also reflects DOS limitations and restricted data memory capabilities. Nonetheless, most systems, whether fixed or freely definable, at least theoretically treat only one concept in any given *entry*. If several *records* are needed to accommodate all data items associated with a concept, they can be linked to form a virtual concept entry. (Figure 1 already exhibits this kind of linkage, where the notation "208" in the lower right-hand corner indicates that there is at least one other card associated with this entry.).

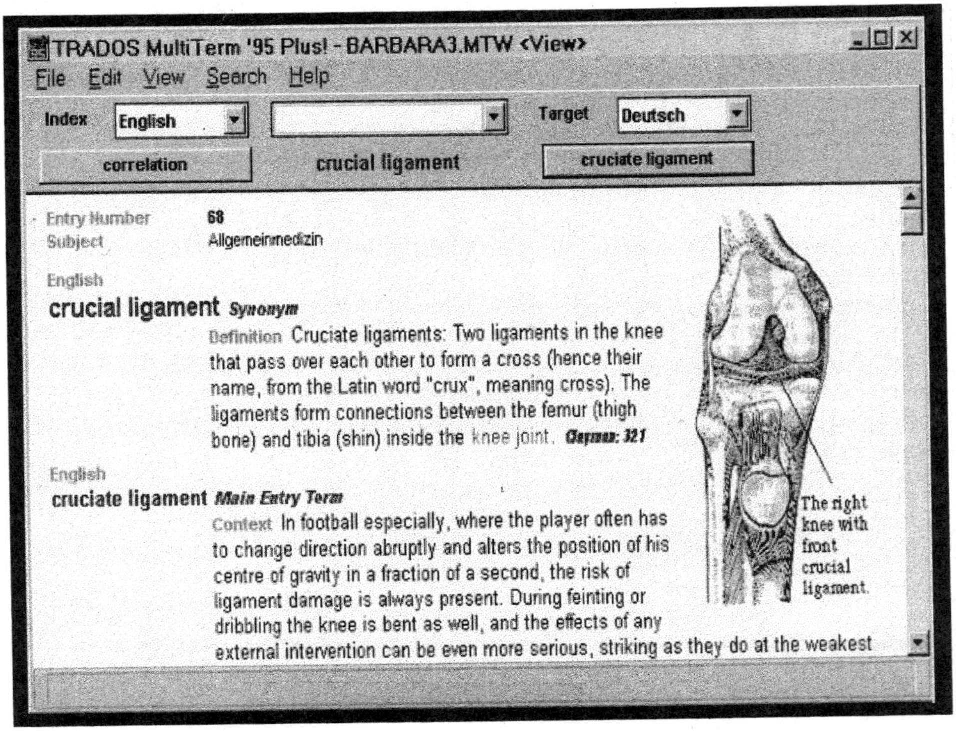

Figure 4: *MultiTerm Display Screen with Graphic*

redundancies

Although multiple records can simulate the effect of a complete concept-oriented entry, they pose a potential data management problem. If a synonym occurs in the initial main entry as well as in its own fully-documented entry, there is the possibility that the term may be deleted from the file in one entry, but remain as a kind of vestigial element in the other, thus creating an inconsistency in the file. Programs generally do not provide any automatic routine for following up on such linked entries—they must be traced and updated manually (Mayer 1993).

SEE *TERMIUM 8.2.2:666*

In the Canadian government's classic TERMIUM entry, division of the entry into English and French sections reflects not only a data management decision, but a political reality as well. Here the English and French portions of the entry appear side-by-side in the user-selected order (Figure 5). This approach gives rise later to the *language set* of the MARTIF entry (see below).

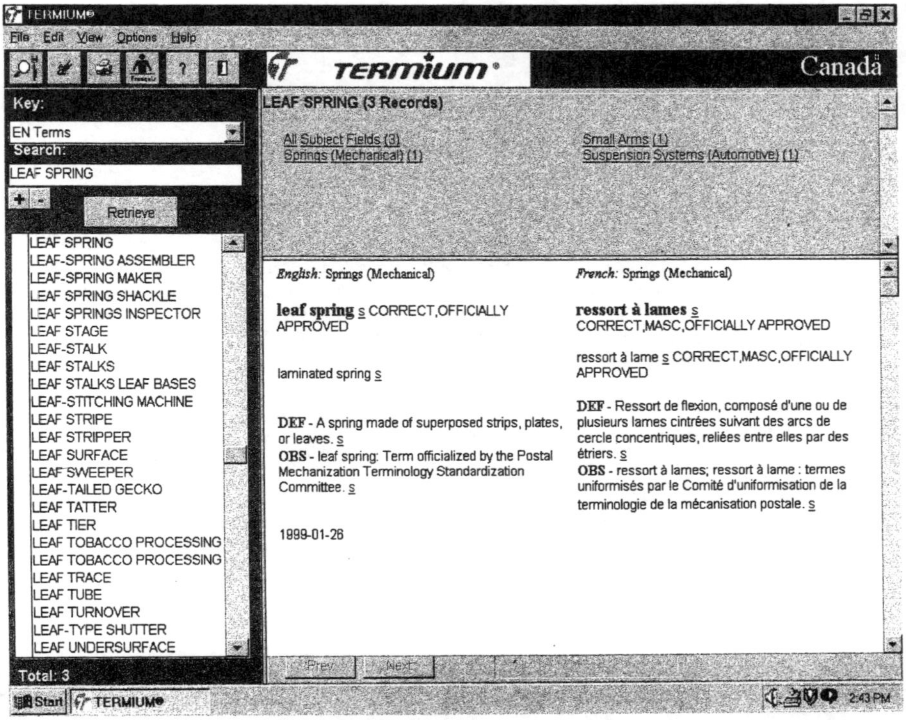

Figure 5: *Parallel Bilingual TERMIUM Entry*

2.3 Data Modeling Variance

term

The data categories listed in ISO 12620 do not exhibit uniform modeling function in all TMSs. There are two fundamental variations that emerge from differing data modeling philosophies and that reflect the needs and goals of different environments. In some data models, certain data categories (for instance, *term*, as well as other data categories such as *main entry term*, *synonym*, and *abbreviated form*) can function as field names that have different kinds of *terms* as their content.

Example 1:

term:	**serializer**
synonym:	**parallel-serial converter**
synonym:	**dynamicizer**

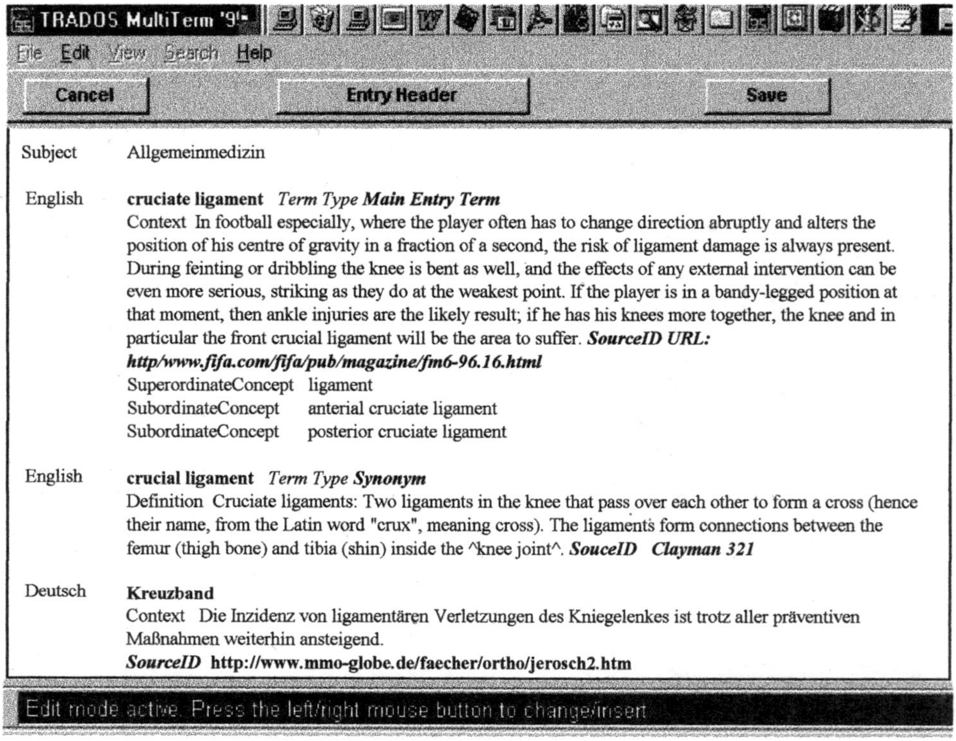

Figure 6: *MultiTerm Data Input Display*

In other databases, only the field name *term* is used for all terms. Some simple databases include no additional information about the term, but more sophisticated environments accommodate the same information as is provided in the first kind of environment, but instead of using distinct field names for the different kinds of terms, this information comprises the content of another field, such as a *term type* field. *term type*

Example 2:

term:	**serializer**
	term type: main entry term
term:	**parallel-serial converter**
	term type: synonym
term:	**dynamicizer**
	term type: synonym

A variation on this approach involves attaching attributes to the primary data categories. An advantage here is that multiple attributes can be easily associated with the same term:

attributes

SEE
Terminology Interchange 8.1.5:613

Example 3:

term:	**serializer**	
	attributes:	main entry term
		preferred term
term:	**dynamicizer**	
	attributes:	synonym
		admitted term

Neither ISO 12200:1999 nor 12620:1999 specifies which data modeling procedure should or must be used in individual databases, but the term type/attribute approach is favored for use in interchange environments.

2.4 Term-oriented Information

term autonomy

As noted above, the second option for data modeling variance is favored in interchange environments. This approach in essence dictates that *all terms are created equal* by reporting each and every term associated with a concept in an independent ***term*** element. Terminologists cite this approach as the principle of *term autonomy* (Schmitz 1998a). These elements can then contain, among other items, subordinate elements such as:

- *part of speech, gender, number*
- *term type*, which can accommodate such information as *synonym, abbreviation*, etc., as attributes of the term
- *definitions, contexts*, and any other descriptive information associated with the term

repeatability
combinability

Term autonomy is closely related to the principles of *repeatability* and *combinability*, which provide for the reuse of data categories and the establishment of links between elements inside entries. *Repeatability* concerns the ability of a particular data category to be used more than once in a terminological entry, and *combinability* allows closely related data categories to be associated with each other, e.g., *part of speech* is combined with *term*. For instance, the data category *entry number* should only occur once in any given entry, but a termbase that allows the inclusion of all terms associated with a concept would allow for the repetition of the data category *term,* as well as of the set of other categories associated with a term (as outlined above).

elemental data items

A further aspect of repeatability involves the need to observe the *elemental nature* of individual data items. This aspect manifests itself in two ways. First of all, it is important that only one kind of data be included in a given category. For instance, a text field such as a *definition* or a *note* can be *linked to*, but should not *contain* bibliographic information about the text. Secondly, if it is desirable to include more than one item of information assigned to a given data category (such as two terms assigned to a concept, or two responsibility codes associated with the same entry), then the data category in question should actually be repeated for each item rather than stringing the items together in a single data field. To be sure, comma-delimitation rules used in some programs allow for this practice, but the consistency of comma notation is highly dependent on the accuracy of data entry, and it can be difficult at times to guarantee that no extraneous commas ever occur as parts of normal text.

data modeling function

Both repeatability and combinability play a role in the Infoterm *fiche* shown in Figure 1. Early efforts to list and categorize terminological data elements attempted to define the potential for repeatability and combinability associated with each data category (Galinski et. al. 1982 and passim; Galinski and Wright 1991, cited in Budin 1996:138-139). The results of these efforts, although interesting, present a degree of complexity that is difficult for even experienced terminologists to appreciate. Consequently, subsequent lists have focused on the basic specification of data category content. Since repeatability and combinability are actually functions of data modeling and data entry design, these features are treated here rather than in the Data Categories chapter, and they are covered in the data interchange standard ISO 12200 instead of in ISO 12620, the data categories standard.

2.5 Presentational Views in Computer Environments

paper orientation

Interestingly, some of the same constraints that were imposed on the term record in paper environments are either inadvertently or intentionally transferred to some computer applications. Perhaps the most significant example of this kind of "print orientation" is the actual representation of the terminological data on-screen in

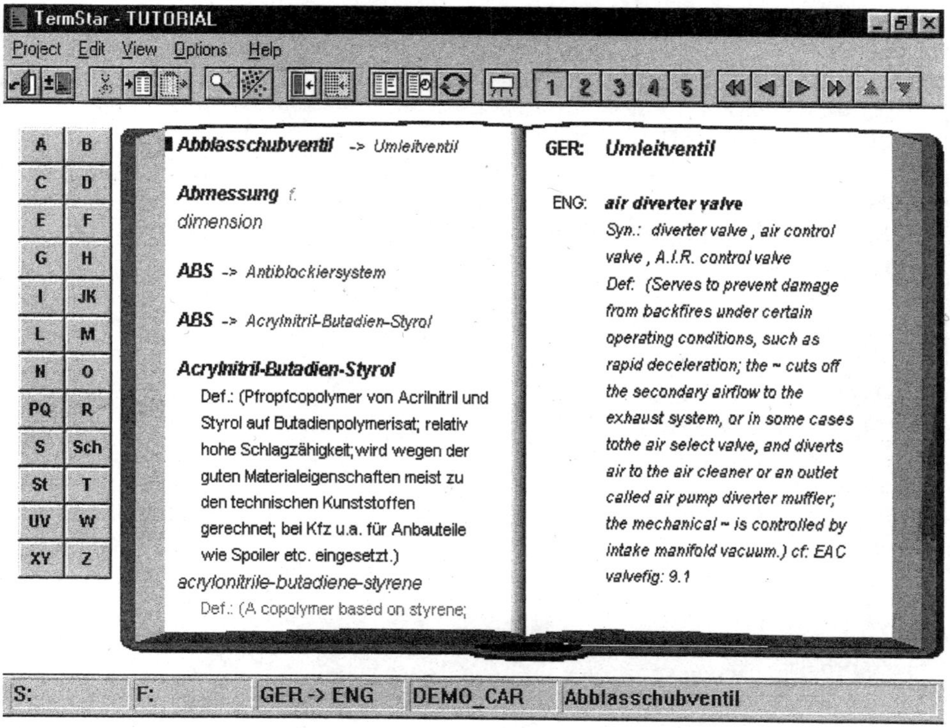

Figure 7: *TermStar Screen Display*.

the form of a "book" display in the TermStar™ terminology management system (Figure 7).

user-oriented structures

In other respects, it can be argued that physical screen dimensions and particularly the limitations of the original DOS environment imposed constraints on the configuration of user interfaces. For instance, both CATS, which is a fixed relational database, and the Termex implementation shown in Figures 2 and 3 accommodate term autonomy. In the screen shots shown, they both present segments of data designed to fill the initial screen display with the most significant information, such as definitions and contexts, is relegated to later screen displays (CATS) or must be accessed by scrolling down into the entry (MTX/Termex) (see Figures 8 and 2, 3, resp.). Where the TermStar layout is a modern Windows design purposefully intended to resemble the presentational features of a printed dictionary, the CATS layout strongly resembles a printed file card.

```
CATS:Suchen        DEEN       Seite 1 von 3          <> Vorgabe   Key.: d
DE-Term:                                              DE:Abk:   DE-Prgm:
Punkt-zu-Mehrpunkt-ISDN-Verbindung________            _________  ...........
                                                  DE-Dskrpt: ________
DE-Sem: ..........................................................
  DE-->: Punkt-zu-Mehrpunkt-Verbindung ......................... DE-Syn. Prgm:
  DE-Syn: .........................................................   .................
          ........................................................   .................
          ........................................................   .................
          ........................................................   .................
EN-Term:                                              EN-Abk:   EN-Prgm:
point-to-multipoint configuration ______________      ________  ............
                                                     EN-Dskrpt: _________
EN-Sem:  ......................................................
  EN-->: point-to-multipoint-configuration ......................... EN-Syn. Prgm:
  EN-Syn: .........................................................   .................
          ........................................................   .................
          ........................................................   .................
          ........................................................   .................
[PgDn]/[PgUp]:Eintrag vor/zurück [F2]:Zum Haupteintrag [F3/F4]:Liste [->] S.2
        [F5]:Druck [F6]: Transf [END]: Ändern [DEL] Lösch [ESC]:Ende
```

Figure 8: *CATS Data Input Screen, Page 1*

medium-related innovations
CD-ROM layouts

Movement into the world of advanced DOS and Windows graphic user interfaces has dramatically changed both the look and function of TMS screen displays. The use of pull-down, pop-up, and right-mouse menus has eliminated the need to resort to excessive screen clutter and abbreviations. As the power and flexibility of the medium expands, developer and user creativity and expectations also undergo a transition, with the result that totally new strategies for presenting the data evolve. For instance, the use of color or reliance on hypertext capability have opened up new ways of representing attributes of the data, of providing for combinability, or of displaying conceptual relationships. Familiarity with new kinds of information layout can lead to dramatically new visualizations of familiar data, for instance along the lines of the now-popular CD-ROM dictionaries (Figure 9). This frame-like functionality holds promise for future TMS interface development.

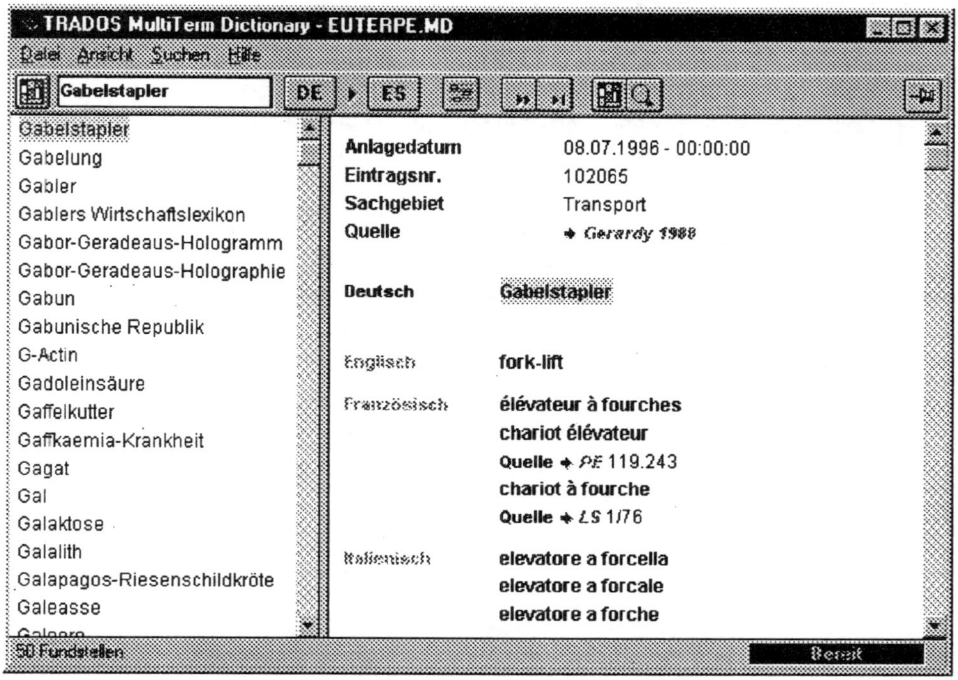

Figure 9: *TRADOS Euterpe CD-ROM Dictionary Layout*

2.6 Directionality

bilingual entries

The number of languages treated in a given TMS frequently reflects the working environment of the original target audience for whom the application was developed. Systems designed for use by individual translators are more likely to accommodate work in two languages, in many cases with a strong sense of directionality from a given language A into a language B. The extreme manifestation of this trend produces term entries that can only be accessed using language A key terms. In other cases, a high prevalence of bilingualism in a particular linguistic setting (such as in Canada) fosters the development of strictly parallel bilingual systems where both languages are given equal weight within the terminological entry.

multilingual entries

Globalization, coupled especially with high-pressure multilingual translation situations that prevail in software localization projects, has expanded the development envelope to include an ever-increasing number of languages in a given TMS. The languages can be treated as totally coequal, in which case a full complement of definitions and contexts are included for each language. As is frequently the case in localization environments, however, one language (generally the corporate "native language") is designated as the pivot language, terms are defined in this language, and quasi-prescriptive equivalents are specified.

SEE
Localization 7.2.5:517

directionality

If absolute equivalence is "legislated" in this fashion, directionality becomes a moot issue because by definition there are no differences between concepts in different languages. In multi-directional, non-prescriptive termbases, however, great care must be taken, not only in reversing the direction of the original entry, but even more so in equating target language terms with each other. In such situations, the *transfer comment* data category from ISO 12620 can provide essential information on differences in concept scope between various language pairs.

2.7 Dimensionality

2.7.1 *Representing Dimensionality*

flatness

Merely listing the data categories used in a TMS actually reveals relatively little about the structure of the database itself and nothing at all about the relationships that exist among those data categories. The paper record (Figure 1) medium is ostensibly a *flat* piece of paper, a two-dimensional object. Computerized media can also be two-dimensional. For instance, a simple relational table or a spreadsheet chart can be two dimensional if the only relations that are possible are those represented by the intersection of the columns and rows (tuple values). As soon as we gain the ability to encode formulas or rules in a given cell, however, we add at least one, if not several, dimensions to the system.

relatedness

2.7.1.1 *Embeddedness.* If we look carefully at the arrangement of the fields on the record in Figure 1, however, we see additional dimensions. While some of the items on the card appear to form a simple list, items relating to the concept description are grouped together in a kind of pick-list fashion that allows the originator of the card to mark (highlight) the relevant data category for the textual material contained in the text box. In this respect, the card reflects the principle of combinability. The note in the third line of the main body of the card refers to the entire record. The note that appears lower down in the list, however, refers specifically to the concept description because it is visually contained inside the concept-description text box. We know that the second note is related to the text because it is *inside* the box containing the text, i.e., it is *embedded* there.

2.7.1.2 *Adjacency.* In terms of explicit embeddedness, the bibliographic notations following the *definition* and *explanation* respectively are coequal with these categories and the other items in the list, but an informed human user will infer logically that *DSR* is dependent on the definition and *ESR* on the explanation. In laying down rules for interpreting the *fiche*, the designer of the card could, however, have simply chosen *SRC* (source) instead of supplying different codes for the two sources. Then he could have dictated that:

togetherness

- SRC following DEF belongs to the definition.
- SRC following EXP belongs to the explanation.

This model reflects the *principle of adjacency*—items next to each other belong together. This rule applies for the *source* references in the Termex entry in Figure 3.

2.7.1.3 *Tagged Text.* The record does not, however, depend solely on adjacency for establishing relationships. Instead, it differentiates between the source of the definition and the source of the explanation by using different tags for marking the text. This practice evolves later in computer environments as a powerful strategy for identifying data items, particularly in indexed textbases. Given this tool, it is possible to state a rule for relating items to one another without their being adjacent. However, if more than one instance of an element, such as a definition, occurs, dependency reverts to the principle of adjacency:

explicit rules

- DSR can only belong to the (preceding) definition (DEF).
- ESR can only belong to the (preceding) explanation (EXP).

In the Termex entry, *Context* and *Context1* rely on the tagged text method as opposed to the adjacency method we saw with *Source*. If these elements were permanently locked in place (like the elements on the hardcopy record), their relationships might remain unambiguous. As noted above, human users of a paper card infer relationships based on their interpretation of the card layout

ambiguity

and their knowledge of the rules set down by the creator of the card. Once we move to a computerized environment, all inferential rules must be explicitated (Holmes-Higgins and Ahmad 1996). The problem arises that the permanent position of the elements is not guaranteed in the electronic entry, especially if we begin to move elements around during data interchange. As a result, adjacency is not a reliable indicator, tagged text is less ambiguous, but embeddedness provides the safest, most robust tool for ensuring that items that belong together always stay together.

2.7.2 *Hierarchical Dimensions*

CLS Framework

Neither ISO 12200 nor 12620 provides guidelines on a framework for structuring termbases, although the layout of the interchange format itself reflects one possible organization of data for those who are working with a structured text approach. In order to guide systems designers, Melby and Wright (1998) have suggested the *CLS Framework*, i.e., *Concept-oriented with Links and Shared references*.

meta-model

We have seen how different kinds of termbases exhibit different structures. The basic idea behind CLS Framework is to create an overall model for termbases that provides a common framework within which a system designer can create either a structured text termbase, a relational database, or even an object-oriented database model. The rational behind this kind of *"meta-model"* is to support automatic bi-directional conversion between systems with little or no loss of critical information.

Within the CLS Framework, a termbase consists of (a) global information about the termbase, (b) a set of concept entries, and (c) a set of references (here called SharedRefs) that can be shared among multiple concept entries or parts of concept entries.

global information

Global information can include administrative information about the entire termbase, such as publication information, the name of the database, the name of the copyright holder, information about the creators of the database, dates, and version numbers. In addition, it can include information about the languages and writing systems used in the termbase, which user group the termbase is intended for, and other administrative information. In an SGML environment, this global information is most likely to reside in the *front* matter (or header) of the SGML document.

concept entries

The *concept entries* make up the *body* of the termbase. CLS Framework does not specify a single layout for concept entries other than to state that they should follow the one concept/one entry criterion already outlined above in this article and should adhere to the principle of term autonomy.

links

SharedRefs

SharedRefs consist of pieces of information that are potentially shared by many concept entries. Examples of SharedRefs might be a complete bibliographic entry or a responsibility entry listing the biodata about a person who created or updated entries. Individual items within entries scattered throughout the termbase can include *links* or pointers to the relevant SharedRefs as needed. SharedRefs can also include graphics, external video or audio resources, charts, tables, and the like. As implied above, the use of SharedRefs is grounded in the well-known principle of database management that redundant information should be avoided in order to ensure economy of storage, consistency, and flexibility. SharedRefs in MARTIF documents as presently configured reside in the *back*, but there is no reason they cannot reside in the *front* as well. In fact, if they do move to the front, then any parsing or processing utilities will find the SharedRefs and have them available for processing from the very beginning of a first pass through an interchange document.

nesting

CLS Framework provides a kind of multi-chambered outer container into which smaller containers for other kinds of information, primarily concept entries, can be packed. The framework is designed to be flexible in that it is capable of accepting a variety of database solutions or entry structures, but one common one that it might accommodate is the deeply nested entry structure shown in Figures 1 and 2 of Chapter 8.1.5 (p. 613), a format that can be roughly mirrored in MultiTerm entries. Another way to represent this hierarchical structure is with a simple outline as shown in Figure 10, where CLS Framework elements are shown in bold italics and concept entry structures in a normal font. Here the premise of multidimensionality is clearly evident in the telescoping shape of the hierarchical system. As noted above, actual detailed entry structures are not specified by CLS Framework.

SEE
Terminology Interchange 8.1.5:613

hierarchical layers

The MultiTerm entry in Figure 11 illustrates a concrete example of such a nested entry. Here the program provides for five hierarchical levels, consisting of the following layers, which can be cross-referenced to the levels outlined in Figure 11:

```
CLS Framework
    Termbase
        Global Termbase Information
            Administrative information

        Concept Entries (Body)
            Concept Entry
                Entry Level Information
                Language Set
                    LangSet Information
                    Term Information Group
                        Term Group
                            Term
                            Term-related info (grammar, term type, etc.)
                        Descriptive Information
                            Definition
                                Source (hypertext link to SharedRef)
                                    Page number
                        Descriptive Information
                            Context
                                Source (hypertext link to SharedRef)
                                    Page number
                        Note
                        Administrative Information
                    Term Information Groups (as needed, etc.)
                Language Set (as needed)
                    Term Information Groups, etc.

        Shared Resources
```

Figure 10: *Nested or Embedded Termbase Structure*

1) The *Concept Entry* itself (*Entry Level Information*), including the entry number, subject field (represented in Figure 11 by *Subject* and *Category* references), and any other top-level information, such as a *note* pertaining to the entire entry

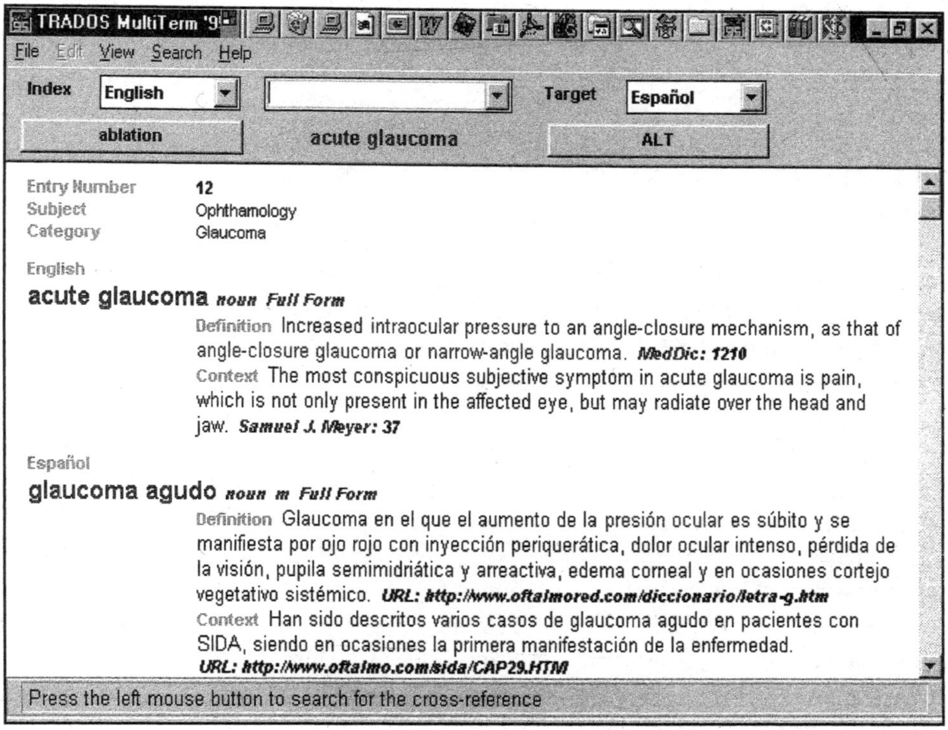

Figure 11: *MultiTerm Entry Model with Deeply Embedded Hierarchical Structure*

2) *Index* fields for terms (and *termGroups*), e.g., *acute glaucoma* and *glaucoma agudo* in Figure 11

3) *Attribute* fields attached directly to the *index* fields, in this case for *part of speech, gender*, and *term type*

3) *Text* fields (*Descriptive Information)*, e.g., **Definition** *Increased interoccular*..., or **Context** *The most conspicuous* ...

4) *Attribute* fields attached to the *text* fields, e.g., *sources* for text fields (e.g., ***MedDic*** for the **Definition**, or ***Samuel J. Meyer*** for the **Context**)

5) *Variant text* fields attached to the *attribute* fields, e.g. for a page number (e.g., ***1210*** or ***37***; this feature is important because if the ever-changing page numbers were included in the *source* field, that field could not be used as a stable and unique linking element.)

The MultiTerm entry in Figure 11 provides for almost all levels of embedding, but the language set level is missing. This is not evident in Figure 11, which only includes one English and one Spanish term. The entry shown in Figures 4 and 6, however, contains two English terms, which will always be grouped together. This represents a sort of language grouping, but there is no provision for assigning any kind of subordinate data that apply to the entire *langSet*. Examples of such information might be *transfer comments* that provide information on the scope of a concept in a given language, or *responsibility* references for the terminologist assigned to documenting a given language.

language level information

2.8 The Abstract Relational View

Citing concerns for redundancy and data integrity, Holmes-Higgins and Ahmad (1996) have recommended creating Entity Relationship (E/R) Diagrams for the purpose of representing relationships within databases. Taking his cue from Holmes-Higgins and Ahmad, Hardman has moved a step further by defining high-level relationships at both what he calls a "meta" level and within standard "non-meta" tables. Figure 12 shows a relational diagram for what he calls the Reltef™ model (Hardman 1997).

entity relationship diagrams

There is insufficient room here to discuss Hardman's model in detail, but we should note that he claims that it resolves many of the shortcomings of the relational database approach while preserving its advantages: "Reltef exploits many strengths of traditional database management systems, including integrity constraints, data normalization, access to data through traditional and extended SQL business rules, and so on. It also provides enormous flexibility and can be customized to address the needs of many data sets in a wide variety of human languages [which is not generally possible in SQL based relational databases]."

relational databases

SQL-based systems

2.9 Further Developments

The concepts outlined above are not the only efforts to provide new options for modeling data. Holms-Higgens and Ahmad, for instance, propose creating object-oriented databases, although no such system exists as yet. Nistrup-Madsen (1998) is exploring new ways to present data in user-friendly, application-oriented relational data models. Sager has supported merging the onomasiological

SEE

Customized Retrieval 8.1.4:600

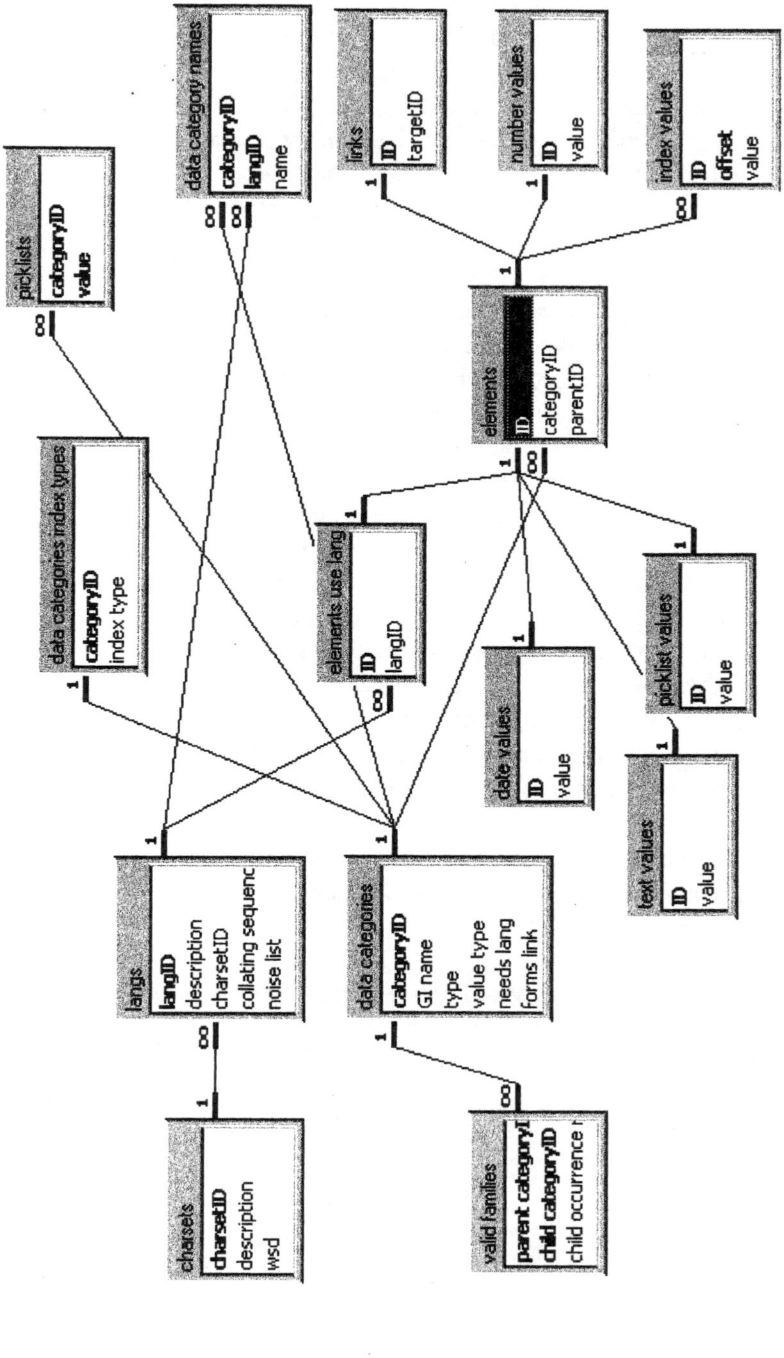

Figure 12: *RELTEF Relational Diagram*

(terminological) and the semasiological (lexicographical) approaches to lexical management (1990). Nkwenti-Azeh's image of multiple user-specific views on a core data collection in Chapter 8.1.4 of this book reflects this position. Hardman implies that Reltef would support such systems, and IBM's *TRANSLEXIS*™ TMS incorporates both views on the core data, but is not yet available for use in the common Windows™ environment.

Ahmad and Rogers have long supported the integration of termbases with dynamic text corpora designed to provide the dimension of living language (Dubuc's *in vivo terminology*, 1978). Their Sebriba™ information retrieval engine serves as an automatic corpus-generation tool. Direct linkage with dynamically spawned, *in-situ* contextual material would indeed change the complexion of the standard terminological entry, where brief contextual material has until now been included in the body of the entry itself.

SEE *Term Extraction 8.4.1:725*

All of these trends underscore the fact that this chapter can only provide a momentary snapshot of the current state of the art in TMS design. Although advances in data memory capacity, computing speed, and interface design have alleviated some of the medium-related constraints outlined above, real-world parameters such as system response time still affect whether some of the theoretical solutions are actually viable in the workplace. Budget considerations in work groups also militate against apparent "ideal" solutions. If, for instance, the core activity of an enterprise focuses on translation, there is no motivation to configure systems that will also meet the needs of technical writers and documentation specialists, even though multi-application termbases may be a highly recommended approach. Only time and experience will show which innovations will survive in practice.

ideal vs. real solutions

1. The availability and appearance of various TMSs undergoes constant change. Any attempt to review specific programs would be quickly outdated in the context of a hardcopy book. The examples used in this article were selected for the purpose of illustrating major points considered in the article. The following information was valid when the book went to press:
Figure 1: The Infoterm fiche was developed based on the work of Eugen Wüster, as reflected in Wüster 1968: 2.28, ff. and terminographical projects dating from 1920 through the 1970s. This course of development is articulated in Felber 1980 and 1981, Galinski et al. 1982 and passim, Galinski 1983.
Figures 2 & 3: Mercury/Termex (MTX, Termex)™ was the first PC-based terminology management system. It was introduced in 1982 by Linguatech International.
Figures 4, 5, and 11: MultiTerm for Windows™ is currently the most widely-used TMS and is

developed and marketed by the TRADOS Corporation.
Figure 6: The TermStar™ TMS is a product of the STAR Corporation.
Figure 7: CATS™ was a DOS-based research and training tool developed at the University of Mainz/Germersheim.
Figure 8: The TRADOS *EUTERPE* dictionary reflects a widespread evolving CD-ROM dictionary format featuring a frame containing a zoom-style representation of the terms treated in a dictionary, with the right-hand window devoted to an individual terminological entry.
Figure 10: TERMIUM™ is published by the Minister of Public Works and Government Services Canada. The format shown here is from the 1999 CD-ROM version of the term bank.
Figure 11: Reltef™ is a proprietary data model developed by Daniel Hardman at CMR Computer It has formed the basis for further development in the field of database modeling.

REFERENCES

Budin, Gerhard. 1996. *Wissensorganisation und Terminologie: Die Komplexität und Dynamik wissenschaftlicher Informations- und Kommunikationsprozesse.* Tübingen: Gunter Narr Verlag.

Dubuc, Robert. 1978. *Manuel Practique de Terminologie*. Quebec: Linguatech and Conseil International de la Langue Française.

Felber, Helmut. 1980. "A Study of Terminological Data Elements on the Basis of the General Theory of Terminology and a Comparative Study of those Data Elements used in Terminological Data Banks all over the World." Vienna: Infoterm.

_______. 1982. "The Terminological Data Elements as Derived from the General Theory of Terminology and their Recording in Machine-Readable Form." *Terminologies for the Eighties*. Munich, New York and Paris: K.G. Saur.

_______. 1984. *Terminology Manual*. Paris: UNESCO and Infoterm.

Fischer, Ingeborg; Freigang, Karl-Heinz; Mayer, Felix; and Reinke, Uwe, eds. 1993. *Akten des Symposiums zum Abschluß des Saarbrücker Modellversuchs*. Saarbrücken: Universität des Saarlandes.

Freigang, Karl-Heinz; Mayer, Felix; and Schmitz, Klaus-Dirk. 1991. Micro- and Minicomputer-based terminology data bases in Europe. TermNet Report 1, Wien: TermNet.

Galinski, Christian, and passim. 1982. *Infoterm Guidelines on Terminography.* Unpublished manuscript. Vienna: Infoterm.

_______. 1983. "Report on the 2nd International Meeting of Experts on 'Guidelines for the Recording of Terminological Data for Machine Processing'." Vienna: Infoterm.

_______. 1984. "Report on the Meeting of an Editing Group on 'Guidelines for the Recording of Terminological Data for machine Processing'." Vienna: Infoterm.

Groß, Rebecca. 1990. *Möglichkeiten des Einsatzes von Microsoft-Works am Übersetzerarbeitsplatz.* Diplomarbeit. Saarbrücken: Universität des Saarlandes.

Hardman, Daniel. 1997. "The Reltef™ Specification." CMR-Termsoft: http ://www.ttt.org/clsframe/reltef.html/. Accessed most recently: 1999-12-05.

Holmes-Higgins, Paul and Ahmad, Khurshid. 1996. "Is Your Terminology in Safe Hands? Data Analysis, Data Modelling and Term Banks." *TKE '96: Terminology and Knowledge*

Engineering. Christian Galinski and Klaus-Dirk Schmitz, eds. Frankfurt/ Main: Indeks Verlag, 215-224.

Hjulstad, Håvard. 1991. "Nordic Terminological Record Format (NTRF)." Oslo: Norwegian Council for Technical Terminology.

ISO 12200:1999. *Computer Applications in Terminology — Machine-readable terminology interchange format (MARTIF) — Negotiated interchange*. Geneva: International Organization for Standardization.

ISO 12620:1999. *Computer applications in terminology — Data categories*. Geneva: International Organization for Standardization.

Mayer, Felix. 1990. "Terminologieverwaltungssysteme für Übersetzer. Ergebnisse einer Untersuchung," *Lebende Sprachen*, 3/90: 106-114.

_______. 1998. *Eintragsmodelle für terminologische Datenbanken: Ein Beitrag zur übersetzungsorientierten Terminographie*. Forum für Fachsprachen-Forschung, Band 44, Hartweg Kalverkämper, ed. Tübingen: Gunter Narr Verlag.

_______. 1993. "Entry Models in Translation-Oriented Computer-Assisted Terminography," *TKE '93. Terminology and Knowledge Engineering. Proceedings Third International Congress on Terminology and Knowledge Engineering*. Klaus-Dirk Schmitz, ed. Frankfurt: Indeks Verlag, 205-214.

Melby, Alan and Wright, Sue Ellen. 1998. "Overview of the CLS Framework." http://www.ttt.org/clsframe/. Accessed most recently 1999-12-04.

Mommenthal, Bernhard.1990. *Entwurf und Implementierung einer einzelübersetzer-orientierten Terminologiebank (D-E-F) auf einem universellen Datenbanksystem - aufbauend auf Terminologie zu einem Teilgebiet der Außenhandelsfinanzierung*. Diplomarbeit. Saarbrücken: Universität des Saarlandes.

Nistrup-Madsen, Bodil. 1999. "The DANTERM Concept." *TAMA '98—Terminology in Advanced Microcomputer Applications*. Proceedings of the 4th TermNet Symposium: Tools for Multilingual Communication. Vienna: TermNet, 55-66.

Reinke, Uwe and Schmitz, Klaus-Dirk. 1998. *Testing the Machine-Readable Terminology Interchange Format (MARTIF)*. Saarbrücker Studien zu Sprachdatenverarbeitung und Übersetzen, Band 13. Saarbrücken: Fachrichtung 8.6, Universität des Saarlandes.

Sager, Juan C. 1990. *A Practical Course in Terminology Processing*. Amsterdam: John Benjamins Publishing Company.

Schmidt, Marie-Therese. 1988. *Nutzung eines integrierten Softwarepaketes auf PC für den Übersetzerarbeitsplatz*. Diplomarbeit, Universität des Saarlandes, Saarbrücken.

Schmitz, Klaus-Dirk. 1990. "Rechnergestützte Terminologieverwaltung am Übersetzerarbeitsplatz," *Terminologie et Traduction*, 3/90: 7-23.

_______. 1996. "Terminology Management Systems. *Il Traduttore Nuovo*, XLVI, 1: 45-60.

Schmitz, Klaus-Dirk. 1998a. "Über wichtige Aspekte bei der Einrichtung einer rechnergestützten Terminologieverwaltung." *Proceedings of the 11th European Symposium on Languages for Special Purposes*, Lita Lundquist, Heribert Picht, and Jacques Qvistgaard, Eds. Copenhagen '97. Copenhagen: Handelshojskolen i Kobenhavn, 383-390.

_______. 1998b. "Terminologiverwaltungssyteme — Einsatzbereiche, Typen und Produkte." ASTTI, ed. *Proceedings of the 1997 Zürich Conference: Theorie und Praxis*. Bern: ASTTI, 171-181.

Wright, Sue Ellen and Melby, Alan. 1993. "TEI-TERM: A Proposed Format for the Interchange of Terminology Data Using Standard Generalized Markup Language." *Standardizing Terminology for Better Communication: Practice, Applied Theory, and Results*. Richard Strehlow and Sue Ellen Wright, eds. ASTM STP 1166. Philadelphia: ASTM.

Wüster, Eugen. 1968. *The Machine Tool—An Interlingual Dictionary of Basic Concepts, comprising an Alphabetical Dictionary and a Classified Vocabulary with Definitions and Illustrations*. Vol. 1, English-French Master Volume. London: Technical Press (The United Nations Organization).

8.1.4 User-Specific Terminological Data Retrieval

BLAISE NKWENTI-AZEH

1 RETRIEVAL REQUIREMENTS OF TERMINOLOGICAL INFORMATION

flexible formats

A main advantage of computer-based terminology over the conventional dictionary lies in the fact that a single database can now accommodate information which was conventionally held in different fixed formats in a number of separate reference tools. The flexibility of organization and retrieval associated with databases allows for various useful information formats to be produced according to specific user needs. Terminology retrieval is heavily dependent on current trends and developments in information and document retrieval systems generally, and, in particular, on the state of the art in compilation and storage techniques.

1.1 Information Categories in Term Banks

INFO

Lexicography vs. Terminology INFB 3:328

Many term banks are term-based, rather than concept-oriented, i.e., the terminological record is focused on the term as a linguistic unit (similar to a lexicographic dictionary entry). Furthermore, term banks have been designed for human users only. Developers have consequently made assumptions about users' threshold linguistic competence and concentrated on certain information categories. This term-oriented approach and the assumptions it implies with respect to user needs are reflected in the information categories used in the term banks.

SEE

Data Categories 8.1.2:553

While there are no limits to the number of categories and/or the amount of information that can be stored in any terminological record (e.g., DANTERM, the Danish national term bank, has defined up to 96 categories), surveys reveal only a small number of recurring information categories, some of which are now considered as the essential minima for the reliable description of a term. Most commercial implementations contain all or an important subset of the following types of information in separate data fields:

— term(s) / phrase	— term number
— grammatical information	— record originator or author
— abbreviation	— date of creation (and updating)
— language - synonyms	— bibliographical source
— subject classification	— context
— definition/explanation	— usage note
— generic term	— quality label
— specific term	— usage variants
— associated term	— foreign language equivalent(s)

The focus of term banks on the human user and the limitations imposed by their dependence on non-dedicated database management system software is highlighted by the fact that even in concept-oriented record designs, the area of conceptual relationships (both intra- and inter-conceptual relationships) is largely unexploited. These relationships are, however, crucial for interrelating linguistic referents of different concepts, and for eventually modeling the knowledge structure of a particular domain. Only by introducing conceptual relationships and defining knowledge structures in terms of these relationships can term banks offer users the opportunity to formulate a diverse range of more interesting, problem-specific queries and, in return, retrieve minimally redundant, needs-oriented responses. In the description of query and output profiles that follows, it is assumed that the full range of terminological information is available, or at least that the technology exists for their eventual representation (Skuce and Meyer 1990). *knowledge structure*

SEE *Concept Analysis 1.4.2:98*

2 ACCESS AND RETRIEVAL OF TERMINOLOGICAL INFORMATION

Users of information retrieval systems have varying degrees of familiarity not only with the system in question but also with information retrieval in general and computer technology as a whole. It is imperative that these systems be user-friendly, i.e., be easy to use, offer some system guidance to infrequent users, and provide "intelligent" responses to otherwise unmatched queries.

SEE *Term Retrieval 6.3:426*

Information retrieval systems can enhance their user-friendliness and improve recall by adopting the principle of user profiles, according to which the user specifies a query (or search) profile, an output profile, or a retrieval profile combining both search and output. The difference between search and output profiles is most *user profiles* evident in online consultation where interaction with the user is possible between search and retrieval. Search profiles require direct access to the information contained in the various data fields. Output profiles are generally defined around information categories or data-field labels.

2.1 Search Profiles

customized search profiles

A serious limitation of printed dictionaries is the predetermined and invariable order which the medium imposes on the information stored and on the possible range of search and retrieval parameters. In alphabetically-ordered dictionaries, the entry term very often constitutes the sole access point to all the information in an entry. Even conceptually structured printed dictionaries, like Wüster (1968) and Holmström et al. (1971), have to be supplemented by alphabetical indexes as the customary method of word-ordering for most languages. Since the information provided by lexical and terminological databases is, however, divided into data fields, multiple search keys for separate data fields can be specified for the database query and individual users or groups of users can take advantage of search profiles to variously restrict their searches to particular categories of information in the database record. For general purposes there is usually a standard search profile that corresponds to the most common user queries—the most basic being a search for the existence of a term record. For specialized users or for particular needs, search profiles can exploit any available boolean operators to build up a set of search parameters and utilize these for all subsequent interrogations of the database, unless they are temporarily overridden.

The search profile thus allows the modification of the standard search strategies by activation or de-activation of less selective search criteria; for example, in a term bank environment:

- automatic performance of free text searches on definitions if no separate entry is found
- upper/lower case neutralization when searching
- removal of the significance of diacritics when searching
- simple morphological rules that search not only for the search string itself, but also for inflected forms
- some sort of spelling approximation algorithm to retrieve entries spelled similarly to the search string

search criteria

- automatic retrieval of all terms containing the search string
- truncated searches for all terms in which the search string occurs in initial, median or final position or as an element of a compound term

2.2 Output Profiles

fixed dictionary output

When consulting printed dictionaries, the user is presented with all available information at once in the sequential order of the dictionary entry; graphical devices (typefaces, indentation, punctuation, etc.) constitute the only means of differentiating between heterogeneous groups of information. On the whole, the dictionary entry may therefore contain too much non-relevant information for the immediate need.

filtering

Output profiles allow users to define a list of information categories they wish to be displayed for all entries retrieved, thereby filtering the results obtained via the search profile. Users may also define the sequence in which the output is to be displayed.

output mode

Database designers generally recognize that queries tend to be directed at subparts of the terminological record. The preferred output profile, which often coincides with the typical or standard requirements of the user-group, can be associated with the username so that it is immediately available on log-in to the system (and thus need not be specified for each consultation). Users normally have the choice of selecting from a range of predefined, system-standard output formats or defining their own personal format from a set of allowable combinations. The predefined system formats may in turn be subcategorized for particular user groups and even linked to user authorization criteria.

Further options that may be provided during retrieval include the definition of different formats for different modes of output (screen consultation, file output, printer output, batch-mode printout to file/printer, etc.).

3 USER-ORIENTED RETRIEVAL OF TERMINOLOGICAL INFORMATION

The main user groups of terminological databases are translators, technical writers, abstractors, teachers, etc. They need to access different subsets of data starting from any one data category as the search term and to combine search parameters and data categories in ways rarely possible in conventional information retrieval software. Their need is for a powerful non-procedural query language which allows the formulation of any logical query commencing from any point in the database. Normal expectations are for a limited number of access paths, as exemplified in the next paragraph.

user groups

- Terms should be accessible
  - via direct query
  - via selection from an index or permuted index
  - and via expansion of the search string to retrieve entries containing the term or entries similar in their orthographic or phonetic form
- Conceptual information should be accessible via direct access using an identification code
  - via conceptual relationships
  - via free text searches on definitions (and possibly contexts)

term accessability

There are even differences in the requirements for any one data category, which term banks of the future should be able to satisfy. For example, different types of definitions, according to the degree of subject specialization of the searcher, could be offered to a scientist, a technical translator, or a student.

SEE *Definitions 1.3.1 INFB 12:340*

Research has permitted the identification of several distinct user groups with their particular retrieval needs.

3.1 Translators and Interpreters

The terminological reference needs of translators underlie the creation of the earliest term banks and translators continue to constitute the largest homogeneous user group. In the routine task of interlingual communication, translators and interpreters require a high degree of accuracy of description, partly because they usually have a less detailed knowledge and understanding of the subject matter than subject specialists do, and partly because they frequently have to paraphrase or coin new terms to fill lexical gaps in the target language. Interpreting needs differ from translation mainly in the mode of consultation (typically batch mode) rather than in the types of information required. The output profile for on-

SEE *Technical Translation 2.1.1:147*

line consultation should assist in comprehension of the source language (SL) text, by e.g., providing access to definitions or explanations of terms. It should also guide translators working as a team to produce texts with uniform terminology. It should be equally possible for the system to perform target language (TL) term substitution on entire texts in online consultation and more particularly in batch retrieval mode.

INFO

Text Designations INFB 22:867

The following templates cover most of the online retrieval requirements for translators and interpreters:

search templates

- term + TL equivalent + source of TL term
- term + TL equivalent + TL source + TL definition
- term + TL equivalent + synonym(s) + quality indicator
- term + definition + variant(s) + source/quality indicator

For cases where the term bank does not contain the TL equivalent sought, an output profile is needed to retrieve the necessary clues for terminologisation or paraphrase. One such package could display matrices of analogous terms, i.e., orthographically related terms (Clas 1990).

For incremental information retrieval, the following sequence would correspond to most requirements:

- term + TL equivalent
  + definition
  + context or usage note
  + quality indicator
  + synonym
  + source
  + subject specification/label or scope note

Batch retrieval of terminological data is mainly required by interpreters in the form of subject-specific, firm-specific or text-oriented bilingual glossaries. The most common form of presentation would be alphabetic lists of term + TL equivalent + usage or scope note. Text-oriented glossaries are also important in that they display terms in the order in which they occur in the SL text. Alphabetical listings may be complemented by systematic monolingual lists. Other combinations are similar to the online packages, i.e. :

terminology for interpreters

- term + definition
  + TL equivalent
- term + TL equivalent + source of TL term

- term + TL equivalent + TL definition + TL context
- term + TL equivalent + synonym(s) + quality indicator
- term + variant(s) + TL equivalents + variants + source/quality indicator

3.2 Technical Writers

SEE *Technical Writing 7.1:445*

Technical writers constitute the second largest group of communication mediators (Sager and McNaught 1981; Sager 1990). This group differs primarily in that their work involves a higher degree of creativity. Abstractors, usually included in this group, combine the task of interlingual translation with that of knowledge reduction. These communication mediators often liaise between specialists in different subjects and between specialists and general readers. For monolingual production or generation, technical writers require the following templates:

- term + variants + usage/scope notes
- term + variants + usage/scope notes + source/quality indicator
- term + variants + context + source/quality indicator
- term + definition + context

abstractors

The retrieval packages for abstractors comprise the same types of information as for monolingual consultation, except that most of this information is associated with the TL equivalent. This implies that terminological databases require comparable information in both SL and TL entries if they are to respond effectively to the needs of this group—a requirement not necessarily crucial for satisfying translator needs, except in environments where members of translation teams are called upon to translate in both or multiple directions at different times.

3.3 Subject Specialists

global retrieval strategies

Subject specialists require terminology support mainly in their monolingual task of term creation and in the multilingual task of understanding the meaning of foreign language terms. This group of users is assumed to have expert knowledge of their respective domains and thus require only occasional reference to the linguistic and semantic information on terms in their primary working language. So far subject specialists have not made extensive use of terminological databases because their needs have not been sufficiently identified and hence some of the information they require is not available. It is evident, nonetheless, that in order for term banks to assist in term creation, these users should be able to retrieve terminological paradigms, term variants, concept families, etc., based, amongst other things, on "what-if" experiments with possible neologisms. Unlike other user groups, subject specialists often require a global view of the term environment rather than a graduated microview. Retrieval packages appropriate for this group of users (particularly for term-creation) would include:

- generic term + definition + coordinate terms
- generic term + variants (including textually-conditioned variants)
- morpho-syntactically related terms
- systematic display of conceptual region showing position of new term.

Retrieval requirements for bi- or multilingual needs are similar to those of translators and other communication mediators, and the same range of output templates would be adequate.

3.4 Terminologists and Lexicographers

KeyWord in Context

Terminologists and specialist lexicographers are the primary creators of term records and require on-line access to the full range of information associated with dictionary-making, e.g., KWIC (KeyWord In Context) citations, documentation of usage, definitions and variant forms.

SEE *TERMIUM* *8.2.2:666*

Online development and maintenance of a reliable database is crucially dependent on access to all the data fields in the collection. The output profile for this group of users should permit systematic verification of concepts and terms. An elaborate framework of terminological relationships, derived from definitions or separately recorded, would provide the often needed support in this direction, e.g., it would enable the terminologist to test the completeness of a conceptual system. Since terminologists also advise on usage and term creation, they need cross-references to lexical, conceptual and pragmatic information in the various records of the database. They may also need to generate dictionaries, glossaries or other lexical reference tools from terminological databases. The output often consists of alphabetically ordered terms and associated information (variants, definition, usage notes, other information as dictated by the particular end-use). In the case where large quantities of information are to be generated, batch-output profiles are most suitable.

3.5 Information and Documentation Specialists

Librarians, information brokers, indexers, etc., use term banks for the reliable identification and description of specialist documents. Their particular needs are for the clear determination of correspondences between the natural language of the ultimate end-user and the artificial language created for identification and retrieval of the required documents or document surrogates. The requirements of this group are very similar to those of the other communication mediators discussed above, and therefore would benefit from similar retrieval packages. Indexing, keyword extraction, and thesaurus construction and searching, however, require quite different combinations of terminological data. Preferred templates include:

SEE *Thesaurus Management* 6.1:381

- term + definition + source
- term + synonym + generic term + specific term
- term + generic term + specific term + TL equivalent
- term + complete systematic display or relations

For graduated retrieval, the most widely favored sequence appears to be:

- term + definition + source
  - + synonym
  - + generic + specific term
  - + TL equivalent

In each case the term would have to be considered in the context of its full conceptual synonymy, i.e., including any spelling variants, shortened or long forms, abbreviations and acronyms. The batch retrieval requirements of this group of users regularly consists of various subject and term listings, including:

conceptual synonymy

- monolingual alphabetical subject field indices
- alphabetical lists of term + definition
- alphabetical lists of term + key related terms
- systematically ordered lists of subject fields

3.6 Language Planners and Standardization Experts

Language planning (e.g., development of a national language) like standardization requires prescription and fixation of usage. The terminological needs of language planners and standardization experts are identical for practical considerations: both types of professionals aim at agreed or unified terminology in their respective communicative situations, and both need an inventory of all available designatory forms before deciding on a preferred form. Terminological databases can facilitate the task of standardization by offering accurate information about all existing designations (including deprecated forms). In return, standardization organizations can contribute to special language communication via the vocabularies and glossaries that result from their work. The work of standardization experts and other language planners can be greatly assisted by online consultation of comprehensive terminological databases and by lists that have been produced off-line of existing standardized and non-standardized vocabulary in systematic or alphabetical order.

SEE *Language Planning 3.2:261*

3.7 Systems for Natural Language Processing (NLP)

natural language processing

Recent investigations into the reusability of term bank information for NLP tasks (McNaught et al. 1991) revealed that there is little directly available information in existing term banks and other machine-readable terminological resources which can be used to support the needs of NLP, and that the terminological information in current term banks requires substantial augmentation in order to become relevant for NLP use. This is not surprising given the main preoccupation of term banks with serving human users. Such augmentation largely concentrates on adding new types, and further differentiating between existing types, of information rather than reformulating existing information. For example:

textually conditioned variants

- Traditionally, little attention has been paid to the recording of textually conditioned variants, even though terms have formal variants in different texts and in different contexts. NLP systems would require specific indication that a form occurring in a text (being parsed, translated, etc.) is indeed a textually conditioned variant of the standard form stored in the term bank.

- Subject codes are of particular interest to NLP as they provide semantic information that can be used to disambiguate senses. Work in the EUROTRA project (Sorensen 1990) demonstrated how subject codes could be used to yield information to help specify the domain-specific information used to fill slots in conceptual frames. In this respect, the subject code (possibly reflecting a fine level of detail) can offer ways of encoding semantic and collocational constraints on terms by limiting the types of fillers a frame can admit. *subject codes*
- Collocations have a particularly important status in NLP. Few term banks store collocational information in explicitly reserved fields. Typically, this material will be included in contexts or explanations. However, in recent years, interest has been growing among terminologists in collocations, and in phraseological elements in general. Work on terminological verbs (L'Homme 1992; Sager 1992) has demonstrated that verb terms have a central role to play in organizing the conceptual mini-systems found in sublanguage sentences. There is therefore a need for term banks to describe the textual environment of terms and for terminologists to treat verbs in much the same light as nouns and adjectives when compiling terminological entries.

SEE *Collocations 8.4.4:788*

If term banks are to respond to the needs of sublanguage- based NLP, they will have to collect and store collocational information in a re-usable manner. Freibott and Heid (1990) describe the construction of a term bank which explicitly incorporates formal descriptions of collocations.

4 MULTI-FUNCTIONAL TERMINOLOGICAL DATA REPRESENTATION

Diversified retrieval of terminological data can supply new information of great value to writers and translators that may also be of use to scientists and information brokers. For example:

- text-oriented glossaries, i.e., lists of terms plus associated information in the order in which they occur in a particular text
- terminological thesauri, i.e., systematically ordered lists of terms in particular subject fields expressing a number of conceptual relationships and facets of the underlying conceptual structure

diversified retrieval

Thus a terminological thesaurus is more than a simple vocabulary as it aims at a coherent representation of the knowledge structure of a subject field. Such a tool has the advantage of showing the full extension of a term and its place within the broader conceptual environment of the overall knowledge structure it belongs to, thus complementing the definition.

Terminological thesauri can be used to provide access to databases or serve as source material for the construction of documentation thesauri. The ROOT Thesaurus, published by the British Standards Institute (BSI 1981), serves such a double function as general scientific vocabulary and information storage and retrieval tool. Among other glossaries that have adopted some kind of conceptual structure, very few can be described as genuine thesauri (see Wüster 1968, Holmström et al. 1971). Many systematic structures consist of groupings or hierarchies of terms based on very arbitrary, questionable and often opaque criteria. This applies even to some glossaries published by standards institutions.

References

Bennett, P.A. et al. 1986. *Multilingual Aspects of Information Technology*. Aldershot: Gower.

BSI . 1981. *BSI ROOT Thesaurus* (in 2 parts). London: British Standards Institution.

Clas, André. 1990. "Les minibanques de terminologie: le prototype BATEM." *Computational Lexicology and Lexicography*. Cignoni, L. and Peters, C., eds. Pisa: Giardini Editori, 219-226.

Freibott, Gerhard and Heid, Ulrich. 1990. "Terminological and Lexical Knowledge from Computer-aided Translation and Technical Writing." *TKE '90: Terminology and Knowledge Engineering*, Vol. 2. Czap, Hans and Nedobity, Wolfgang, eds. Frankfurt: Indeks Verlag, 522-535.

Holmstrom, John E.; Fickelson, Maurice; and Jejcic, Danilo. 1971. *Trilingual Dictionary for Materials and Structures, English/French/German*. International Union of Testing and Research Laboratories for Materials Structures (RILEM). Oxford: Pergamon Press.

L'Homme, Marie-Claude. 1992. "Les unites phraseologiques verbales et leur representation en terminographie." *Terminologie et Traduction*, 2/3, 493-505.

McNaught, John; Nkwenti-Azeh, Blaise; Martin, W.; and ten Pas, E. 1991. *Eurotra-7 Study, Doc-11: Feasibility of standards for terminological description of lexical items*, Final version. Luxembourg: CEC.

Sager, Juan C. 1990. *A Practical Course in Terminology Processing*. Amsterdam: Benjamins.

________. 1992. "Future developments and research in phraseology and terminology related to translation," *Terminologie & Traduction*, 2/3, 583-588.

________. and McNaught John. 1981. *Specifications of a Linguistic Data Bank for the UK*. British Library Research and Development Report No. 5644. Manchester: CCL.

Skuce, Douglas and Meyer, Ingrid. 1990. "Concept analysis and terminology: a knowledge-based approach to documentation." *Proceedings of COLING 90*. Helsinki, 56-58.

Sorensen, Henrik S. 1990. "The Use of Knowledge-Based Frames for Terms in EUROTRA." *TKE '90, Terminology and Knowledge Engineering: Applications*. Frankfurt: Indeks Verlag, 300-309.

Wüster, Eugen. 1968. *The Machine Tool, An Interlingual Dictionary of Basic Concepts*. London: Technical Press.

8.1.5 Terminology Interchange

ALAN MELBY, KLAUS-DIRK SCHMITZ, AND SUE ELLEN WRIGHT

1 INTRODUCTION

electronic data exchange

MARTIF

This chapter will describe a method of exchanging terminological data electronically, that is, in machine-readable form such as on a diskette or other storage medium, via electronic mail on a company Intranet, or on the World Wide Web. First, we will discuss why there is a need to store, transmit, and exchange terminology in electronic form. We will review several approaches to electronic terminology interchange and focus on one solution called the Machine-Readable Terminology Interchange Format, abbreviated *MARTIF* (ISO 12200:1999). (Originally called *TIF*, the *Terminology Interchange Format*, the name was then changed to *E-TIF*, the *Electronic Terminology Interchange Format*, to avoid confusion with *TIFF*, the *Tagged Image File Format*, which also uses the common file extension *tif*. The current name is the result of much discussion in ISO Technical Committee 37, *Terminology (principles and coordination)*, Sub-Committee 3 for Computer Applications, Working Group 3 (ISO TC37/SC 3/WG 3).

supplemental information

This chapter serves as an introduction for terminologists who are considering using interchange formats in their work, but it does not provide all the information necessary for implementing the interchange format. For this degree of detail, we recommend that users acquire the standard itself from their national standards bodies and that they consult other support literature cited in the references. Readers will also benefit the most from this chapter if they have first read chapters 8.1.2 on data categories and 8.1.3 on terminological entries. An interchange format can also be used for dissemination and data analysis, but this chapter will treat interchange, which involves transmission of information between two software applications via an intermediate format.

2 The Need to Supply Terminology in Electronic Form

sharing information

Increasing numbers of technical writers, standardizers and translators today are sharing terminology because they are working together on common projects under severe time constraints in localization projects where they must also exchange data with their clients, frequently all over the globe. They also know they can conserve labor-intensive research and data-entry time by sharing information, thus reducing the overall cost of terminology management.

integrated environments

It goes without saying that today texts are almost exclusively created and subsequently transmitted in electronic form. Furthermore, translators are using terminology management systems (TMS), translation-oriented text databases (textbases), often called translation memories (TM), machine translation programs (MT), or various localization-related tools and utilities. More and more of these high-level users are striving to integrate a variety of tools into a seamless work environment and to leverage (i.e., reuse) terminological information found in termbases and MT lexicons (Melby 1982, 1994; Wright 1998; Melby and Wright 1999), which requires that all components of the system interact electronically.

leveraging paper resources

In the past, if terminology was shared at all, it was customary to supply terminological information on paper. This practice was satisfactory as long as human translators only used this material to manually look up terms that were then incorporated into texts. Today, however, paper as a medium has become problematic for several reasons. If the translator is given a terminology file on paper and still wants to use computer-aided translation tools, the terminological data must either be keyboarded into a termbase or must be scanned using an optical character reader (OCR). Both these alternatives are inefficient and error-prone.

updating

Another problem with terminology on paper is that new terms appear so often that documents must be reprinted very frequently, which is not only expensive, but ecologically unsound as well. If only the changes are printed, the user must consult many updates in addition to the main (and outdated) version. Closely related to updating, merging files generated by different individuals in different environments requires reformatting of paper versions.

The prevalence of electronic environments coupled with the disadvantages of paper reflect the obvious need to create, maintain, and exchange terminological data in electronic form. In such situations, decisions must be made with respect to the format of these resources. Therefore, we must consider the format of electronic terminology files and make adjustments for inevitable variations and even more troublesome inconsistencies.

resource format

3 The Need for a Terminology Interchange Format

If everyone used the same terminology management software on the same type of computer in the same operating environment with the same objectives, all terminology files could be maintained in a format appropriate to these uniform conditions (Stellbrink 1993). Despite the attraction of a uniform format, there are many aspects of diversity that characterize the terminology management scene:

incompatible formats

SEE

Multilingualism, character sets 3.1:251 ff

- Many competing packages have their own internal file formats designed to meet the needs of different kinds of users.
- Work groups use a variety of platforms, even sometimes within the same enterprise.
- Files that contain non-English characters, especially non-alphabetical CJKV (i.e., Chinese, Japanese, Korean, and Vietnamese) ideographs, require special handling.
- Different applications (TMS functions as opposed to TM or MT) require additional or varying degrees of information, and lexicographical approaches can (but do not necessarily have to) be at odds with terminological structures.

One way to allow interchange among all these termbases would be to write two conversion utilities for each pair of formats, one for importing data and one for exporting it. If, for instance, there were only thirty-five different formats (there are probably many more than that), this approach would require over two thousand separate conversion routines. If the prospect of this daunting task is cause for despair, however, let us consider that if there were a common intermediate format for terminology files, the number of conversion programs required is reduced from over two thousand to just seventy (one import and one export routine to convert data back and forth between each program and the interchange format). Moreover, instead of individuals being concerned with at least (theoretically) thirty-five such sets of conversion routines, they would only need to worry about the pair of utilities that facilitates

mathematical permutations

conversion between their system and the interchange format. This scenario is very much like being able to use ASCII, ANSI, or RTF conversion capability to transport text files instead of dealing with many different word-processing conversions.

document production chain

In the document production chain, terminologists, technical writers and translators may all use the same resources in a homogenous environment, but frequently, product developers and in-house system users also require access to the same resources, despite their different needs. In global markets and international organizations, these diverse user groups may be scattered around the globe and work on a variety of platforms, not to mention the need to interact with outside suppliers, package designers, typesetters, etc.

inter-company terminology harmonization

Consistency between enterprises is also essential. Computer software is a prime example. Microsoft, for instance, makes its glossaries available on the Internet in 20-some languages so that everyone programming for the Microsoft Windows™ interface can use consistent terminology (Microsoft 1999). These resources are available online in comma-delimited form. Simple search and display programs are available for viewing the files and searching for individual strings. In many respects, the entries are more reminiscent of translation memory segments than of standard glossaries or formal terminologies. Although the text chunks in the glossaries might be used as contextual references in a terminology file, they would have to be incorporated into some sort of exchange format in order to actually integrate them automatically into another resource with a more complex data model.

translation management scenarios

The following scenarios illustrate situations in which the electronic exchange of terminology would be needed.

- A professional translator might receive a terminology file from a client along with a source text to be translated. In this case the file would probably be bilingual and include client-specific target-language terms that are to be used consistently when translating that client's texts. The translator would need to import the terms into his or her TMS so that the appropriate target-language terms can be quickly accessed and inserted into texts during translation, but if the file format differs from that used by the translator's TMS, interchange utility becomes necessary. This also applies if the translator wants to leverage the information for MT or TM applications as well.

- In another scenario, a creator of source text (for instance, a technical writer) receives a product description file from designers in the home office of the company. In this case, the file might be monolingual and include technical terms used in the company in accordance with company-specific definitions. The technical writers would be under the obligation to ensure that standard company terms are used consistently in the text. They might also hand the finished documentation on to a translation project manager, who would be responsible for seeing to it that standard company target language equivalents are identified and documented for each language into which text is to be translated. They might have to convert the resulting terminology files to several different database formats before distributing them as just-in-time (JIT) resources for their outside translation team in order to guarantee near-simultaneous delivery of their product to a variety of locales *(simship)*.

authoring tools

INFO

Controlled Language
INFB 25:872

- In a third situation, a company with a significant existing termbase might find that it was no longer desirable to continue using the existing TMS or the company wishes to shift to a different platform, etc. In this case, the technical writers and translators need to convert data from the old to the new database formats in order to leverage legacy data in the new system.

leveraging legacy data

Admittedly, scenarios like those described above may not yet be everyday occurrences in all companies, but current trends indicate that rapid interchange of terminology in the form of electronic files between various offices around the world using different hardware and software platforms is needed to improve both productivity and quality for both translators and monolingual authors who are required to use preferred terminology. Preservation of all or at least most of the information in the source termbase is assumed to be an important factor in the choice of an intermediate format.

4 The Evolution of Terminology Interchange Formats

4.1 MATER

MATER

Given the need for an electronic format for interchanging terminology files, it is not surprising that several such formats have already been developed. Prior to the development of the current generation of exchange formats, perhaps the best-known terminology interchange format was the International Standard ISO 6156:1986, usually called by its short name, MATER. This standard was a pioneering effort that began in the 1970s and culminated in approval as an International Standard in 1986. It is intended mainly for the interchange of data between mainframe-based term banks via 9-track magnetic tape. This technology was a reasonable one at the time that work on MATER began. Microcomputers were relatively unknown and very limited in power and application. By the time MATER became a standard, however, the IBM PC and the Apple Macintosh were taking over the market. Terminology files were no longer restricted to mainframe computers, and 9-track large-reel magnetic tape was no longer the primary medium for storing and exchanging data. Furthermore, the format and the selection of available data categories in the standard were inflexible, not to mention that the treatment of both terminological and lexicographical data in the same standard proved unwieldy at the time. Aside from the exchange of data between such systems as the Siemens TEAM system and EURODICAUTOM, the term bank of the European Union, MATER was seldom actually used in practice.

4.2 MicroMATER

MicroMATER

In the early 1980s, work began on a proposal for a new terminology interchange format called MicroMATER (implying an adaptation of MATER suitable for use with microcomputers). This work was even carried out with the cooperation of some of the designers of MATER. The principal aims of MicroMATER were to provide a flexible interchange format that would allow terminology files to be transmitted among various computer platforms (mainframes, minis, and micros) without loss of information, with a fully open choice of data categories and flexible data category lengths. MicroMATER files were also somewhat easier to edit and view in a standard text editor than were MATER files..

The results of the MicroMATER project were published (Melby 1991), but by then, those involved in the project had already started to develop a new terminology interchange format in cooperation with the Text Encoding Initiative (TEI).

4.3 Nordic Terminology Record Format

NTRF

Another project developed a terminology interchange format for use among several term banks in Norway, Denmark, and other Nordic countries, appropriately called the Nordic Terminology Record Format (NTRF, Hjulstad 1991; Nilsson 1999;605; 607; rtt 1999). The format, which is still actively used in the Nordterm environment, was the first serious SGML-like interchange format.

4.4 The Text Encoding Initiative

INFO

SGML
INFB 21:863

The Text Encoding Initiative (TEI) had a tremendous appeal to the members of the MicroMATER project. One reason was that the TEI was based on SGML. The MicroMATER format, although not a true SGML application, had incorporated many SGML-like features. Although it seems obvious today, the decision to use SGML to develop an interchange format came as the result of a serious debate conducted at the time among theorists and practitioners, who agreed that SGML represented the logical medium for this purpose (see Heid 1991 and Reinke 1991). When the Terminology Workgroup was formed within the TEI in March of 1991, the group decided to totally reevaluate the design rather than to adapt any existing format to the then-current TEI P1 Guidelines (Sperberg-McQueen and Burnard 1990; Melby et al. 1994).

text-related issues

The main reason for joining the efforts of the TEI, as opposed to pursuing independent development, was that many of the solutions for problems encountered in encoding other texts were also needed for terminological data, such as dealing with character-set problems and facilitating interchange across 7-bit paths, coding cross-references, including illustrations, and referencing other external data objects. As a work group of the TEI, the MicroMATER group was able to interact with others dealing with similar problems and to create an environment that would encourage additional external review and validation. Not insignificant was the additional consideration that those who need a terminology interchange format would also frequently have to deal with significant quantities of text and would be able to utilize the TEI environment for assistance in the world-wide shift from presentational to descriptive markup.

4.5 Global Standards

ISO

Although TEI constituted a fertile environment within which to develop the early stages of the standard, it appeared that the best venue for achieving universal acceptance was the international standards community. In 1992, the draft stages of the TEI proposal were adopted as a work item by ISO TC 37/SC 3/WG 3, where it eventually acquired its final form and title. Migration to the ISO environment further expanded the involvement of experts from a wide variety of countries, as well as the inclusion of diverse theoretical and methodological approaches. The following discussion of the standard reflects the work of this broader group, together with future efforts to link the work of the ISO group to various other projects related to terminology interchange. This effort, which is described in Section 9, is called the SALT project (Standards-based Access to Multilingual Lexicons and Terminologies).

5 Design Considerations for a Terminology Interchange Format

5.1 Basic Principles

theory-neutrality

flexibility

simplicity

A terminology interchange format is only useful if it is widely used. In light of the fact that no previously proposed format had enjoyed widespread use, it was decided within TEI and then in the ISO group that it was above all necessary to develop a format that had some likelihood of broad acceptance and implementation. One primary stumbling block would be the perception that the format was biased toward any particular school of terminological theory. Therefore, it must appear reasonably theory-neutral. Another obvious stumbling block would be if it could not adequately represent the information in any particular terminology database. Thus, it must be flexible. Finally, it must not look so different from the way terminologists are used to looking at their data that it appears entirely too abstract and complex. Thus, it must appear as simple as possible, while at the same time providing the power to express a wide variety of data models.

5.2 Generic Identifiers and Data Category Classification

Before going any further, the work group decided to examine the structure and content of as many existing termbases as possible (Wright and Budin 1994). After looking at over thirty of them, it was observed, as predicted, that although many field types (called "data categories" in termbases) were used identically across numerous termbases, other data categories were particular to just one or a few termbases. One potential problem was the sheer number of different data categories, which amounted to several hundred, plus, in many cases, the enumeration of permissible instances, i.e., allowed values.

SEE *Data Categories 8.1.2:552*

Furthermore, it became apparent that as different applications evolve, new data categories become necessary. Although it is possible to devise a unique SGML generic identifier (GI) for each field type, the diversity of field names and their tendency to proliferate poses an ongoing problem. All GIs used to mark up a document type must be declared in the Document Type Definition (DTD) for that document type. Once a DTD is published and becomes widely used, adding new GIs becomes difficult because any additions entail changes in the DTD. A stable DTD is not as much an issue when only general-purpose SGML tools are used, but in many industrial settings, the structural constraints of a DTD are encoded into import and export routines using the programming language found in the rest of the software application, in which case a stable DTD becomes very important to commercial acceptance.

constant evolution

The solution to this concern is to define a minimum number of broad (i.e., truly generic) identifiers for the types of elements appearing in the document type. In the case of terminology, these GIs include *term*, *termNote* (for term-related data such as grammatical information or information about term types, restrictions, etc.), *descrip* for definitions and other descriptive information, *admin* for administrative information, and *note* for additional information of any kind. These GIs, along with a number of others designed to meet technical and administrative needs within the MARTIF format, are defined in the DTD.

generic identifiers

Term contains only terms and terminological units, such as phrases, whereas *termNote*, *descrip*, and *admin* function as *meta data categories*. In order to accommodate all the many data categories found in terminology databases, each individual category is

values of "type"

assigned to one of the meta data categories, i.e., to one of the GIs, and differentiated in SGML as a value of the attribute *type*. For instance, if *<descrip>* is a GI, then a data field containing a *definition* can be identified as:

<descrip type='definition'>

Because the values of the attribute *type* do not have to be declared in the DTD (Style 1), they can be unlimited in number and new ones can be added at any time without changing the DTD. The emerging XML environment provides powerful tools for implementing this principle. For some, however, it is preferable to define more specific GIs than to have a stable, concise DTD. For those groups an alternative representation has been defined that corresponds directly to the primary representation but that generates a distinct GI for each data category (DTD Style 2). The complex DTD needed to define this alternative representation ends up being quite long, but it can be generated by an algorithm and does not have to be worked out "by hand". Style 2 may be more appropriate for the dissemination of data, and Style 1 definitely more effective for the lossless exchange of data.

constrained MARTIF

Although the Style 1 DTD does not list all the data categories, it is important for conversion routines and general practice to recognize a standard list of all available data categories that could serve as a guide both to anyone who is using MARTIF and anyone who is designing a new terminology database. This effort was developed from a project based at the Kent State University Institute for Applied Linguistics (Wright 1991) that began as a part of the MicroMATER effort and subsequently became ISO 12620: 1999. Together, the MARTIF DTD, representing the core structure of a terminological data collection, and the inventory of data categories form a framework within which various interchange formats can be defined. Throughout the development of the MARTIF project, it was anticipated that after the MARTIF framework, sometimes called Negotiated MARTIF, became a standard, one particular application of this standard, or possibly a family of applications, defined primarily by a particular (constrained) subset(s) of data categories and their permissible instance, would be standardized as well during a second phase of the MARTIF project. This second phase is underway as of this writing. This effort will also make MARTIF XML compliant.

5.3 Format Structure

normalized structure

Early in the development of MARTIF, it seemed politically attractive to develop a structure for the interchange format that would be flexible enough to visibly mirror the structure of any termbase. This approach would have resulted in a highly variable interchange layout closely resembling the source termbase for a given data set. The advantage of this approach would have been that system developers would immediately recognize the way that their data related to the original format. This would make export routines very easy to write, but import routines would become much more difficult because they would have had to cope with a wide range of variables, necessitating complex mechanisms for reorganizing data. Consequently, the ISO Working Group decided to create a single "normalized" structure into which the data from any terminology database could be "poured" after a certain amount of manipulation.

<termEntry>

The normalized MARTIF structure takes advantage of SGML's ability to embed subordinate logical elements inside higher-level elements, resulting in "nested" structures rather like children's nesting toys, where small elements fit inside larger ones (see Figures 1 and 2). This principle is based on the assumption that all terminological data is made up of identifiable parts that can be fit together into a logical, hierarchical structure. The most important component of the MARTIF structure reflects the assumption that concept-oriented termbases are by definition organized into terminological entries, consequently the fundamental element of a MARTIF document is the *<termEntry>*.

one concept, one entry

SEE

Entry Structures 8.1.3: 572

Although the terminological entries appear to be more or less universal, examination of existing TMSs reveals great variety within entry structures. As noted previously in this book, some multilingual termbases divide entries by language sections. Others treat similar data categories in two or more languages as a set, for instance grouping all terms, all definitions, etc., together. Some termbases place language sections in separate, but linked, records, while others combine them into one entry. MARTIF addresses this problem by grouping all terms assigned to the same concept together into a single terminological entry, a principle that reflects widely-held theoretical views on terminological entries.

<langSet> <ntig>

Many databases are rather "flat", meaning that there is no explicit hierarchical nesting of information. Detailed examination of the organization of information in many systems reveals that data in terminological entries usually pertain either to the entire entry (entry-level information), to one of the language sections, or to one of the terms treated in the entry (see Figure 2). Based on this observation, the structural elements representing the next subordinate levels of information in the terminological entry are the *language set*, represented by the SGML GI *<langSet>*, and the *term information group (tig)*, represented in ISO 12200 by the *<ntig>* GI. In the evolving new standards, simple term information groups can be represented by *tig*, but *tigs* with nested data categories use *ntig*. Any mention of *ntig* in this chapter can be understood to include *tig* unless otherwise indicated. Synonyms, variants, and equivalents in other languages are each treated in its own *<ntig>*, thus preserving the principle of *term autonomy*, which means that they may each be documented by a full set of subordinate data categories. The *<ntig>* element is neutral with respect to whether an entry is monolingual or multilingual and does not imply any particular theoretical bias since all entries contain terms and each term may—and almost always does—be associated with additional information. Entry-level information that applies equally to all languages and terms in the entry appears at the beginning of the entry prior to the first *<langSet>*. Any information that pertains to all terms used in a given language, such as differences that occur in conceptual relationships in that language, can be included at the *<langSet>* level.

group tags

Within the individual *<ntig>*, group tags can be used to bundle data segments containing associated information such as notes or pointers together with descriptive or administrative information (*<termGrp>*, *<descripGrp>*, and *<adminGrp>*). These group tags, along with *term*, comprise meta data categories: term-related information (*termNote*), *descrip* and *admin*, together with links that point to related units of information. MARTIF uses SGML's powerful pointer mechanisms to forge these links to other entries, back-matter (or front-matter) items such as bibliographical entries treated as shared references within the SGML document, or even to external data, such as graphical illustrations, tabular material, etc., residing in the same system, accessible on an Intranet, or on the WWW. It is also possible to link within entries, such as pointing back and forth between abbreviated forms and their respective expanded forms.

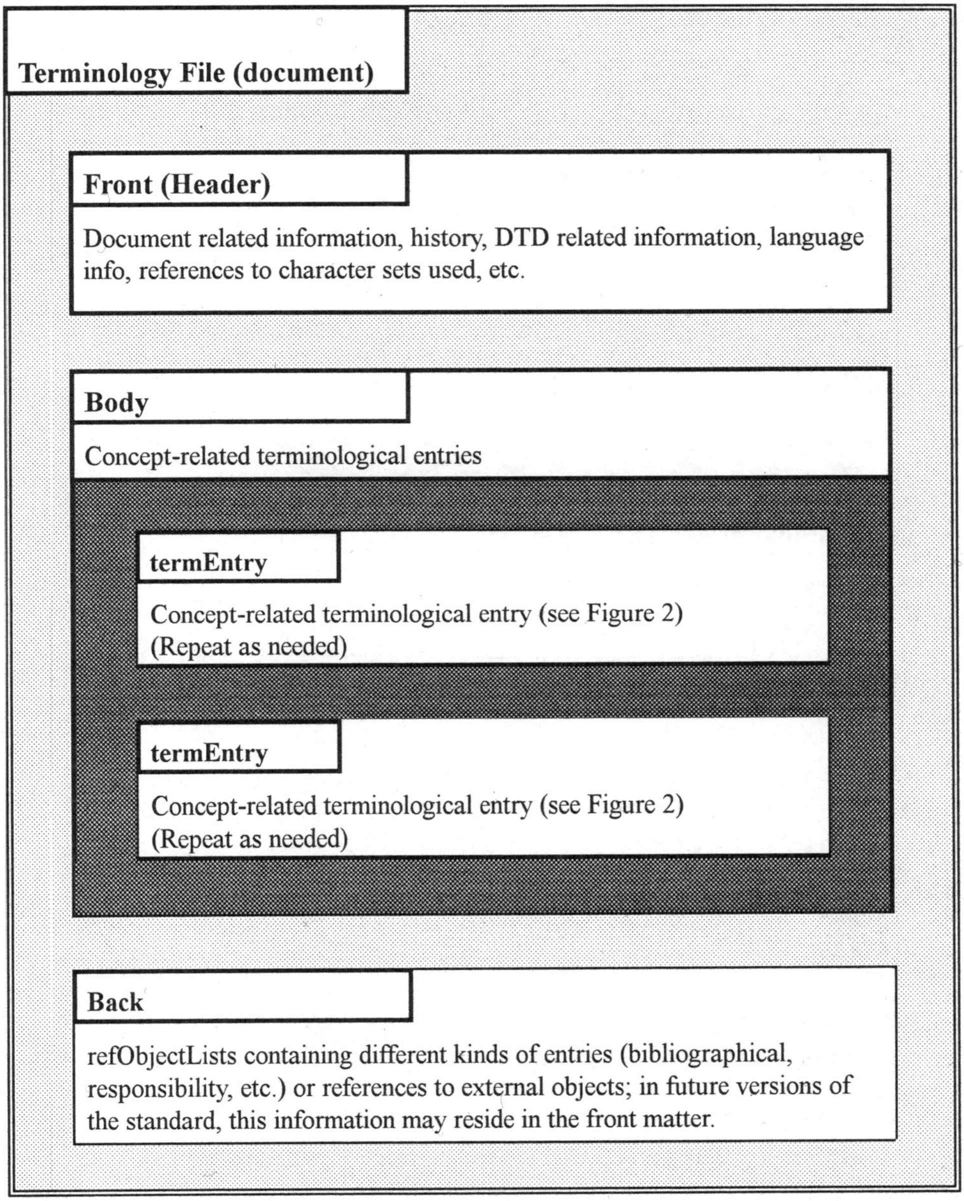

Figure 1: *MARTIF Document Structure*

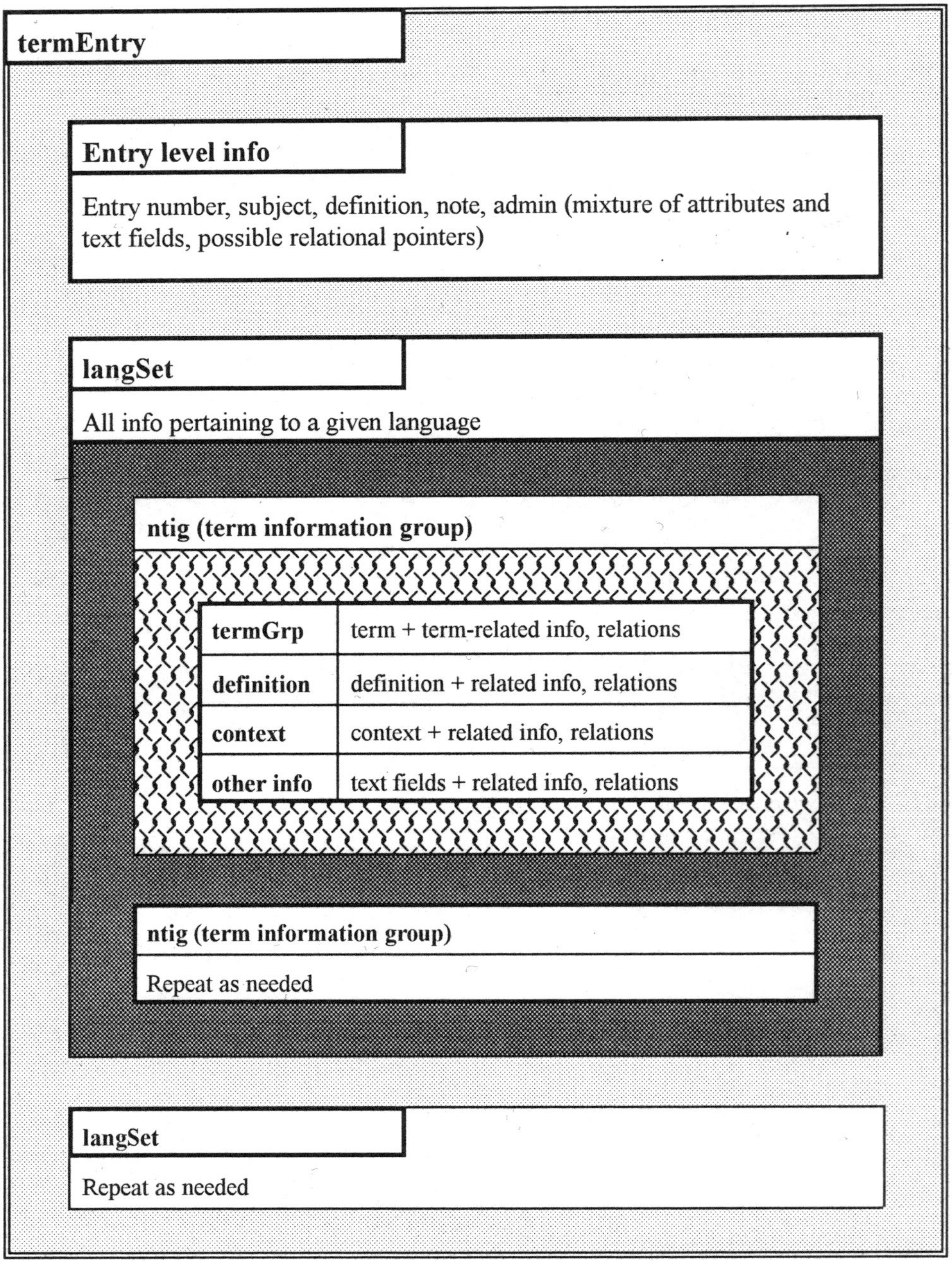

Figure 2: *MARTIF <termEntry> Structure*

presentational vs. logical representation

As implied above, a fully normalized structure features one *<termEntry>* per concept and one *<ntig>* per term. Figures 3 and 4 show a sample terminological entry illustrating the difference between conventional print formats and MARTIF. The data they contain comes from ISO 472:1988, *Plastics—Vocabulary*. Figure 3 reflects ISO presentational format (ISO 10241), in which standard typographical and layout features (boldface fonts, indentation, blank lines, parallel language presentation) represent the terminological entry structure. Figure 4 shows the same data represented in the MARTIF format, which explicitly marks the hierarchical relationships among data segments. Both the presentational version, with its implicit relationships, and the MARTIF version, with its explicit relationships, represent the same terminological information. In both cases, the underlying conceptual data model reflects the position that a terminological entry consists of terms logically grouped by language, with descriptive, administrative, and other associated information potentially appearing at any of three levels: entry, language, and term.

MARTIF markup

INFO

Entailed Terms INFB 14:342

Unfortunately, space does not permit a complete description of all the aspects of Figure 4, but ISO 12200 does indeed treat these issues in depth. In brief, the information inside the *<termGrp>* tags documents the term as a lexical unit, while the *<descripGrp>* data provides concept-related information (in this case a definition). The *<ref>* elements point to other entries cross-referenced in the original terminology standard (*entailed terms*). The diacritics used in the original French text have been converted here to SGML entities consisting of 7-bit ASCII characters that can be read and manipulated in all known computer systems and across Internet gateways. Implementation of MARTIF in XML, however, enables the use of Unicode, which provides the capability to handle the characters for most modern languages that are likely to occur in terminology management environments.

MARTIF Structure

We have focused so far on an explanation of the general aspects of the terminological entry itself as the fundamental information element contained in a termbase. Figure 1 illustrates how these terminological entries fit into the overall structure of the MARTIF document. Here, in keeping with the original TEI flavor of SGML, the document consists of three major containers: a front, a body,

thermal degradation
the entirety of all deleterious chemical modifications of plastic at elevated temperature
NOTE—It is essential to report the temperature and other environmental conditions at which the phenomenon is studied.
See also **aging**, **degradation** and **deterioration**.

décomposition thermique
ensemble de toutes les modifications chimiques nuisibles d'un plastique a température élevée
NOTE—Il est essentiel d'indiquer la température et les autres conditions d'environnement dans lesquelles le phénomène est étudié.
Voir aussi **viellissement**, **dégradation** et **détérioration**.

Figure 3: *ISO Data Presentation*

front

and a back. In the case of a MARTIF document, the header contains declarations such as authoring information and history, information on languages, and character sets. It might include a declaration of the specific subset of ISO 12620 data categories that are used in this document.

back

The body contains the terminological entries themselves. In the back (or possibly in the front), we find shared resources that can be referenced from anywhere in the document. These take the form of organized lists. Bibliographic and responsibility entries are each collated in their respective *refObjectLists*. Other kinds of information that resides on the computer system, but outside the immediate file (external foreign objects) might include graphics, audio, video, or virtual reality files, charts and graphs, or even other termbases. The argument for placing these items in the front instead of in the back is that this would facilitate their being processed first, thus making them immediately accessible when they are referenced later by links in the body.

references

For more information on the MARTIF format, readers who are seriously interested in the internal structures and functions of the standard should consult ISO 12200 itself, as well as other articles on the subject (Schmitz 1995, 1998; Melby et al. 1996). Potential users of the standard should bear in mind, however, that as more user-friendly processing utilities emerge, many people will be

```
<termEntry>
   <adminGrp> <admin type='responsibility'> ISO TC 61, Plastics </admin>
   </adminGrp>
   <langSet lang=en>
   <ntig>
      <termGrp><term>thermal degradation</term>
      <termNote type='partOfSpeech'>n</termNote></termGrp>
      <descripGrp><descrip type='definition'>the entirety of all deleterious
      chemical modifications of plastic at elevated temperature </descrip><note>It
      is essential to report the temperature and other environmental conditions at
      which the phenomenon is studied.</note>
      </descripGrp>
      <descrip type='relatedTerm' target='en47222'>aging</ref>*
      <descrip type='relatedTerm' target='en47248'>degradation</ref>
      <descrip type='relatedTerm' target='en47250'>deterioration</ref>
   </ntig></langSet>
   <langSet lang=fr>
   <ntig>
      <termGrp><term>d&eacute;composition thermique</term>
      <termNote type='partOfSpeech'>n</termNote>
      <termNote type='gender'>f</termNote></termGrp>
      <descripGrp><descrip type='definition'>ensemble de toutes les modifications
      chimiques nuisibles d'un plastique a temp&eacute;rature
      &eacute;lev&eacute;e</descrip>
      <note>Il est essentiel d'indiquer la temp&eacute;rature et les autres condi-
      tions d'environnement dans lesquelles le ph&eacute;nom&egrave;ne est
      &eacute;tudi&eacute;.</note>
      </descripGrp>
      <descrip type='relatedTerm' target='fr47222'>viellissement</ref>
      <descrip type='relatedTerm' target='fr47248'>d&eacute;gradation</ref>
      <descrip type='relatedTerm' target='fr47250'>
      d&eacute;t&eacute;rioration</ref>
   </ntig></langSet>
</termEntry>
```

Figure 4: *MARTIF Data Representation (* This example implements changes introduced as part of an amendment to ISO 12200.)*

able to use the standard without worrying much about its internal workings, just as many people create RTF files by converting from advanced word-processing or other applications without ever looking at the raw RTF code.

black-box solutions

6 Interchange Procedures

6.1 Validation Procedures

MARTIF software

What users will need to know is how to apply the procedures for processing a MARTIF document. Let us assume that a terminology file has been received as a MARTIF document. To be able to use MARTIF effectively to import these terminological data, a user needs access to a software package with at least the following three functions:

- validation of the MARTIF format
- conversion to some text viewing and editing format
- conversion to the import format of the local TMS

Needless to say, few users will have the programming sophistication or the time to create this kind of software themselves. As the use of MARTIF becomes more prevalent, developers of commercial language technology, and possibly independent programmers, will provide user-friendly conversion and validation routines designed to simplify these procedures.

validation

First, users must ensure that the file is a formally valid MARTIF file. Invalid files can result from transmission errors or from misunderstandings on the part of the party supplying the data. Formal validation only checks the structure of the file. It does not guarantee that the translation equivalents in the file are well-chosen or verify any other content-related aspects of the file.

quick inspection

File validation should consist of three logical steps, which ideally may be combined into a single software application:

- First, a "quick-inspection" software utility should be run on the file to determine whether it is indeed a well-formed MARTIF file. On occasion, one might receive an incomplete file that must be embedded in a standard MARTIF framework before it can be used to import data.
- Second, the file should be checked for conformance to the MARTIF core structure. This can be done using a general-

purpose SGML or XML parser. Typical errors at this stage would be missing required end tags, invalid generic identifiers, and invalid (that is, unresolved or ambiguous) global ID references. Of course, any errors detected by the parser must be corrected and the parser must run without errors before the user goes on to step three.

SGML/XML parsing

- The third step is to run a MARTIF-specific validation program on the file to check for conformance to MARTIF requirements that are stated in the data category module, such as the values of the "type" attribute used to qualify MARTIF generic identifiers. This validation routine can also check other MARTIF requirements, such as the format of dates and the content of any other type of element that has a well-defined format or list of permissible values. The second and third steps can be combined and checked using a general-purpose XML parser if the user has chosen to generate the complete data category framework based on the XML schema representation of the individual data categories. This validation can be performed using freeware XML tools.

MARTIF validation

6.1.1 *Conversion to Easily-Viewed Format*

After completing the validation step, users may want to at least spot check the data in the file using an electronic viewer. Since a MARTIF file is in a platform-independent, descriptive format, it is easier to view if it is first converted to a form that can be read by the locally-used word processor, web browser, or text editor. This will enable human users to view it and, if desired, to select entries to be imported into the local TMS.

viewing the data

MARTIF files contain only seven-bit ASCII characters (technically, ISO 646 characters). Any non-ASCII characters, such as *é* are represented as SGML entities, e.g., the ***é*** notation that makes reading the French segment of Figure 4 a challenge. Anyone who has experience looking at HTML code in non-English languages will be familiar with these entities. While they can be irritating for standard western European languages (BFIGS), they are even completely unintelligible for humans trying to read Arabic or CJK languages. Furthermore, if the alternate hexidecimal notation is used, even standard European languages become unreadable for a human. By converting the file to some sort of viewer, users can see real diacritics or characters instead of thinking about entities or hexidecimals. In addition to displaying true characters, a conversion routine could also provide for user-friendly font and layout conventions. In in-house environments or when working with familiar partners, however, this viewing step may in many cases be eliminated and the user goes directly to the next step.

6.1.2 *Conversion to Import Format*

local termbase requirements

Once a MARTIF file has been validated and viewed (which need only take a few minutes per step if there are no problems with the file), the entries can be converted to the format of the local termbase. Of course, the conversion routine must also include some character-entity conversion. The exact conversion depends on what is expected by the import format of the local termbase.

table of correspondence

The heart of the import conversion is a table of correspondences between the data categories used in MARTIF and those used in the local import format. ISO 12200 makes the assumption that the interchange partners are aware of each other and that they can agree among themselves on a set of descriptive and administrative tags, values of the attribute *type* to be used with these tags, and permissible instances. (See the discussion of *blind interchange* below for situations where interchange partners do not "know" each other.)

conversion routine

Since the file has already been validated, the conversion routine can assume that its structure conforms to the MARTIF DTD and the standard data categories. The conversion routine will usually process one entry at a time. Essentially, conversion routines can use the ordering and embedding rules that govern the MARTIF *<termEntry>* to sort out information and to reassign it to the slots provided in the target TMS. For instance, the one-entry-per-concept and the one-*ntig*-per-term rules enable the import routine to interpret the various *ntigs* from the current entry and then to reformat them appropriately. The *lang* attribute associated with each *langSet* facilitates straightforward division by language of the information included in the *langSets*, etc.

testing MARTIF

So far, case studies show that the fully normalized MARTIF DTD is sufficiently flexible to represent the information in any TMS, and routines have been written to perform interchange between various formats. An additional tool that might be useful is a "pre-merge compare" utility designed to check a local database to see whether terms from an interchange file already exist in the local TMS. One particularly detailed experiment in MARTIF conversion is described in detail by Reinke and Schmitz (1998), who found that their major problems lay not with the limitations of the interchange format, but rather with inherent difficulties presented by rich vs. impoverished data models and the shortcomings of proprietary import routines.

6.1.3 *Exporting Terminological Entries*

export routines

The process of writing an export routine from a local TMS to MARTIF is essentially the reverse of the process of writing an import routine. The first step is to select a subset of the local TMS that is to be exported. It is necessary to prepare a table of correspondences between data categories used in the source TMS and the interchange data categories used for MARTIF. Difficulties can arise if a source data category is broader than the interchange data category. The worst case is that human interaction would be needed to restructure data in order to harmonize data category content with the defined content of the standardized data categories. These concerns underscore the importance of term autonomy, granularity, the elemental nature of data categories, and other factors treated in Chapter 8.1.3. Of course, local non-English characters must be mapped into standard entity names, and the information from the source TMS must be reformatted to match the normalized MARTIF format.

conversion effort

After renaming data categories, reconfiguring permissible instances, and reformatting special characters, the exporter runs a conversion program to create a MARTIF file and validates the file using the same tools described above for checking an import file. This process may sound formidable, and indeed it will entail a certain degree of effort the first time it is implemented, but the goal is to fully automate the process so that users do not have to give attention to details during subsequent conversion transactions.

7 BLIND INTERCHANGE

negotiated interchange

As described in Sections 5 and 6 and illustrated in several scenarios, the MARTIF standard provides an open, flexible format for exchanging terminological data between different terminology management systems. Used together, ISO 12200 and 12620 provide a fairly precise definition of data for interchange purposes.

Nevertheless, the standards make the assumption that interchange partners will probably engage in supplemental negotiation in order to ensure exchange of data without information loss or corruption.

blind interchange

As a result of these concerns, various researchers have conducted empirical studies using concrete terminology resources with the goal of determining the feasibility of using the MARTIF standard to exchange terminological data between TMSs with different entry structures and different data categories, but without involving a separate negotiation between each pair of partners. The primary objective of these projects has been to ascertain whether so-called *blind interchange* is possible, i.e., whether the negotiation process can be sufficiently formalized that a "pre-negotiation" can be expressed in a machine-processable format and multiple exchange partners can use the resulting blind format without "seeing" each other (see Hardman 1996; Wright 1996; Melby, Schmitz, and Wright 1996; Melby and Alder 1997; Melby and Wright 1998; and Reinke and Schmitz 1998).

This work resulted in draft documents that describe blind interchange using MARTIF (Alder et. al 1998; Alder 1998). Based on these efforts, ISO TC 37/SC 3/WG 3 launched a project to expand upon the capabilities of MARTIF in order to create an environment where blind interchange would be ensured. The blind format defined in this project is called MARTIF with Specified Constraints (MSC).

7.1 Character Set Solutions

SGML character entities

As implied above, there are at least two solutions to the problem of representing characters not defined in ISO 646. Umlauted and accented Latin letters and other special characters can be represented using the character entities defined in Annex D of the SGML standard, ISO 8879 (see the examples Figure 4). As noted above, this solution is almost totally inadequate for languages that do not use the Latin alphabet.

XML

INFO

Unicode INFB 20:861

Another solution is to use the coding specified for XML (Extensible Markup Language), which comprises a subset of SGML that has been defined under the aegis of the World Wide Web (see World Wide Web Consortium 1997 und Microsoft Corporation 1998). This modification brings MARTIF and related standards in line with the latest developments in markup language.

Moreover, it offers a reasonable solution to the character set problem because XML is based on the 16-Bit Unicode character set (Unicode Consortium 1996; ISO 10646:1993), which defines 38,885 characters. This comprises almost all the characters that are used in the various modern languages of the world, including Arabic, Chinese, Japanese and Thai.

7.2 Data Modeling Variance

data category ambiguity

Annex A of ISO 12620 defines over two hundred terminological data categories, and Annex A of ISO 12000 specifies the corresponding MARTIF representation for these data categories. Nevertheless, depending on the design of a given TMS, some kinds of closely related information can be documented in different data categories without violating the specifications of these standards. For instance, one termbase might store information on *part of speech* in a *grammar* data category, whereas another might code it with the more granular designation *part of speech*.

data category subset

This kind of data modeling variance is inappropriate for blind interchange because it requires that interchange partners clarify specifically how a given kind of data is to be represented in the interchange format. Consequently, any blind interchange format must define an unambiguous set of data categories selected from ISO 12620. (This does not rule out the possibility that new data categories may have to be added to 12620 in the future.)

supplemental validation

As noted in Section 5.2 above, the MARTIF DTD defines only structural and group elements (meta-data elements) as generic identifiers and specifies the actual data categories as values of the *type* attribute. These values are not checked against the DTD during the standard parsing stage. In a blind interchange format the data categories are limited to those explicitly defined as a subset of ISO 12620. They are still treated as values of the attribute *type*, but they are also explicitly included in data category specification (DCS) module that can be validated using an XML schema.

7.3 Data Category Content

optional content

Another aspect of data categories that is not specified in the standards is their content. ISO 12620 and 12200 only provide sample recommendations for content values that *could* be used. Consequently it would be possible to code "*part of speech* =

noun" as *noun, n., sub, sub.* or *substantive* without violating the standard.

registered classification systems

For blind interchange, however, the content of many data categories must be unambiguously defined. A blind interchange format will specify data category content wherever this is feasible. Certainly it does not make sense to try to check the content of textual data categories such as *term, definition, context*, or *note*. The content of categories with specified permissible values, however, such as *part of speech, gender,* or *term type*, can indeed be validated. Specifying the content of a category such as *subject field* is problematic, however. To be sure, it would be possible to standardize a set of permissible values, but real variations among individual users and subject fields are too great to make it reasonable to anticipate any kind of global solution. Experts in individual subject fields, however, can agree to sets of classifications for their domains and can record this information with a central registration authority to create a family of related formats. Interchange documents could then indicate which classification system was being used and the document could be validated against this information.

autonomous XML validator

As implied above, the Blind MARTIF strategy does not propose that the data categories or their content be listed in the DTD or that these values should be validated by the SGML parser. Instead, an autonomous validation tool (with an XML conformant control apparatus) has been developed specifically for the purpose of checking data categories (values of the attribute *type*) and their content. Thus a standard XML parser will be used together with a validation tool to determine whether a document conforms to a particular blind interchange format. Another advantage of using a separate validator to check the more restrictive elements associated with blind MARTIF (MSC) is that these documents remain compliant with ISO 12200 and can be parsed using the standard MARTIF DTD, even though they are in effect more restricted than a standard negotiated MARTIF document. Thus standard MARTIF and blind MARTIF will comprise a seamless sequence of compatible standards.

7.4 User Subsets

A group of researchers working in a given discipline could want to exchange data with each other without serious loss of information or significant editing prior to import. They could agree ahead of time to use Blind MARTIF, but also to restrict their work to a specified subset of the data categories in ISO 12620. This could include using an industry-specific documentary language or thesaurus to classify concepts in the termbase. Terminological data that are interchanged within this (closed) user group would follow the Blind MARTIF specifications and could therefore be parsed and validated by the standard tools, but the validator could also be used to check the restriction to the agreed subset of data categories by modifying the XML rule set for the validator.

group-specific restrictions

Another very similar scenario might involve a group of individuals, such as professional translators, who want a very simple, "quick and dirty" format in which they can reliably document a very small subset of categories and quickly disseminate their data to numerous colleagues (such as in a localization team). Bonthrone has proposed such a solution under the rubric of "MARTIF-lite" (Bonthrone 1996).

MARTIF-lite

8 APPLYING MARTIF

Readers who have made it successfully to this point are well aware that data interchange involves a fair degree of complexity. Melby et al. (1996) have attempted to address these issues. The fact remains, however, that many terminologists and localization (L10N) specialists in particular have informed themselves of the basic issues involved in interchange and would like to be able to use proposed solutions, but this is not feasible because no practical tools and utilities are readily available yet. Endorsements from professional organizations and research institutions, e.g., the American Translators Association (Committees for Terminology and Translation and Computers), FIT (the International Federation of Translators), or Infoterm (the International Information Centre for Terminology in Vienna), may look impressive on paper, but they don't help individuals apply the standards.

tool availability

In language engineering and L10N environments, the availability of in-house or consultant-based programming talent has facilitated the use of the standards on a daily basis in real industrial environments to exchange and disseminate terminological information to

industrial implementation

external translators and writers either by direct data transfer or on in-house intranets (Walmer 1999).

expert committees

The "group of experts" scenario outlined above was originally visualized by Sager as a network of collaborative specialists as early as 1990, before the hardware, software, and inter-networking requirements for such a system were readily available to a wide range of users (Sager 1990:140). Today MARTIF is also the foundation for terminology management in ASTM Technical Committee E-31 for Healthcare Informatics (ASTM E-31, 1998) and its companion international activities. This group, however, also represents an example of a committee that has access to special computing expertise in order to implement their solution.

variations

MARTIF and data categories specified in 12620 contributed significantly to the development of the Virtual Hypertext Glossary (VHG), which provides a distributed environment for protein chemists to share their terminological data on the WWW (Murray-Rust and West 2000). Although the core VHG engine reflects a variation on the original MARTIF theme, the basic design remains compatible with the MARTIF structure.

black box solutions

Many software publishers and system designers want to be able to incorporate the conversion directly into their products, however, and to offer the average individual or industrial user "black-box" solution to the data conversion problem that will be as easy to use as rtf conversions should ideally be. The developers of popular TMSs such as TRADOS MultiTerm for Windows™, EP™, and TermStar™ are currently creating the kinds of conversion utilities that should make it feasible for average users to achieve successful interchange without expensive outside consulting assistance. Companies and broad-based multilingual terminology resources (MTRs) have chosen to use the MARTIF format and its spin-off formats in their data sharing environments.

9 Integration with Other Interchange Formats

integrated environments

A major initiative, code-named SALT (Standards-based Access to Multilingual Lexicons and Terminologies) began early in 2000 with the intent of refining and testing a particular MARTIF blind interchange family of formats called XLT. Over thirty universities, institutes, translation technology developers, and translation and localization service providers from Europe and North America have agreed to participate. The goal of this project will be to link the existing and future development of the MARTIF standard with other exchange formats, such as OLIF (Open Lexicon Interchange Format) and TBX (TermBase eXchange). OLIF has been defined by the EU-Commission-supported Otelo Project (Open Translation Environment for LOcalization) and is designed to facilitate exchange of data between lexicons used as dictionaries for different machine translation systems (see also Thurmair, Ritzke, and McCormick 1998). A Special Interest Group (SIG) within LISA (the Localisation Industry Standards Association) was founded with the name OSCAR (the Open Standards for Container/Content Allowing Re-use). In a record eighteen months, this group created the TMX (Translation Memory eXchange) exchange format for data resources in translation memory textbases. Work continues on the parallel development of the TBX format for termbase resources that reside in integrated environments. Other projects with which the SALT group is collaborating include OLIF's TQPro (Translation Quality for Professionals) project and the EAGLES ISLE (Expert Advisory Group for Language Engineering Standards / International Standards in Language Engineering) project, which is focusing on the creation of meta-descriptions for multimodal/multimedia language resources (MMLR), specifically in the area of language corpora. SALT, along with a number of these projects, is funded by the EU Human Language Technologies (HLT) program. There is close collaboration among the developers of the various formats under the umbrella of the SALT project (see also Melby 1998). In addition, discussion continues in ISO TC 37/SC 3/WGs 3 and 4 to create a metadata model that will facilitate the interoperability of interchange data from the two different styles of data representation described in Section 5.2.

INFO

Meta Model INFB 29:879

Accessing MTRs INFB 30:884

Human Language Technologies INFB 31:887

References

Ahmad, Khurshid; Ogonowski, Antoine; Dauphin, Eva; Sta, Jean-David; and Arppe, Antti. 1996. "Engineering Terminology — A Case for a Linguistically-informed Terminology Database." *TKE '96: Terminology and Knowledge Engineering.* Frankfurt am Main: Indeks Verlag, 166-178.

ASTM E-31. 1998. http://www.astm.org/. (Search Committee listings for E.31.) Accessed most recently 2000-07-20.

Alder, Aaron D.; Corrodini, Ryan A.; Melby, Alan K.; Schmitz, Klaus-Dirk; and Wright, Sue Ellen. 1998. "Blind MARTIF." *CLS Framework*. http://www.ttt.org/clsframe/. Accessed most recently 2000-07-20.

Bonthrone, Robin. 1996. "The Business of Terminology: A European Perspective." *TermNet News*. 54/55: 1-3.

Budin, Gerhard; Melby, Alan; and Wright, Sue Ellen. 1993. "Terminology Interchange Format (TIF): A Tutorial." *TermNet News*, 40, 5-63.

Budin, Gerhard; Lonsdale, Daryle; Melby, Alan; and Wright, Sue Ellen. 1999. "Integrating Translation Technologies Using SALT." In: *Proceedings of ASLIB*. London: ASLI, unnumbered pages.

Hardman, Daniel. 1996. *A Practical Proposal for the Blind Interchange of Terminological Data*. Unpublished Masters Thesis. Provo, Utah: Brigham Young University.

Heid, Ulrich. 1991. "Eurotra-7: Feasibility and Project Definition Study on the Reusability of Lexical and Terminological Resources in Computerised Applications. Intermediate Report—A guide to the documents of Phase 1."

Hjulstad, Håvard. 1991. "Nordic Terminological Record Format (NTRF)." Oslo: Norwegian Council for Technical Terminology.

ISO 472:1988. *Plastics—Vocabulary*. Geneva: International Organization for Standardization.

ISO 646:1991. *Information technology — ISO 7-bit coded character set for information interchange*. Geneva: International Organization for Standardization.

ISO 6156:1987. *Magnetic Tape Exchange for Terminological/Lexicographical Records (MATER)*. Geneva: International Organization for Standardization.

ISO 8879:1986. *Information processing—Text and office systems—Standard Generalized Markup Language (SGML)*. Geneva: International Organization for Standardization.

ISO 10241:1992. *Terminology standards—Preparation and layout*. Geneva: International Organization for Standardization.

ISO 10646-1:1993. *Information technology — Universal Multiple-Octet Coded Character Set (UCS) — Part 1: Architecture and Basic Multilingual Plane*. Geneva: International Organization for Standardization.

ISO 12200:1999. *Computer Applications in Terminology—Machine-Readable Terminology Interchange Format (MARTIF)—Negotiated Interchange*. Geneva: International Organization for Standardization.

ISO 12620:1999. *Computer Applications in Terminology—Data Categories*. Geneva: International Organization for Standardization.

LeMeur, André. 1998. "GENETER : un format générique pour la distribution et la réutilisation de données terminologiques hétérogènes." http://www.uhb.fr/Langues/Craie/balneo/demo_geneter.pl. Accessed most recently 2000-07-20.

LISA. 1998. "Translation Memory eXchange." http://www.lisa.unige.ch/tmx/index.html. Accessed most recently on 2000-07-20.

Melby, Alan K. 1982. "Multi-level Translation Aids in a Distributed System." *North Holland Linguistic Series # 47 — Proceedings of COLING '82*. Jan Horecky, ed. Amsterdam: North Holland Publishing Company, 215-220.

Melby, Alan K. 1991. "MicroMATER: A Proposed Format for Exchanging Lexical/Terminological Data Files," *META: Translators Journal*, Vol. 36, No. 1.

Melby, Alan. K. 1994. "The CAT is on the Multi-level MAT." *The ATA Chronicle*, XXIII, 2, 14-15: 35.

Melby, Alan K.; Budin, Gerhard; Shreve, Gregory M.; and Wright, Sue Ellen. 1994. "Base Tag Set for Terminological Data." *Guidelines for the Encoding and Interchange of Machine Readable Texts, TEI P2*, C. Michael Sperberg-McQueen and Lou Burnard (eds). Chicago and Oxford: ACH, ACL, ALLC, 3771-390.

Melby, Alan K.; Schmitz, Klaus-Dirk; and Wright, Sue Ellen. 1996. "The Machine-Readable Interchange Format (MARTIF): Putting Complexity in Perspective." *TermNet News*. 54/55: 11-20.

Melby, Alan K. and Alder, Aaron D. 1997. "MARTIF I." *CLS Framework*. http://www.ttt.org/clsframe/. Accessed most recently 1999-10/11.

Melby, Alan and Wright, Sue Ellen. 1998. "Overview of the CLS Framework." http://www.ttt.org/clsframework/. Accessed most recently 2000-07-20.

Melby, Alan and Wright, Sue Ellen. 1999. "Leveraging Terminological Data for Use in Conjunction with Lexicographical Resources," *Proceedings of the Fifth International Congress on Terminology and Knowledge Engineering, August 23-27, Innsbruck, Austria (TKE'99)*. TermNet: Vienna, 544-569.

Microsoft. 1993. *The GUI Guide—International Terminology for the Windows™ Interface*. Redmond, Washington: Microsoft Press.

Microsoft. 1999. [English Computer Terminology, accessed 1999-10-09] http://microsoft.com/globaldev/reference/glossary.asp; [Multilingual Glossaries, accessed 1999-10-09] ftp://ftp.microsoft.com/developr/MSDN/NewUp/Glossary/ .

Microsoft Corporation. 1998. "XML Online Workshop". http://msdn.microsoft.com/xml/index.asp. Accessed most recently 2000-07-20.

Murray-Rust, Peter and West, Leslie. 2000. "Virtual Hypertext Glossary (VHG)." http://www.vhg.org.uk/index.html. Accessed most recently 2000-07-17.

Nilsson, Henrik. 1999. "Terminology [ɪn saɪt]—Nordic Termbank Services via Internet." In: Terminology and Knowledge Engineering '99. Peter Sandrini, Ed. Vienna: TermNet, 603-609.

OTELO. 1998. "LE-2703 OTELO: Common Access to Translation Services." http://www.otelo.lu/seite2.htm. Accessed most recently 2000-07-20.

Reinke, Uwe. 1991. *Der Austausch Terminographischer Daten. Diplomarbeit zur Erlangung des Grades eines Diplom-Übersetzers*. Saarbrücken: Universität des Saarlandes, Fachbereich 8: Neuere Sprach- und Literaturwissenschaften; Fachrichtung 8.6: Angewandte Sprachwissenschaft sowie Übersetzen und Dolmetschen.

Reinke, Uwe and Schmitz, Klaus-Dirk. 1998. *Testing the Machine Readable Terminology Interchange Format (MARTIF)*. Saarbrücker Studien zu Sprachdatenverarbeitung und

Übersetzen. Saarbrücken: Fachrichtung 8.6—Angewandte Sprachwissenschaft sowie Übersetzen und Dolmetschen—Universität des Saarlandes.

rtt [Rådet for Teknisk terminologi]. 1999. http://www.rtt.org/htrf/htrf/htm. Accessed 1999-12-09.

Sager, Juan C. 1990. *A Practical Course in Terminology Processing*. Amsterdam and Philadelphia: John Benjamins Publishing Company.

Schmitz, Klaus-Dirk. 1996. "MARTIF: A New ISO Standard for the Interchange of Terminological Data." Translated by Sue Ellen Wright. *TermNet News*. 50/51: 4-5.

Schmitz, Klaus-Dirk. 1998. "MARTIF—Ein SGML-basiertes Austauschformat für terminologische Daten." *SGML-Anwendungen und Perspektiven*. Wiebke Möhr und Ingrid Schmitt, Hrsg. Berlin: Springer-Verlag, 109-121.

Sperberg-McQueen, C. Michael and Burnard, Lou, eds. 1990. *Guidelines for the Encoding and Interchange of Machine Readable Texts. TEI P1, Draft Version 1.1.* Chicago and Oxford: ACH, ACL, ALLC.

Stellbrink, Hans-Jürgen. 1993. "Selling Terminology at a Price—The Wrong Approach Price." *Proceedings of TKE '93: Terminology and Knowledge Engineering*. Klaus-Dirk Schmitz and Christian Galinski, eds. Frankfurt/Main: Indeks Verlag: 292-299.

Thurmair, Gregor; Ritzke, Johannes; and McCormick, Susan. 1998. "The Open Lexicon Interchange Format OLIF." *Proceedings of the 4th International Symposium on Terminology in Advanced Microcomputer Applications*. Vienna: TermNet: 237-262.

Unicode Consortium. 1996. *The Unicode Standard*, Version 2.0. Reading, Massachusetts: Addison Wesley.

Walmer, Daphne. 1999. "One Company's Efforts to Improve Translation and Localization."*Technical Communication: Journal of the Society for Technical Communication*. 46/2: 230-237.

World Wide Web Consortium. 1997. "Extensible Markup Language (XML)." http://www.w3.org/XML. Accessed most recently 2000-07-20.

Wright, Sue Ellen. 1991. "The MicroMATER Tagset: Proposed Data Categories for the Exchange of Terminological Data between Terminological Database Systems," *Proceedings of the NISKO International Conference on Knowledge Organization, Terminology and Information Access Management.* Bratislava: NISKO: 147-159.

Wright, Sue Ellen and Budin, Gerhard. 1994. "Data Elements in Terminological Entries: An Empirical Approach." *Terminology*. Vol. 1, No. 1. Amsterdam: John Benjamins Publishing Company.

Wright, Sue Ellen. 1996. "Blind Interchange of Terminological Data: Problems and Possibilities," *Proceedings of the 10th European LSP Symposium*, Vienna, 29-Aug.-1 Sept., 1995. Gerhard Budin, ed. Vienna: IITF-Infoterm-TermNet: 1131-1154.

Wright, Sue Ellen. 1998. "Trends in Language Engineering." *Proceedings of the '98 TAMA Conference.* Vienna: TermNet: 3-22.

Section 8.2
Database Applications and Products

The three articles in this section all represent case studies documenting specific projects or "snapshots" taken at a given point in time during the development of an organization.

8.2.1 Preparing Multi-Volume Illustrated Terminological Dictionaries

JOCELYNE LE NÉAL

1 INTRODUCTION

A terminological dictionary, also known as a vocabulary, differs from a standard language dictionary in that it provides the words and expressions, in one or more languages, used to designate the concepts relating to a particular subject field, as well as the definitions and, in some cases, illustrations of these terms. Perhaps the question "What is the best way of compiling such a dictionary?" can best be answered by another question: "Why do you wish to do so?" Indeed, the most suitable methods to adopt will depend not only on available resources, such as staff, documentation, computer resources and time, but also on the audience being addressed, as well as on the objective being pursued—whether it be to standardize the terminology in a given organization, for example, or to produce a translation tool or a reference for explaining specialized vocabulary. This chapter describes how the terminologists at Teleglobe Canada Inc. went about defining and carrying out their project of producing a multi-volume English-French dictionary entitled the *Bilingual Dictionary of International Telecommunications* (BDIT).

terminological dictionaries

INFO

Terminology Collections INFB 1:325

2 DEFINING THE PROJECT

Teleglobe is Canada's international telecommunications carrier. Headquartered in Montréal, it operates switching centers in Québec, Ontario and British Columbia, as well as submarine cable stations on the east and west coasts and satellite earth stations located in different parts of the country. Approximately 55% of its employees are Francophone and 45% are Anglophone. Until its privatization in 1984, Teleglobe was a Crown Corporation and, as such, had to provide service in Canada's two official languages.

official bilingualism

SEE

Francophone Experience 3.2:261

Although it now falls under Québec's *Charter of the French language* establishing French as the official language of work in the province, Teleglobe is allowed—and indeed forced—to deal with carriers and customers outside Québec in other languages, mainly English. Against this backdrop, one can easily understand the need to communicate in both French and English throughout the corporation.

SEE *Multilingualism 3.1:245*

Given the status of English as the *lingua franca* of telecommunications, one might be led to believe that the terminology of the field is well established in that language, such that the interests of communication between Teleglobe's employees, and between the corporation and its customers, correspondents, and other partners would be adequately served by producing glossaries of English technical terms and their French equivalents.

status of English

Such an assessment, however, would fail to take into account the scope and complexity of the subject field itself. Indeed, since the invention of electric telegraphy in 1837 and of telephony some forty years later, and especially since the advent of electronic switching in the early 1960s and the progressive digitization of networks over the last twenty years, telecommunications has diversified to include such vastly different offerings as data transmission, facsimile and E-mail, video and multimedia conferencing, intelligent network services, land, maritime and aeronautical mobile communications and advanced space research services, to name but a few. Technology has evolved from manual exchanges to asynchronous transfer mode systems and switchboards in the sky, from telephone wires to fiber-optic cables and VSAT (very small aperture terminal) satellite links—all of which are based on highly sophisticated techniques like image compression and adaptive differential pulse code modulation, and which integrate the latest advances in computer science and electronics.

digitization

Standards organizations such as the International Telecommunication Union's International Telegraph and Telephone Consultative Committee (CCITT) and International Radio Consultative Committee (CCIR), the International Electrotechnical Commission (IEC), the International Organization for Standardization (ISO), the Institute of Electrical and Electronics Engineers (IEEE), and the American National Standards Institute (ANSI) lost no time in setting up terminology committees to provide terms and definitions for this avalanche of new concepts. However, the publications of

terminology standards

SEE *International Standardization 2.2.3:209*

these organizations tend to be expensive and not all of them are readily available. More importantly, the Teleglobe team felt a more descriptive reference was needed as a complement to the vocabularies they produced, both to account for the multiple terms and expressions found in the literature and to explain the new realities in greater detail to a broad audience including the following groups: technical and nontechnical staff at Teleglobe, members of international telecommunications organizations, domestic and foreign carriers, telecommunications managers and users in other companies, translators, and other language specialists.

SEE *Descriptive Terminology 2.1.4:184*

3 Description of the Dictionary

end product

The decision was made that the best way to meet the needs of all these groups would be to compile a dictionary that could be used as either a bilingual or a unilingual French or English reference. Before explaining how it was produced, a brief description of the end product is in order.

The *Bilingual Dictionary of International Telecommunications* (BDIT) consists of a series of four volumes published by Teleglobe over an eight-year period. Volume 1 (*Antenna Theory*) dates back to 1983; Volume 2 (*Transmission Equipment*) appeared in 1985; Volume 3 (*Switching Equipment*) was released in 1988 and Volume 4 (*Telecommunication Services*), in 1991. In total, these define some 12,000 English and French terms representing approximately 3,000 concepts. The series was to include four additional volumes dealing with other subfields, but it was abandoned due to cutbacks in staff and a redeployment of the corporation's resources in its core sectors of activity.

dictionary layout

Each volume of the BDIT contains definitions in both English and French as well as technical and/or linguistic notes when warranted. In addition, Volumes 2 and 3 are illustrated. Entries are structured like a terminology record in that terms are grouped by concept (see Appendix A), and the concepts are numbered and arranged in sections, each covering a distinct subfield. Access to the entries is provided through two indexes listing the English and French terms separately and referring users to the appropriate concept number. Dictionary entries also include cross-references to related or opposite concepts defined elsewhere in the same

INFO *Lexicography vs. Terminology INFB 3:328*

volume. While this structure may appear somewhat cumbersome at first, it offers two major advantages:

information access

- It allows users to consult the dictionary in the same way in either language and to find all the information pertaining to a given concept in the same entry.
- The structure of each volume reflects that of the subject field that it covers. Thus, users can get a good idea of the scope and ramifications of the field simply from the table of contents. Moreover, they can use the extensive cross-reference system to read through the volume in hypertext mode, as it were, in order to gain some insight into the relationships between concepts.

4 RESOURCES

The two resources used to produce the BDIT represent the kinds of computer applications that are indispensable in conducting any extensive terminological research project. The first of these is a proper classification scheme for the subject field under study. Sufficient time must be taken at the outset to establish a solid framework for the entire project, since a piecemeal approach is totally inadequate for studying the elaborate system of concepts that exists in most complex disciplines, and could result in major inconsistencies. On a more practical level, it is also very important to define clearly the scope and direction of the research, given the extended time frame involved and the very real possibility that it might be carried out by several different collaborators.

INFO *Classification INFB 9:334*

The resource used by the Teleglobe team for this purpose was a terminology thesaurus, an outline of which is shown in Appendix B. It was developed by a member of the team—who happened to be a telecommunications engineer—as part of a project for the Translation Bureau of the Department of the then Secretary of State of Canada. Preparation of this thesaurus took about six months. It involved extensive reading, defining the chosen descriptors, drawing up a list of at least 100 terms falling under each, as well as a list of other candidate descriptors and a structured list of descriptors and non-descriptors, and compiling a bibliography of the thesauri, classification systems, dictionaries, etc., that were consulted. Such a formal approach may not be required in establishing every classification scheme, but the importance of this kind

SEE *Thesaurus Management 6.1:381*

of preparatory work—and of the involvement at this stage of specialists in the field—can have a profound effect on the quality and utility of the final outcome of the project.

project planning

Initially, the BDIT was to comprise a total of nine volumes, seven of which would correspond to the primary or first-level descriptors of the thesaurus. The eighth volume, entitled *Telecommunications*, would deal with very general concepts or those which, for any other reason, defied classification under the chosen descriptors, while the ninth volume would contain the complete thesaurus as well as general English and French indexes of all the terms appearing in the BDIT. The order of treatment of the different subfields was not predetermined. It was simply decided for the engineer to start out by handling the most complex topic, i.e., antenna theory, and to allow the nonspecialists in the team to introduce themselves to telecommunications by first studying concrete concepts, namely those relating to the various types of equipment. The second idea proved to be a very good one, although in retrospect, the first volume should have covered a topic of greater general interest for the target audience.

project evolution

The findings and experience gained during this initial phase led the team to redefine the structure of the BDIT as they went along, with most reassessments taking place at the start of work on a new volume. For example, it quickly became apparent that the number of terms relating to all the equipment used in the industry and the components thereof far outnumbered what the group had intended to cover in a single volume. Moreover, it was important to publish the volumes at regular intervals of two to three years so as not to fall into the trap of going on forever on any single topic in an attempt to cover it completely or to keep pace with rapidly evolving technology. These considerations led to a new strategy for the terms under the "telecommunication equipment" descriptor: the decision was made to devote one volume to switching equipment, another to transmission equipment and its two subfields, radio transmission equipment and transmission media, and not to treat general terms under the broad descriptor, since a good part of that terminology (telephone sets and other terminal equipment, structures such as conduits, telephone poles, etc.) was deemed to be more the domain of manufacturers than of an international carrier such as Teleglobe.

mainframe data collection

It is important to note, however, that the changes were applied to the structure of the dictionary and *not* to the thesaurus itself, except for minor refinements to accommodate new concepts. Thus the thesaurus constituted a solid foundation on which to build the second major resource, the database, which was initially set up on a central computer using a relational database management program called STAIRS™. Because of STAIRS' limited word processing capabilities, this database only included terms and their formal definitions, which were extracted from standards, specialized vocabularies, and glossaries in monographs, technical reports, and periodicals. This database was later converted into EDIBASE™ (see Acknowledgments) format designed for operation in a PC environment. At the time this article was written, the database contained some 30,000 terminology records, the majority of which are unilingual English records. This database served as the starting point for Volumes 3 and 4 of the BDIT. It served as a vital resource in planning each volume, defining its scope, drawing up the base list, choosing the corpus, etc., and provided an initial batch of records with which to start actual terminological research.

PC-based environment

Once in a PC environment, a separate database was created for each new volume, containing thousands of additional records as researchers scanned through the literature in search of additional terms, as well as contexts, explanations on usage, idiomatic expressions employed in conjunction with the terms, etc. The same record format was adapted for this database as for the other (see Appendix C) so as to be able to transfer the records with "lasting" value to the permanent collection before purging the rest and beginning anew with the next volume. In today's workplace environments, where each terminologist would probably be equipped with a PC, it would have made more sense to have written the dictionary entries directly into a third database configured for this purpose. However, it is not necessarily desirable to do away with paper completely, since the mass of information collected in each case was too large to manipulate and assimilate on screen.

5 Methodology

The core team involved in the production of the three largest volumes consisted of three people: two bilingual terminologists, of whom one was an Anglophone and the other a Francophone, and one assistant terminologist, who was also fluent in both working languages. Contrary to what one might anticipate, the terminologists' work was not divided on the basis of language, since this would have required both to research all the subfields. Moreover—and this raises a much more fundamental issue—the English or French part of the dictionary would in all likelihood have become a translation of the other, partially defeating the whole purpose of the project and of terminology as a whole, and making it considerably more difficult to respect and highlight the different systems of concepts at play in the two languages.

team composition

Consequently, once the outline of a given volume had been established, each terminologist assumed responsibility for a certain number of subfields and was allocated a period of time typically in the order of 18 months to cover them as thoroughly as possible in light of their relative importance in the industry and/or for the corporation. As for the assistant terminologist, her role was essentially to input all the data that was collected—which involved choosing the contexts for terms scanned by both terminologists in the course of their readings—and to compile the indexes as the work progressed so as to keep each terminologist informed of the concepts already defined by the other. Recent advances in computer support tools for this kind of work can simplify much of this effort. Neverthelesss, involving students or other assistants with some background in terminology is an extremely effective way of increasing productivity in such a project.

division of labor

Others played a key role once the definitions were written. Among these were experts from Teleglobe and other companies or organizations, who were asked to check the accuracy of the technical information contained in the entries and to comment on usage or on any terms which we had had to propose. Some of these specialists were also consulted during the planning process or in the course of the research, e.g., for technical explanations or help in obtaining privileged information on particular topics. Other collaborators included the graphic artists, colleagues in English and French translation, who revised the entries from a linguistic point of view, and the department director who, in addition to assuming the role of project leader, coordinated the publication aspects and took an active role in revising the manuscripts.

subject field experts

A first aspect in the treatment of each subfield involved the choice of concepts to retain. In an attempt to meet the needs of a majority of users, we applied the three following criteria in making our selection:

concept selection

1. All basic concepts in the field as well as those which, at a more detailed level of classification, play a pivotal role in the system of concepts
2. Concepts that are difficult to designate or translate, or that are denoted by synonymous or polysemic terms
3. Concepts that are widely talked about without being clearly understood, and that consequently give rise to buzzwords which are often misused or abused. Broad concepts such as "electronic switching" and "E-mail" fall under this category.

Treatment of synonymy was inclusive rather than exclusive in the choice of terms. This is consistent with the aim of reflecting usage rather than promoting the use of some terms to the detriment of others, an activity which was felt should be left to competent standards or inter-corporate committees formed for this purpose. The policy, then, was to list all terms and expressions that were sufficiently documented in the sources so as to facilitate the decoding as well as the encoding of technical documents and provide access to the entries through as many key terms as possible. Terms that were deprecated by the authors consulted, or that struck the team as being either misnomers or improperly formed, but whose use is nonetheless widespread, are mentioned in a note and emphasized in the index. Thus, in Volume 2, *récepteur hétérodyne* is included in the last French note of entry 442 and appears in regular rather than boldface type in the index (see Appendix A). To avoid overburdening the entries, however, not all the attested orthographic and syntactic variants of a term were listed. The team relied on rules of grammar and standard dictionaries to resolve such issues as hyphenation in complex English terms. Finally, it should be noted that entry terms in each language are presented in alphabetical order, followed by their abbreviations and by geographic or frequency labels where appropriate. This practice (rather than grouping terms according to their position in the concept system) reflects a strong general user preference for alphabetical arrangement.

INFO

Descriptive & Prescriptive Terminology INFB 4:329

The content of the entries was determined by two main objectives. First, as in the treatment of synonymy, maximum use was made of the information gathered. Thus, an entry always begins with a formal definition presenting the essential characteristics of the concept: nature, form, composition, purpose, means, cause or effect, as the case may be. Wherever possible, the definition begins with a generic term defined in the same volume. The formal definition is followed by examples or by a trailer sentence providing nonessential semantic traits if these are useful in understanding the concept. Thus, in entry 217 of Volume 3, shown in Appendix D, the *line concentrator* is defined by its nature (*concentrator*), its purpose (*reduces the number of wire pairs in the local loop*) and the means by which this is achieved (*by connecting a number of distant subscriber lines to a smaller number of shared analog lines*), whereas the mechanisms that are involved in the process are further described in the follow-on sentence. The main terminology records on which the entry is based are reproduced in Appendix C.

INFO

Standard Definitions INFB 12:340

If warranted, further technical or linguistic information is included in the form of notes. In general, the technical notes contain facts that are not as closely associated with the definition as those in the trailer sentence, as is the case in entry 217, although the line is not always clearly drawn between the two. As for the linguistic notes, they serve multiple purposes: to warn users against certain terms, as was previously stated, and to explain why they are deprecated; to comment on usage, particularly when the concept is denoted by a large number of terms; to add a useful related term in either language for which no equivalent was found in the other; or to provide more idiomatic means of expressing the same idea from another perspective, as in the second English and French notes of entry 97 of Volume 3, also reproduced in Appendix D. *Hot stand-by system*, *parallel synchronous system* and the other entry terms listed in this example were included in the dictionary mainly because of their key role in the system of concepts. Although a few occurrences of these terms were found in the literature, we encountered expressions such as *parallel synchronous mode*, *parallel synchronous arrangement*, and *synchronous duplication of processors* much more often, and felt users should be informed of this fact. This same line of reasoning also led us to word the entries in such a way as to reflect the language of the subject field to the

notes

greatest extent possible without resorting to jargonistic or stilted means of expression.

distinguishing features

The second major objective pursued in writing the entries was to show the structure of the field. The last two sample entries in Appendix D (Nos. 662 and 650 of Volume 2) reveal how similar concepts were defined—in this case different types of optical fibers—so as to make it easy for the user to draw a parallel between them and note the differences at first reading. The illustrations were also used to highlight the distinguishing features of each concept and thereby facilitate comparison. Similarly, the entries were worded corresponding to quasi-synonyms or relating to the various components of a given system in such a way as to allow the user to situate the concept without difficulty in relation to those immediately surrounding it in the classification scheme.

illustrations & presentation

INFO
Entailed Terms INFB 14:342

A great deal of information can be conveyed to users through illustrations, obviously, as well as through the judicious and consistent use of a few typographic codes, and this is easier than ever to achieve with the computer-aided drafting, desktop publishing and advanced word processing software available today. In preparing the BDIT, the team gathered as many illustrations as possible of a given concept during the term extraction phase, and then asked the graphic artists to produce a composite drawing based on the examples we had collected and on our own instructions concerning labeling and the highlighting of significant features. With regard to typography, as shown in Appendix A, entry terms appear in boldface characters both in the text itself, to make them stand out, and in the index, to distinguish them from terms and expressions that are only mentioned in a note. Illustrated terms are marked in the index by an asterisk following the concept number, and terms occurring within a definition or note that are defined elsewhere in the same volume appear in italics, followed by the appropriate concept number. For practical reasons, no cross-references were made between volumes.

SEE
Terminology Extraction 8.4.1.:725

In retrospect, of course, it is clear that a number of new tools could have helped speed up some of the more tedious aspects of large-scale terminology projects like the BDIT, particularly term extraction. One long-awaited instrument would be a hand-held digitizer or electronic highlighter that would allow the terminologist to capture selected passages of a text in graphic mode while reading through it. The passages could then be processed by a

new tools

character recognition system to allow for subsequent manipulations using a word processor. At the time this article was written, the slow speed, high error rate and cost, and limited adaptability of existing models prevented their use.

scanning tools

Other avenues, however, hold more promise in the short term. First, an increasing number of dictionaries, encyclopedias and other publications are available in electronic form, either on the Web or CD-ROM, making it much easier to access and process the information. Many of these, notably those produced in hypertext mode, provide navigational aids that can help the terminologist piece together the elements of a complex concept or make associations between related concepts. In-house documents are largely produced using word processors, such that they too can be manipulated easily. Second, a few software systems have been released on the market that can scan through documents in computer-readable form and be used to different ends by the terminologist. Key word in context (KWIC) systems, as their name implies, can be employed in conjunction with a stop list to produce alphabetical listings of words in context, i.e., preceded and/or followed by a predetermined number of other words, with an indication of the page and line number of each occurrence (see 6.1, Thesaurus Management, Figure 4:403). Other systems go even further. Typical applications scan documents and draw up a number of lists: an alphabetical list of all words with an indication of frequency; a list of words by descending order of frequency; an alphabetical list of collocations, i.e., word strings that recur in the text; a list of collocations by descending order of length (determined by the number of words) with an indication of frequency; and an alphabetical list of collocations, sorted by length. Furthermore, linked with commercial or in-house data banks, programs can perform a search of existing records, with or without human intervention, and generate a list of the collocations in the source text, along with the equivalents it has found for each in the banks and their sources. Terminologists can use these lists to isolate the terms and determine which warrant the most research.

These aids notwithstanding, the production of large-scale terminological dictionaries remains an arduous and labor-intensive, albeit extremely rewarding, task.

Appendix A

Dictionary Format

Term Entry

442 **double-detection receiver** (infrequent)
superheterodyne receiver; superhet

A *radio receiver* (455) that mixes (heterodynes) the incoming radio-frequency carrier signal with a locally generated frequency in order to convert the RF carrier down to a fixed intermediate frequency (IF) at which the signal is selectively amplified before it is fed into the *detector* (154) for demodulation. See *frequency converter* (32).

Notes: 1. Exceptionally, the intermediate frequency may be higher than the received radio frequency.

2. Some superheterodyne receivers have more than one *local oscillator* (383) and shift the carrier down through successive intermediate frequencies. See *double superheterodyne receiver* (441) and *triple superheterodyne receiver* (466).

3. Reception through intermediate-frequency stages above the limit of audibility was originally called "supersonic heterodyne reception". See *heterodyne receiver* (447).

récepteur à changement de fréquence
récepteur superhétérodyne

Récepteur radioélectrique (455) qui, à l'aide d'un *oscillateur local* (383), transpose une porteuse radiofréquence en une fréquence fixe dite fréquence intermédiaire, puis effectue une amplification sélective à cette fréquence avant que le signal ne soit démodulé par le *détecteur* (154). Voir *changeur de fréquence* (32).

Notes : 1. Le plus souvent, la fréquence intermédiaire est inférieure à la radiofréquence captée par l'antenne.

2. Le récepteur peut comporter plusieurs oscillateurs produisant plus d'une transposition en fréquence intermédiaire du signal radiofréquence capté. Voir *récepteur à double changement de fréquence* (441) et *récepteur à triple changement de fréquence* (466).

3. Le terme «récepteur hétérodyne», parfois utilisé dans ce sens, est déconseillé : le principe de la réception hétérodyne suppose une transposition directement aux fréquences audibles, sans fréquence intermédiaire. Voir *récepteur hétérodyne* (447).

SINGLE SUPERHETERODYNE RECEIVER
RÉCEPTEUR À SIMPLE CHANGEMENT DE FRÉQUENCE

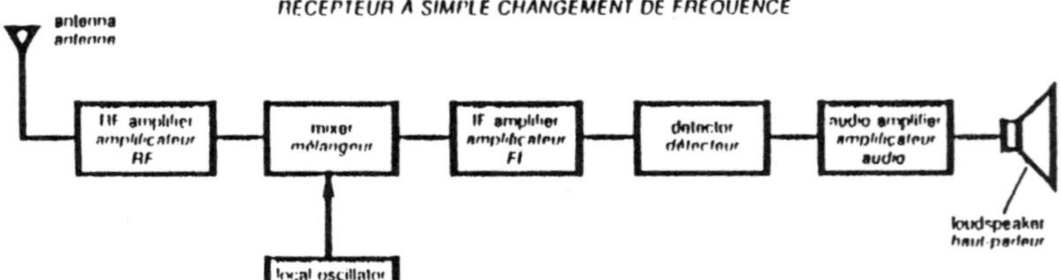

Index

Appendix B
Thesaurus

TERMI Thesaurus; Subject field: Telecommunications
Marcel Nouvel; Working document, September 1981

Telecommunications
- antenna theory
  - antennas
- switching systems
- technical management (telecommunications)
  - (telecommunication) traffic
- telecommunication equipment
  - switching equipment
  - transmission equipment
    - radio transmission equipment
    - transmission media
- telecommunication networks
- telecommunication services
- transmission systems
  - data transmission systems
  - radio transmission systems

Appendix C
Sample Terminology Records*

*These database records correspond to dictionary entry no. 217 of Volume 3, reproduced in Appendix D. See legend for an explanation of the field labels and contents. The complete set of data categories is presented in the first record. In subsequent records, empty categories are indicated by the "…" symbol.

NO	F001765
ESTERM	
ECTERM	line concentrator //a; stand alone concentrator //a
EDEF	DE. A switching device which concentrates traffic from a number of circuits or subscriber's lines onto a smaller number of circuits to a parent local exchange, where a similar switching device deconcentrates the traffic to the original number of lines. In the case of subscribers' lines, the correspondence of the lines before concentration and after deconcentration must be maintained. The system is both-way working, i.e., traffic from the exchange is concentrated onto the same circuits and deconcentrated to the subscribers as well. //a OB. See Fig. 1/Q.9, p. 309. //b
FSTERM	
FCTERM	concentrateur de lignes //c; concentrateur autonome //c
FDEF	DE. Organe de commutation qui concentre le trafic fourni par un certain nombre de circuits ou de lignes d'abonné sur un nombre plus restreint de circuits aboutissant au central urbain dont dépend cet organe et dans lequel un organe de commutation similaire déconcentre le trafic sur un nombre de lignes égal au nombre de circuits ou de lignes ayant fait l'objet de la concentration. S'il s'agit de lignes d'abonné, la correspondance des lignes avant la concentration et après la déconcentration doit être préservée. Ce système fonctionne dans les deux sens, c'est-à-dire que le trafic en provenance du central est concentré sur les mêmes circuits et déconcentré ensuite sur les lignes d'abonné. //c
SOURCE	a: CCITT-6E, 1981, vol. X.1, p. 138 b: 2Yv c: CCITT-6F, 1981, vol. X.1, p. 54
DOMAINE	support de transmission

AUT eg

DATE 841003

TYPE BLDD

FIN

NO F010883

ESTERM concentrator //a

ECTERM line concentrator //a

EDEF DE. A device that connects several subscriber lines which are not all likely to be used at the same time to a smaller group of trunks for more economical transmission. In telecommunications, concentration or sharing is achieved through the use of a circuit-switching mechanism or electronic subsystem. //a

FSTERM, FCTERM, FDEF ...

SOURCE a: SMICO 1982, p. 27

DOMAINE support de transmission

AUT eg

DATE 861115

TYPE ENDE

FIN

NO D005046

ESTERM concentrator //a

ECTERM line concentrator //a

EDEF CO. Loops are busy (i.e. connected to trunks or other loops) only a small percentage of the time (...) Where suitable calling patterns exist, this has led to the consideration of line concentrators for introduction between station sets and the central office. A concentrator is a small switching machine which allows a number of loops to be connected to the central office over a small number of shared lines which are, in effect, trunks. //a

FSTERM, FCTERM, FDEF ...

SOURCE a: BSCTE-2, 1977, vol. 1, p. 5

DOMAINE support de transmission

AUT eg
DATE 860812
TYPE ENCO
FIN

NO D000774
ESTERM, ECTERM, EDEF, FSTERM …
FCTERM concentrateur de lignes //a
FDEF CO. Les concentrateurs de lignes ont pour but d'économiser des lignes d'abonnés en leur substituant, sur une fraction de leur longueur, des jonctions en nombre moins important. Les opérations de concentration et de déconcentration sont effectuées par deux organes de commutation, l'un appelé passif ou distant, relié par une ligne individuelle à chaque abonné, l'autre, appelé actif ou local, relié par une ligne individuelle à l'équipement d'abonné de l'autocommutateur. //a
SOURCE a: FPT-7, 1982, p. 641
DOMAINE support de transmission
AUT eg
DATE 870309
TYPE FRCO
FIN

NO D006985
ESTERM, ECTERM, EDEF, FSTERM …
FCTERM concentrateur d'abonné //a
FDEF OB. Fig. 2 - Concentrateur d'abonné pour 80 à 100 abonnés. //a
SOURCE a: ONELE 1981, vol. 61, no. 5, p. 19
DOMAINE support de transmission
AUT eg
DATE 861217
TYPE FROB
FIN

Legend:

NO	Record number. The initial letter identifies the database (F designates the permanent record collection and D, the volume-specific one) and is followed by a unique six-digit number which is automatically generated by the prevalidation program.
ESTERM	English simple term(s)—i.e., terms that consist of a single word—including abbreviations. Entry terms, listed in the order in which they are found, are separated by semicolons and are each followed by a source reference letter (e.g., //a). Simple and complex terms are listed separately in both languages because the only way to retrieve just the former is by specifying the field in the query.
ECTERM	English complex terms—i.e., terms that consist of a multiword expression.
EDEF	English supporting text, which may be either a formal definition (DE.), a context (CO.), an observation (OB.) or, less frequently, a passage in support of a proposed term coined by the terminologist by inference (IN.) or analogy (AN.). The supporting text is followed by a source reference letter.
FSTERM	French simple term.
FCTERM	French complex term.
FDEF	French supporting text.
SOURCE	Coded references (may be either publications or resource people).
DOMAINE	Subject field, which corresponds to one of the descriptors of the thesaurus.
AUT	Initials of the terminologist who produced the record.
DATE	Date the record was produced.
TYPE	Record type, which is generated automatically by the prevalidation program for statistical purposes. The first two letters indicate whether the record is bilingual (BL) or unilingual English (EN) or French (FR); the last two indicate the type of supporting texts provided in each language (definitions, contexts, observations, texts substantiating an inference, or lack of supporting text).
FIN	End of record indicator.

Appendix D: Dictionary Entries

Example 1:

217 **line concentrator**
line connector (British)
remote line concentrator
stand-alone concentrator (infrequent)
A *concentrator* (206) which reduces the number of wire pairs in the local loop by connecting a number of distant subscriber lines, not all of which are likely to be used at the same time, to a smaller number of shared analog lines effectively used as trunks These extend to a complementary circuit switching device at the central office where the original number of lines is restored. See *concentrator* (207), *line concentrator* (218) and *remote concentrator* (226)
Note Such concentrators can be used with either *electromechanical* or *electronic switching systems* (62, 76) The concentration ratio is inversely proportional to expected average line usage
concentrateur autonome (peu utilisé)
concentrateur d'abonnés
concentrateur de lignes (très utilisé)
concentrateur de lignes d'abonné
Concentrateur (206) qui diminue le nombre de paires téléphoniques nécessaires pour desservir un groupe d'abonnés distants en raccordant sur un petit nombre de lignes analogiques, qui deviennent alors des jonctions, un nombre plus grand de lignes d'abonné À une partie dite « passive » ou « distante » correspond une autre partie dite « active » ou « locale » constituée d'un second organe de commutation qui effectue la déconcentration au central. Voir *concentrateur* (207), *concentrateur de lignes* (218) et *concentrateur distant* (226)
Note Ce type de concentrateur peut être raccordé à un *autocommutateur électromécanique* (62) ou *électronique* (76) Le rapport de concentration est déterminé en fonction de la probabilité d'occupation des lignes d'abonné

LINE CONCENTRATOR IN THE LOCAL LOOP
CONCENTRATEUR DE LIGNES DANS LE RÉSEAU LOCAL

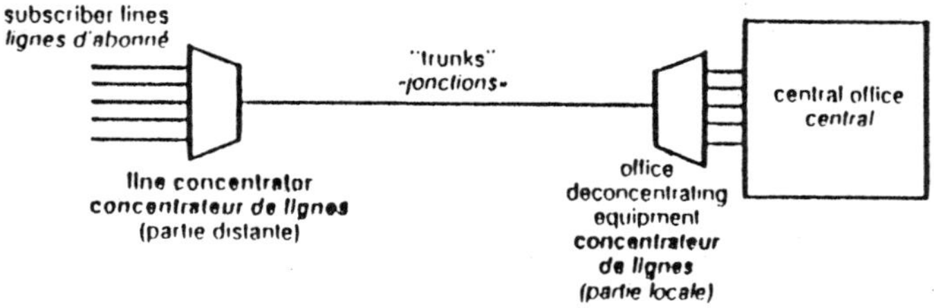

Example 2:

97 **hot standby system**
parallel synchronous system
synchronized reserve system

A *dual processor system* (92) in which the two *central processors* (398) simultaneously perform the same operations on the same calls, the results of the second processor serving only as a basis of comparison with those of the first, which actually controls the system. Any divergence due to error or failure initiates immediate diagnostic procedures and the shutdown of the faulty unit. See *load-sharing system* (99) and *cold standby system* (87).

Notes: 1. In such an arrangement, the second processor serves to increase the reliability but not the capacity of the system.
2. Expressions such as "parallel synchronous mode", "parallel synchronous arrangement" or "synchronous duplication of processors", which refer to the operation of the processors rather than to the switching system itself, are more frequently encountered than the above terms.

système en microsynchronisme
système en réplique synchrone

Système doublé (92) dans lequel les deux *calculateurs centraux* (398), l'un commandant le fonctionnement du système et l'autre travaillant à blanc, effectuent simultanément les mêmes opérations pour les mêmes appels afin de permettre une comparaison systématique de leurs résultats. Toute divergence apparaît alors comme une erreur, déclenchant l'exécution quasi instantanée d'un programme de test et, éventuellement, la mise hors service du calculateur défaillant. Voir *système en partage de trafic* (99) et *système en mode actif-réserve* (87)

Notes: 1. La duplication des calculateurs sert à augmenter la fiabilité mais non pas la capacité du système.
2. Le plus souvent, on se réfère non pas à l'autocommutateur mais au mode d'exploitation des calculateurs que l'on désigne par les expressions « calculateurs fonctionnant en microsynchronisme », « paire de calculateurs travaillant en synchronisme », « commande en microsynchronisme », « microsynchronisme » ou « duplication par réplique synchrone ».

SYNCHRONIZED RESERVE SYSTEM
SYSTÈME EN MICROSYNCHRONISME

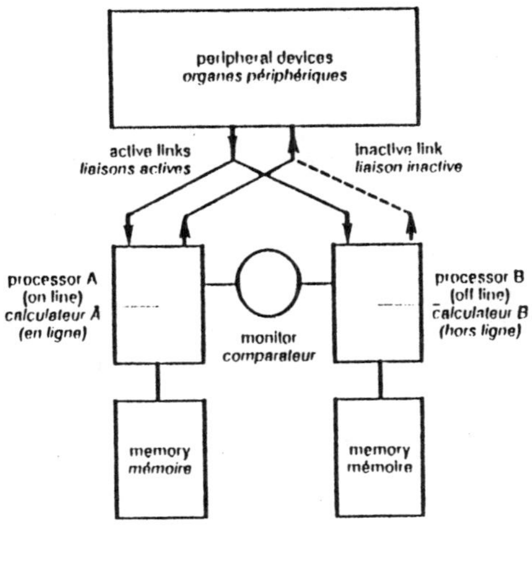

Example 3:

662 step-index fiber
step-index optical waveguide
An *optical fiber* (646) characterized by a uniform refractive index within the *core* (641) and a sharp decrease in refractive index at the *core-cladding interface* (642). As opposed to *graded-index fiber* (650).
fibre à échelon d'indice
fibre à saut d'indice (très utilisé)
Fibre optique (646) dont l'indice de réfraction est constant dans le *coeur* (641) et diminue brusquement à l'*interface coeur-gaine* (642). Par opposition à *fibre à gradient d'indice* (650).

STEP-INDEX MULTIMODE FIBER
FIBRE MULTIMODE À SAUT D'INDICE

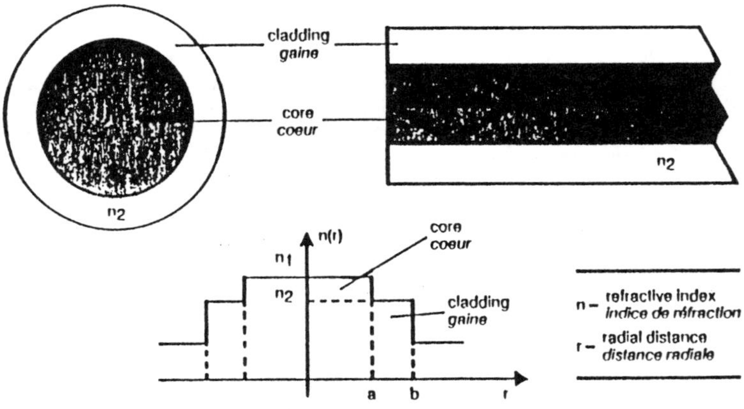

Index profile
profil d'indice

ACKNOWLEDGMENTS:

1) This article is an adaptation of a paper entitled "Expérience d'un mégaprojet", which was presented in Montréal at the second conference of the Canadian Translators and Interpreters Council (CTIC) on June 1, 1990, and published in the Conference Proceedings. It is being published with the permission of the CTIC.

2) *Bilingual Dictionary of International Telecommunications*, Volume 1, *Antenna Theory*, by Marcel Nouvel (1983); Volume 2, *Transmission Equipment*, by Andy Lauriston and Jocelyne Le Néal (1985); Volume 3, *Switching Equipment*, by Andy Lauriston and Jocelyne Le Néal (1988); Volume 4, *Telecommunication Services*, by Andy Lauriston and Jocelyne Le Néal (1991). Distributed by Teleglobe Canada Inc., Language Services, 1000 de La Gauchetière Street West, MONTRÉAL, QUÉBEC H3B 4X5, CANADA; telephone (514) 868-7464; fax (514) 868-8143.

cont. next page

Example 4:

650 **graded-index fiber**
graded-index optical waveguide
An *optical fiber* (646) whose *core* (641) has a refractive index that varies as a function of the radial distance from the fiber axis. The refractive index of the core decreases from the center of the fiber to the *core-cladding interface* (642), such that light rays propagating within the core are continually refocused by refraction toward the axis. As opposed to *step-index fiber* (662). See illustration, page 334.

fibre à gradient d'indice
Fibre optique (646) dont le *coeur* (641) a un indice de réfraction qui varie en fonction de la distance radiale. L'indice de réfraction du coeur décroît progressivement de l'axe vers la *gaine optique* (639), de sorte que la trajectoire des rayons lumineux qui se propagent dans le coeur est une sinusoide qui se refocalise périodiquement sur l'axe. Par opposition à *fibre à saut d'indice* (662).

GRADED-INDEX MULTIMODE FIBER
FIBRE MULTIMODE À GRADIENT D'INDICE

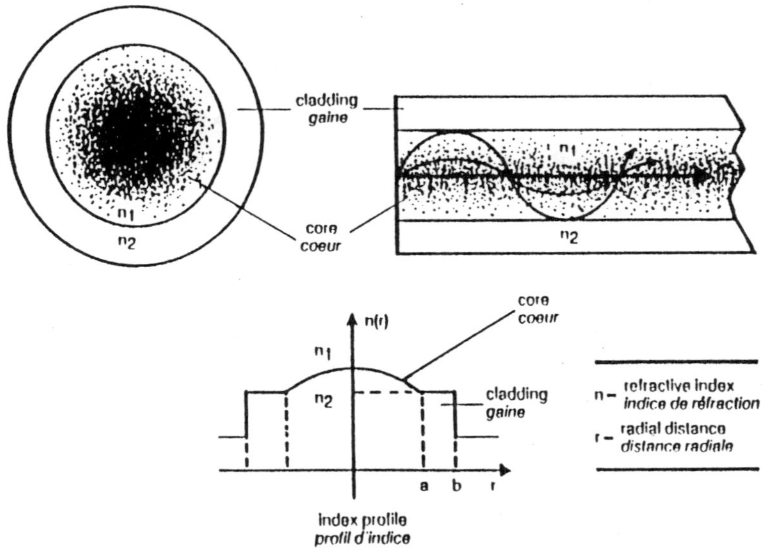

ACKNOWLEDGMENTS, cont.

3) EDIBASE is a text database management software produced by INFORM II Microfor of Montréal and widely used in Canada for terminology bank applications. Additional information may be obtained from INFORM II Microfor, 801 Sherbrooke St. East, Suite 615, MONTRÉAL, QUÉBEC H2L 1K7, CANADA; telephone (514) 524-7722; fax (514) 524-5441.

4) Further information can be obtained from the developer, CLC Ltd., 239 Champlain St., HULL QUÉBEC J8X 3R7, CANADA; telephone (819) 777-5779; fax (819) 777-0239.

8.2.2 Practical Considerations for a Term Bank: Termium

HELEN HUTCHESON

1 OVERVIEW

Canadian bilingualism

INFO

Termbase Terminology INFB 28:878

Canada is a bilingual country with two official languages, French and English. Promotion of correct terminology usage is of utmost importance in the Canadian context. The Government of Canada established the TERMIUM® data bank as a means of distributing and standardizing the terminology used throughout the federal public service, encouraging the use of Canada's official languages, and promoting the bilingual nature of Canada. TERMIUM®, which contains over a million bilingual records, is operated by the Terminology and Standardization Directorate (TSD) of the Canadian Government Translation Bureau. Maintaining and updating such a large bank are enormous tasks for terminologists, analysts, and managers alike. This article highlights some aspects of the professional management of the bilingual terminological data. It attempts to provide answers to such basic questions as how to organize terminological data, how to classify and retrieve information, how to maintain and control quality, and how to ensure that the data meet client needs. In today's world, these issues must be examined in the light of decreasing human and financial resources. This paper also outlines solutions which have been retained by the TSD. It is hoped that this information will help other organizations in the development and maintenance of their terminology data banks.

2 HISTORICAL BACKGROUND

TERMIUM®

The TSD has acquired a great deal of expertise in terminology management over the years as TERMIUM® has gradually evolved. TERMIUM® was acquired from the University of Montreal by the Translation Bureau of the Secretary of State Department in 1975 and established as a term bank for Bureau translators. (The name

TERMIUM® was created by appending the University's initials to the first part of the word "terminology".) Hundreds of thousands of records, accumulated over the years by the Bureau's translators, were added to the records in the University of Montreal Terminology Bank (BTUM). The original BTUM software was modified and installed as TERMIUM® II in 1977. By 1985, TERMIUM® III was implemented and, in 1990, another redesign of the data bank was undertaken using a new version of its underlying software. TERMIUM® IV was released in 1994.

It is from TERMIUM® IV that data is extracted for TERMIUM® on CD-ROM, which was first commercialized in 1989. In 1999, TERMIUM Plus® was made available to clients via the Internet.

3 Issues

3.1 Data Bank Structure

A comparison of some aspects of data organization in TERMIUM® III and TERMIUM® IV reveals certain challenges involved in structuring terminological information and shows how the TSD overcame these difficulties. Figure 1 illustrates TERMIUM® III's rather complex file structure.

file structure

The original Terminology Data Collection three-file structure was intended to meet the requirements of both clients and terminology managers. Over time, it became evident that the structure was difficult to manage, in part because of the number of terminologists working in the Directorate (about 100 in 1985 when this file structure was first introduced, and presently about 45). Consequently, when technological change made it possible to redevelop the online version of TERMIUM®, the file structure was reevaluated. This reorganized structure is illustrated in Figure 2. It should be noted that this file structure is used for data management purposes and is not extracted and distributed to clients.

TERMIUM® IV

As can be seen, the Multilingual Data Collection has been eliminated. After implementation of TERMIUM® IV, all entry terms were grouped on a record dealing with one concept. The TSD now directly manages non-French/English terminology, rather than simply storing it, as it had done with TERMIUM® III. In

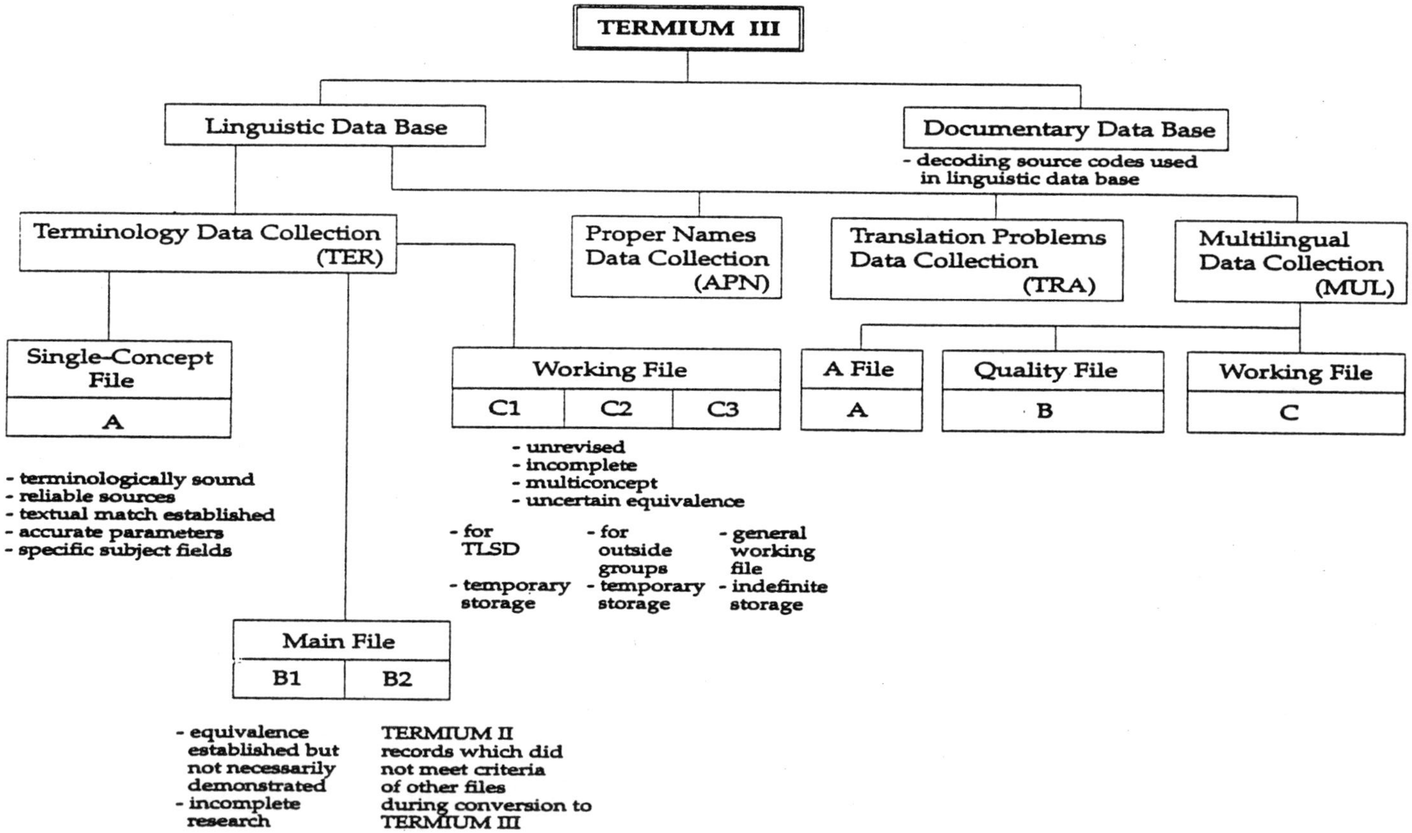

Figure 1: *File Structure, TERMIUM III*

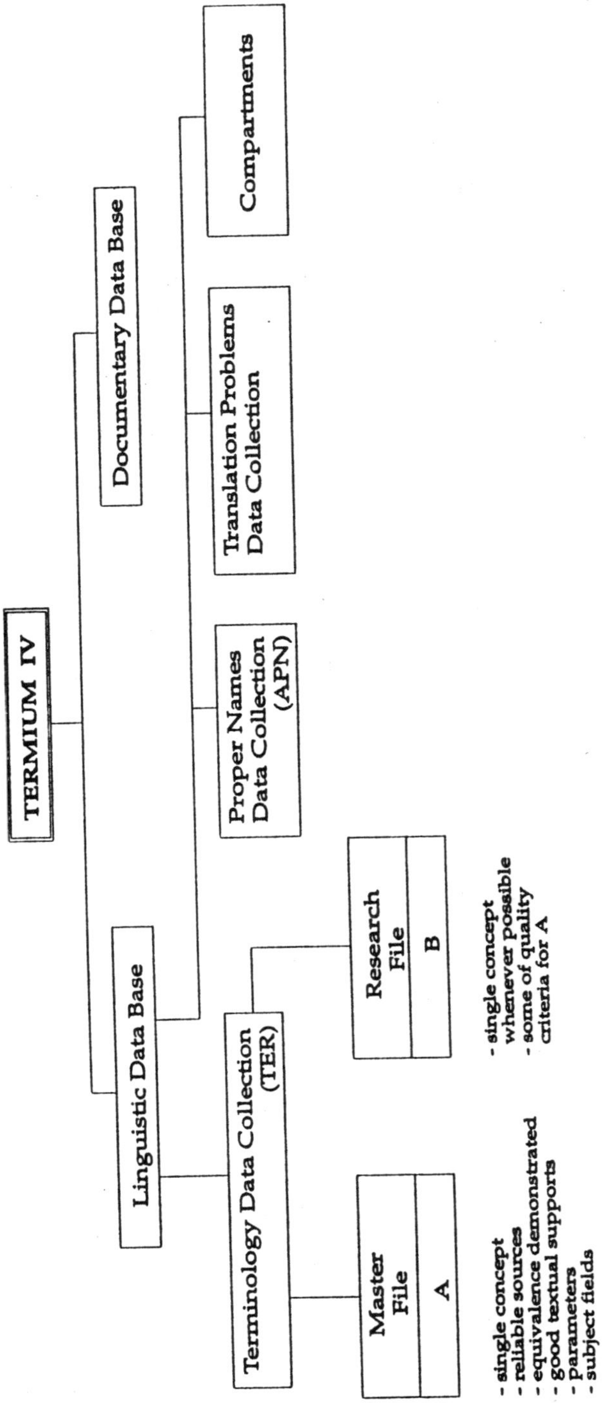

Figure 2: *File Structure, TERMIUM IV*

order to supplement its multilingual file and to reflect global tendencies, the Spanish component is being further developed. TERMIUM Plus® on the Internet includes about 60,000 records with a Spanish module.

simplified structure

The file structure has been simplified significantly. The *A* file, or Master File, now includes all records in the *A* file and most records in the *B1* file of TERMIUM® III's Terminology Data Collection and in the *B* file of the Multilingual Data Collection. The new *B* file contains all records stored in the *B2*, *C1* and *C3* files, and some records in the *B1* file of TERMIUM® III's Terminology Data Collection, as well as records in the *C* file of the Multilingual Data Collection.

single concept principle

SEE

Entry Structures 8.1.3:572

A file structure is an indispensable management tool for any large term bank. It provides terminologists with an indication of the record's reliability, of the originator's assessment of equivalence, and of adherence to the single-concept principle. (According to this principle, the database should contain only one record for any given concept, and any given record should deal with only one concept.) The file structure is also a valuable tool for terminology managers, allowing them to trace the evolution of the database's contents, evaluate the quality of records over time, organize work plans to ensure updating of research-level records, and monitor adherence to screening guidelines. The file structure can also be used as a means of data extraction. Terminology managers must therefore consider file structure very carefully, in the light of client and data-manager needs, when they are establishing term banks.

3.2 Subject-Field Classification

INFO

Classification INFB 9:334

Subject-field classification is another issue which must be meticulously analyzed by term-bank managers. Accurate identification of subject fields for each record is vital, as they are primary retrieval tools. The subject-field classification system used in TERMIUM® was created at the University of Montreal (the BTUM classification). In this system, three-character codes are used to indicate the three levels of the hierarchical structure created for the classification of terminological information. For example:

- H - The Economy (class)
  - HE - Finance (division)
    - HEG - Banking (section)

subject-field summary

Because the same term may be used in different contexts to refer to different realities, there may be several records in the bank for any given expression. A subject-field summary allows users to select the record(s) pertinent to their contexts during querying. Users may also obtain a printout of all records in a specific subject field. Therefore, assigning precise, accurate subject fields to records is of utmost importance.

field interdependency

Classifying terms in this classification scheme is perhaps the most problematic part of terminology management, in part because of the interdependency of fields. For example, terminologists must decide whether terms in pharmacology should also be classed under chemistry, whether a certain mineral term belongs to mineralogy or petrology, or both, and whether "smelting" of nickel has a particular meaning distinct from that of the "smelting" of metals in general.

interdisciplinary classification

A further challenge involves the assignment of primary subject fields (i.e., the fields with which the terms are naturally associated) and fields of application. For example, there are many terms in statistics, graphics, general theory and mathematics that are used in several subject fields without perceptible change in meaning. A single record indicating the primary subject field and the various fields of application suffices in such a case. If a single concept occurs in more than one subject field, but is designated by different equivalents, separate records must be drafted. Terminologists must determine whether they will classify a term in all the fields in which it may be used, or whether they will indicate only those fields representing concept networks with which the term has the most direct connection. The first option improves results when sets of records are extracted by subject field. However, during querying, subject-field summaries become lengthy, perhaps unnecessarily.

classification gaps

In the TSD, fundamental subject-field classification difficulties were compounded by gaps in the classification scheme. These sometimes prevented terminologists from classifying records accurately. Various methods have been employed to circumvent this problem, including the classification of records in the closest possible field, in several related fields, or in "general" sections. The system is constantly undergoing revision. With TERMIUM® III, a new section was added to each division to classify "non-classifiable terms". These records are to be reviewed periodically in order to identify gaps in the structure that could be eliminated by updating the classification system.

subject-field familiarity

The classification system itself presented terminology managers with another challenge: the scope of the various headings had not been explicitly defined. The TSD overcame this situation by giving terminologists responsibility for specific subject fields. The more familiar terminologists are with a field, the greater the probability that they will assign accurate subject codes consistently. Familiarity with a subject field also facilitates updating of the classification scheme.

provisional classification

Despite the principle of subject-field responsibility, limited resources often make it impossible for terminologists to do in-depth research for each record prepared. When the terminologist possesses insufficient information for classifying records precisely and accurately, TERMIUM® provides a means of provisionally classifying records. A temporary field may be assigned by replacing the section level (e.g., HE-) or the division level (e.g., H--) with hyphens. This type of classification is also useful for rapidly processing batches of records. All provisionally classified records are intended to be properly classified when additional information is available, and the contents of the provisional fields are reviewed periodically.

modification

The term bank and the classification scheme should both be flexible enough to allow revisions that reflect changing realities. The TERMIUM® subject-field classification provides managers with the flexibility to create new subject-field codes relatively easily. However, whole blocks of records may in some cases have to be reclassified, either manually or by using global commands. Modifications to the classification system are extremely important for upgrading data bank contents.

3.3 Updating

Both term-bank structure and subject-field classification are key elements which must be considered by terminology managers when they are planning screening and updating.

screening

In the TSD, screening is the procedure whereby a terminologist analyzes a concept, consolidates information concerning one concept on one record, cancels inaccurate records, corrects errors, and updates data. Consequently, all terms expressing a given concept—synonyms, spelling and syntactic variants, abbreviations, etc.—are consolidated on a single record. It is through this process that TERMIUM®'s contents are constantly upgraded.

upgrading

Upgrading the database is the responsibility of all TERMIUM® users. TERMIUM® has always been a shared tool resulting from the co-operation of all its users. The data bank's clients are not only users of a service, but also suppliers of terminology records. Collaboration agreements establishing data exchanges have resulted in the creation of a shared collection of terminological information from which all TERMIUM® users can benefit by avoiding duplication of research efforts and limiting the number of terms used to designate the same concept.

incomplete entries

It has been determined that all terminology records, no matter how incomplete, contain potentially useful information. Relatively incomplete data may consequently be stored in TERMIUM® IV as long as they are loaded into the appropriate file. However, before any new information is loaded into the database, a thorough screening of the bank's contents must be carried out.

planning

For this continuous process to occur rationally and systematically, managers must establish screening and updating plans by analyzing client needs and identifying priority fields. Quality and an improved response rate must be used as indicators of success in these fields.

statistical tracking

The system itself provides an excellent means by which managers can monitor the content and quality of the database. TERMIUM® is capable of generating statistical reports on the number of records per term, per subject field, per file, etc. TERMIUM® IV includes a counter that will register the number of times a record is displayed and by whom. When combined with user feedback concerning the distributed versions of TERMIUM®, these statistics are useful in the preparation of work plans.

access enhancement

It is essential that user needs be considered by terminology managers. They must ensure that clients, who often query incorrect terms, have access to the records in the data bank. TERMIUM® IV facilitates this task by indexing key terms (e.g., irregular plurals, syntactic variants) with entry terms. Records with such terms will be accessed when queried, but the terms will not appear as entry terms on the records. The display of TERMIUM®'s index of entry terms also greatly improves access to the data bank's records. Response rates are further improved by the addition of a word-search option in TERMIUM® IV. Users are able to access all records containing a word or group of words in the entry terms or supporting texts.

target audience

Once accessibility has been ensured, terminologists must make certain that the record's contents meet client expectations. The TSD's main target audience includes translators, for whom contexts are very important. It is recommended that contexts and examples be systematically added during the updating process. Terminologists are also encouraged to draft their own definitions and observations to improve the overall quality of the term bank.

terminologist's workstation

Practical tools have been designed by the TSD to help terminologists in this task. A terminologist's workstation (LATTER©) provides terminologists with a means of managing, storing and analyzing terminological data. It was developed, in conjunction with a desktop publishing system (Publiciel©), to avoid duplication of effort within the TSD. Terminologists can now use the workstation to prepare records for loading into TERMIUM® or for transfer to Publiciel©. This economizes valuable resources, and eliminates rekeying and repeated proofreading.

spelling and grammar

Although technological advances have now eliminated much of the need for simple proofreading, it still remains an important task that cannot be abandoned. Terminology managers are continuously working to facilitate this task. A major spell-check project was undertaken. However, not all errors can be detected (e.g., "manger/ manager", "un/e société"). Proofreading then remains an essential activity.

validation

Many of the basic inconsistencies which found their way into TERMIUM® II were eliminated by a set of validation rules in subsequent versions of TERMIUM®. Such data elements as subject-field codes, project codes, originator codes and source codes are checked against control lists of valid codes. Records with errors are not allowed into the database.

automatic entry generation

Other discrepancies found in the database are the direct result of actions taken during the course of TERMIUM®'s history. In 1977, e.g., the decision was made to load thousands of translator records into TERMIUM® using optical character recognition. Numerous errors resulted as many of these records were handwritten. Also, because the records had been prepared for personal manual files, rather than for a central electronic database, many inconsistencies resulted when all the files were merged into one large system. Although these records have proved extremely useful, terminologists are constantly updating them to improve the overall quality of the database.

duplicate entries

There are cases in TERMIUM® where several records on the same concept may exist. This situation also occurs when two or more records dealing with the same concept are loaded into the data bank during the same time interval. In-depth research is required to determine whether the data on these records can be consolidated. A thorough screening of the database must, however, be weighted by such factors as time and available documentation.

subject-field screening

Because TSD terminologists are responsible for screening given subject fields, duplication of effort is reduced to a minimum. However, terminologists must not only focus only on the fields for which they are responsible, but should also study terms as a function of their primary subject fields. Failure to recognize the relationship between primary subject fields and fields of application leads to the creation of multiple records, all referring to the same concept, but with different fields assigned.

polysemy

This situation is particularly evident for polysemous terms (e.g., "charge", "load"). Special attention must be paid to the single-concept principle in order to avoid a proliferation of records that would make consultation unnecessarily lengthy and inefficient and lead to confusion on the part of users. For these cases, there needs to be some central control by data managers to prevent an avalanche of records from being displayed upon querying.

In order to ensure that screening is carried out effectively and economically, the process must be well-documented and guidelines must be strictly adhered to in a uniform manner. Follow-up should be provided in the form of revision and training when needed. Screening is a costly operation which must be completed on a continual basis to improve quality, respond to client needs, and ensure adherence to the single-concept principle.

4 CONCLUSION

As can be seen, terminology management, in the context of a large bilingual term bank, is a complex issue involving careful situational analysis to determine client needs and to attain and maintain quality. The bank's overall structure must be easy to understand for terminologists and users alike. User guides and handbooks should be well-prepared so that uniform work methods are ensured. Subject-field classification schemes should be flexible enough to allow for updating. All records to be loaded into the system should be analyzed to determine the effect on the database as a whole. New technologies must be investigated as a means of refining the term bank.

expanded data types

The future holds many more challenges for terminology managers at the TSD. Presently, analysts are examining the storage and retrieval of linguistic data; the storage and retrieval of images; and the storage, management and retrieval of long-text data.

In its efforts to keep up with rapidly changing technologies and to meet the requirements and expectations of clients, the TSD is continuing to develop and refine applications for the benefit of language specialists. By improving the tools used by terminologists and by encouraging the participation of partner organizations, the TSD can ensure that the contents of its term bank are as up-to-date and as useful as possible.

REFERENCES

Department of the Secretary of State of Canada. 1984. *Terminologist's Handbook* (Draft). Ottawa.

Department of the Secretary of State. 1993. *TERMIUM®*. Ottawa.

Department of the Secretary of State of Canada. 1986. *TERMIUM® Record Completion Guide*. Ottawa.

Department of the Secretary of State of Canada. 1986. *TERMIUM® Subject-field Guide*. Ottawa.

Williams, Malcolm. 1993. "TERMIUM® and LATTER©. An Update." In: Schmitz, Klaus.-Dirk, ed. *TKE 93. Proceedings of the Third International Congress on Terminology and Knowledge Engineering, Cologne*. Frankfurt am Main: Indeks Verlag, 412-419.

8.2.3 Globalization and Terminology Management

KARA WARBURTON[1]

1 GLOBALIZATION AS A CORPORATE STRATEGY

As a large international corporation with a customer base that spans the globe, IBM takes globalization seriously. Globalization is a corporate strategy that is a key to IBM's competitiveness in international markets. For IBM, globalization means designing products that support the language, culture, and character encoding elements of the communities where they are marketed. This need must be met consistently, comprehensively, and quickly. Because time to market is a major factor in a product's competitiveness, closing the gap between its release in English and in other markets is critical, particularly with the rapid growth of some non-English markets in the high-tech industry. Through the deployment of standard corporate tools and processes, IBM has successfully narrowed this gap in recent years, and has even achieved worldwide simultaneous general availability for some products. In addition to better serving our customers in various countries, this achievement allows our large multinational customers to use our products to improve their business processes wherever they are applied.

SEE

Technical Translation 2.1.1:147

Localization 7.2.5:516

INFO

Globalization Defined INFB 27:877

This chapter describes the role that terminology management plays within IBM in the development and translation of products for global markets. It also presents some of the processes and tools that support terminology management and translation, and describes the challenges that the company currently faces in harmonizing these processes and tools.

[1]Due to the many recent advancements in information technology and marketplace globalization, this article is a substantial rewrite of the one originally written for this collection by Mary S. Neff, Leonard Cantor, and Evelyn Obermaier. Parts of the earlier article have been incorporated with permission.

2 THE SCOPE OF IBM TRANSLATION AND TERMINOLOGY

simship

As one of the world's largest publishers, IBM Corporation is also one of the world's largest customers for technical translations. Before any new hardware or software product can reach the world market, its user interface and documentation must be available in the national languages of the intended market. Some thirty IBM translation centers (TCs) worldwide manage the translation of several hundred thousand pages of documentation and other product assets annually from English into up to thirty languages. Large customers that operate globally look to IBM precisely because the company can offer them versions of a product in several languages, often simultaneously.

consistency

The high volume of technical material, the worldwide distribution of the company's development laboratories, and the rapid rate of change in the computing domain present major challenges to achieving the consistency in terminology necessary for high quality texts. Because IBM products are developed in different laboratories, similar or identical concepts or objects have come to be associated with different terms. For instance, the terms *shadow cursor*, *scale line cursor*, *grid cursor*, and *scale cursor* all refer to the same concept, but in different products. There is a natural tendency for individuals involved in creative aspects of product development to use terms that distinguish and embellish, even to the point of inventing new terms for concepts that already have acceptable designations. For product suites that have multiple components, often developed at different sites or by different teams, the problem can even occur within the same product. As Dubuc explains, such a plethora of synonyms impedes effective communication (Dubuc 1992).

SEE *Term Formation 1.2.1:25*

IBM development laboratories have made a significant contribution to the English terminology of computing. Many objects and concepts, along with their English designations, continue to be invented in the development laboratories around the world. Consider *virtual storage*, *virtual memory*, and other *virtual* compounds. Even when a product is developed in a site where English is not the spoken language, the source language is still English, in which case factors of foreign-language influence in the terminology can come into play.

The translation centers have played a significant role in the development and dissemination of computing terminology in their respective languages. Many objects and concepts born in the various development laboratories were first given translated names by IBM translators. A famous example is the French word for computer, *ordinateur*, coined by a French scholar on request from IBM. The German word *virtuell* was invented to translate terms like *virtual storage*, *virtueller Speicher.* No one discounts anecdotes about the Académie française calling up the IBM TC in Paris with questions on correct French computing terminology.

SEE *Language Planning 3.2:261*

terminology assets

In the development labs, the total quantity of terminology in glossaries, local databases, Web sites, and so forth, is unknown. According to conservative estimates there are more than 40,000 terms. The published version of the *IBM Dictionary of Computing* contains some 22,000 entries, and the *AS/400 Glossary* includes nearly 6,000.

regional collections

Most translation centers have developed a local bilingual terminology database to serve their translation needs. The larger collections require full-time staff to maintain them. Many TCs use an IBM terminology management system, TransLexis, to manage their collections, while others use its precursor, TMS, or an alternative, such as a Lotus Notes database, ASCII lists, or DB2, a proven IBM relational database management system. Some databases can be queried from a Web browser. The approximate size of the main databases, by number of entries, is: Germany - 35,000, Canada - 43,000, France - 36,000, Japan - 42,000, Taiwan - 16,000, China - 124,000, Korea, - 2,000, Poland - 17,000, Russia - 8,000, Spain - 2,500, Norway - 26,000, Hungary - 23,000, Brazil - 8,000. Spain is also currently developing an English-Catalan termbase.

resource incompatibility

Thus, in total, IBM's terminology assets exceed 440,000 entries maintained in multiple separate collections. These collections can have different properties and structures, making them incompatible with each other and with certain text-processing tools. The resources needed to manage such a multitude of collections and to remedy incompatibilities that affect production can be overwhelming. The magnitude of the differences in processes and tools for managing terminology is sufficiently large that IBM is now looking at consolidation.

3 PROCESSES AND TOOLS

Development labs and translation centers are focused on delivering a quality work product as efficiently and quickly as possible. This often means that many different innovative practices are developed locally. IBM constantly evaluates those practices to leverage its size and global reach, with the aim to achieve cost savings by using the same or similar processes and tools, where appropriate. This "One Team, One Tool" approach seeks to reduce inefficiencies or duplication by harmonizing processes and tools across the corporation. Both translation and terminology are areas where this approach can improve productivity.

tool harmonization

3.1 The Translation Process

The translation process starts with the development of source text, usually in English. Multiple developers and writers may be involved in creating the English version of a product, making it difficult to maintain consistent terminology and cohesive style in the source text. Even minor inconsistencies such as in capitalization, hyphenation ("hot link", "hot-link"), morphology ("open map" and "opened map") and syntax ("session request initiation" and "session initiation request") are a major challenge for translators.

SEE

Technical Writing 7.1:445

Without a macro view of the product texts, translators cannot detect terminology inconsistencies in the source text. Therefore, unless these inconsistencies are eliminated at the source, they must be caught at each translation center, either by human verification, which is labor-intensive and error-prone, or by the use of consistency-checking tools. Otherwise, each inconsistent term in English will become an inconsistent term in the translation, with negative consequences for product quality and usability. Stylistic problems can produce ambiguities that can lead to mistranslation. Many errors and terminology problems in the translated language can thus be traced to the source language.

inconsistencies

Many text processing and software development tools in the marketplace produce information that cannot be translated easily. The translation process can be complicated by the multitude of different file formats, programming languages, and markup. Consequently, IBM developed and implemented its own "translation friendly" tools. For English text production, this is the Information

translation friendly tools

Development Workbench (IDWB), while for translation, it is the Translation Workbench (TWB) and the TranslationManager (TM). Other tools complement these key production environments.

The IDWB provides a Standardized General Markup Language (SGML) authoring environment for all IBM writers. SGML (ISO-8879) is a standard, cross-platform, meta-language for text markup. Documents written in SGML can be used to produce multiple output formats, such as HTML, Postscript, and IPF, IBM's Information Presentation Facility, which is an OS/2-based online documentation delivery system. Authoring in the IDWB standard format and "transforming" to alternate formats, rather than authoring in multiple source formats, simplifies file handling and text processing.

INFO

SGML
INFB 21:863

A strategic component of the IDWB is the Easy English Analyzer (EEA). This tool analyzes the syntax, grammar, consistency, and vocabulary of document files against a set of criteria for text quality and translatability. It points out ambiguity and complexity caused by such structures as double negatives, the passive voice, long noun strings, or excessively long sentences, and suggests clearer and simpler alternatives.

authoring tools

Using a base English vocabulary of some 80,000 words, and the technical terms from the *IBM Dictionary of Computing* and major product glossaries, EEA also performs a vocabulary and terminology check. It points out inconsistent terms caused by alternating singular and plural forms, hyphenation, capitalization, or spelling, and it identifies any new terms. Some of these may be newly coined terms which can then be analyzed to determine if they are needed or if an existing term should be used. The writer can then use this list as a basis for developing a product glossary for customers and can also provide it to TCs for the English portion of their bilingual project dictionary.

term validation

EEA can be used with various file types. A numeric quality measure is provided at the end of the EEA analysis. Running the analysis several times enables writers to set measurable goals, track their progress, see the results of their changes, and hone their skills. Seeing the new term list early in a project also alerts them to potential terminology problems which can then be corrected before they proliferate.

quality metrics

Messages and text that appear on software panels or windows are called Machine Readable Information / Program Integrated In-

translatable

formation (MRI/PII). This text is not produced in the IDWB because it is typically written by the software developers themselves and integrated into the program source files. Writers, usability experts, and editors may review the text. Conventions to maximally divorce display text from programming instructions, or code, are clearly outlined in corporate software enablement guidelines. MRI/PII files can be of many different formats, reflecting the multitude of programming languages, platforms, interfaces, and development tools. The Translation Workbench recognizes these file types, and, using IBM-specific markup tables, separates the translatable material from the non-translatable markup. This makes it easier to translate the text and reduces the chance of machine-readable markup being inadvertently translated or corrupted.

resources

Another IBM tool, called CHKPII (pronounced "check P.I.I."), is used to perform additional verification at both the development labs and the translation centers. CHKPII is a multi-platform program that validates MRI/PII files before and after translation. It is available as part of the IDWB, the TWB, and as a stand-alone program. CHKPII can be run against virtually any file type in use in IBM products, and it is updated to support new file types as required. At the development labs, it is used to verify that the files conform to certain translation requirements and can be seamlessly imported into the TWB. After translation, it is used by the translation centers to compare translated files with the originals and verify that the code and markup match perfectly. All translatable text is excluded from the comparison. CHKPII can also extract the translatable text from the file, generating a markup-free output file, which can then be used for further processes such as term extraction.

SEE *Term Extraction 8.4.1:725*

After lab developers have produced their text and run EEA and CHKPII, they use another proprietary tool to package the files, including a term list and any product glossary, and send them to a Central File Management facility, or CFM. The CFM does some additional verification and then distributes the files to the appropriate translation centers. Using a central hub for file verification and handling ensures that the distribution process is consistent and efficient.

file verification

At the translation centers, the files are imported into the Translation Workbench, a workstation-based environment for preparing and processing translation material. This tool set contains functions

Translation Workbench

for file management and verification, translation memory creation and analysis, word counting, IDWB transform processing, dictionary creation, and project folder creation, among others.

translation memory

The translation memory is a database of bilingual sentences or sentence units. These units are recorded in memory when the translator completes translating each segment of text. The previously-translated segments are automatically shown to translators when they later come across the same or a similar segment. It has been estimated that more than twenty percent of technical text is repeated. This is particularly true for repeated releases of products and for product families. Reusing existing translations for this repeated material saves resources and maintains translation consistency between products and product releases.

Translation Manager

The Translation Manager (TM) is the IBM state-of-the-art translation editor product. This tool is used to do the actual translation. It is available externally as a fully supported and complete translation editor. The IBM internal version provides specific enhancements for use in translating IBM materials. TM features dual-window editing for source text and translation, automatic display of entries from multiple dictionaries, transfer of source layout information to the target, automatic generation of translated segments from a translation memory file, and customizable dictionaries. This machine-assisted translation environment automates repetitive tasks, freeing the translator to focus on the finer points of translation.

glossary tools

INFO

Unicode INFB 20:861

Glossaries and indexes, which are arranged alphabetically, can be processed by yet another IBM tool, called SORTGLOS. This multi-platform tool sorts lists of terms alphabetically after translation, and it supports the sorting sequence of English and European languages. Another IBM tool, TWBDBSRT, was developed for this purpose for Asian languages, which have unique double-byte properties. These tools eliminate much repetitive manual work. Custom parameters can be specified to handle many different list structures.

archiving

When translation is complete, the translation center performs a quality check before returning the files to the labs. The labs coordinate the testing of the translated product, often using the services of a translation tester from the TC. The files are then archived so that they can be used to generate a new translation memory for any future version of the product.

3.2 Terminology

Throughout the product development and translation process, the management of terminology plays an important role. Problems in the source are often repeated in the target language, so a strategy for terminology management that includes both source and target languages is essential.

3.2.1 *Source Language Terminology*

terminology coordination

Product developers and writers must coordinate their terminology usage. Terms that appear on the user interface must match those used in the supporting documentation or online help, especially such product-specific terms as the names of windows or options; otherwise, the user will be confused.

editors

Editors often act as a bridge between product developers and writers, since they typically see both sets of files and therefore may have a more complete view of the texts. Editors gather and distribute appropriate terminology resources: specialized dictionaries, previous or related product glossaries, Web site addresses, and so forth. They also prepare guidelines about the use of terminology and the development of product glossaries. Through their linguistic training and experience with writing styles and audiences, editors are generally sensitive to translation challenges presented by certain writing styles, culturally sensitive terminology, or unnecessary synonyms. They keep the translation process constantly at the forefront of their minds during editing, and they advise writers of potential problems.

centralized resources

For English terminology use, several centralized resources are widely referred to, such as the *IBM Dictionary of Computing* and other major glossaries. The *IBM Dictionary of Computing* was published in 1993 and is a standard manual on the desks of IBM writers. The version available on the Internet contains approximately 4,500 new or updated terms to complement the published manual. Any employee can propose a term to be included in this resource, using a Lotus Notes Database on the intranet. Terms and definitions are reviewed by experts before being included in the dictionary.

terminology tools

In summary, writers use various tools and processes to collect and document English terminology: EEA, editor support, corporate terminology guidelines, and existing terminology resources.

They generate term lists from EEA, which they can send to translation centers. They produce product glossaries using the IDWB in standard SGML format. They must accurately designate and describe new technologies by using existing terms or carefully creating new ones, while ensuring terminology consistency across large numbers of files.

3.2.2. *Target Language Terminology*

hybrid environments

In response to translation volumes and market conditions, some translation work is done by off-site translators. In recent years, translation operations have migrated from the mainframe to workstations, and from less to more fully supported machine-aided translation. Within this hybrid environment of people and technology, the database of bilingual terminology has emerged as an essential tool to promote cohesion and consistency.

in-house terminology management

Achieving consistent terminology in translations is a major challenge since multiple translators may work on portions of a given product, often in isolation from each other. To control terminology among multiple translators, terminology management has remained in-house and is usually the responsibility of a staff terminologist or translation manager. Thus there are two distinct terminology users at TCs: the coordinating terminologist and the translators.

project dictionaries

The key task of the coordinating terminologist is to produce bilingual project dictionaries for the TranslationManager tool, which is used by the translators. These project dictionaries are displayed in a graphical user interface that is integrated with the translation editor. Where there is more than one possible equivalent for a given English term, sufficient information, such as a usage note, is provided to enable the various translators to consistently select the right one for the context. The project dictionary includes all key words in the source English text for the product being translated: technical and non-technical terms, such as the labels on user interface objects and the names of product features and functions.

product glossaries

Since the English product glossary is intended for customers whose mother tongue is generally English, it does not completely meet the requirements of a project dictionary intended to assist translation. Corporate guidelines specify that product glossaries include only technical terms or other product-specific terms that users may not be familiar with. They do not need to include more common terms like *numeric value* or *trailing blank*, or window names, whose meanings are largely compositional, like *Data File Utility* or *Create and Select Field Definitions*. Such terms, however, must be translated consistently and ought to be included in the TM project dictionary. A term such as *cover page* is not a computing term and so it is absent from the product glossaries. A term like *hexadecimal collating sequence* is not included either, since it is already covered by *hexadecimal* and *collating sequence*. But compound terms in English may not be composed the same way in other languages, or perhaps multiple compositional structures are possible, so to ensure consistency, the terms should be included in a translation project dictionary.

translation terminology

Some of the necessary but missing terminology needed for translation can be identified if the writer submits the term lists generated by EEA. Invariably, there are always terms that slip through the identification process and are found by either the TC terminologist or the translator.

dictionary stages

To produce TM project dictionaries, the terminologist requires knowledge about the product and technology, standardized terminologies, syntax, semantics, morphology, and other fields of linguistics. The following are the key stages in this process:

1. Collect the terms in the supplied product glossary and EEA term files.
2. Identify other key words from the publications, help files, and screens, messages, and any other product assets.
3. Research and document the source terms.
4. Determine translations and usage information from any available terminology database and lexicographic resources.
5. For terms that are new or unknown, consult product experts in the development laboratory or in the target language community.
6. For new objects or technologies, create new terms (*neologisms*) in the target language, as needed.
7. Enter the information into a central bilingual database at the TC.
8. Export the project terms from the database and import the output into a TM project dictionary.

TranslationManager project dictionaries are designed with translators in mind. They display the English term found in the source text and all possible translation equivalents, with other basic information, and allow terms to be easily inserted into the translated text. They look and behave like a conventional dictionary by organizing data in a linear fashion with the term as focal point. Indeed, TM dictionaries can be created outside TM in flat SGML format and then imported into TM. The coordinating terminologist at the TC, however, manages terminology across multiple projects. The added dimension and scope of this terminology activity introduces the challenges of documenting synonyms and multiple translations for a concept, which may be distinguishable by product area, operating system, or other usage factors. This level of terminology management requires a more sophisticated approach that adopts a basic principle: concept orientation.

concept orientation

4 Lexicographic and Terminographic Approaches

Concept orientation refers to the principle of using the concept as the focal point for terminology data. This principle is a recognized international standard for terminology databases. Since multiple terms, in multiple languages, can refer to the same concept, the concept is the shared element that must be used to link the terms together in a multidimensional database. Effective and proactive management of synonyms and equivalents requires linking these terms through the concept.

INFO
Lexicography vs. Terminology
INFB 3:328

The importance of this requirement becomes obvious when term-oriented collections are compared to concept-oriented ones. Like conventional dictionaries, term-oriented, or *lexicographic*, collections are lists of words, arranged linearly in alphabetical order, with all the meanings of each word described below it. Concept-oriented, or *terminographic*, collections consist of multiple records associated by concept, and usually stored in a multidimensional structure such as a database, with multiple access points to each record. IBM uses both types of collections to serve different end users and needs. A product glossary produced in IDWB by a writer is a lexicographical work arranged alphabetically by term with cross-references to related terms that are also defined. In order to get a picture of all the synonyms and related concepts for a given term, the user needs to look up each cross-reference separately, possibly a three, four, or five step process. Most pro-

lexicographic collections

duct users, as *passive* users of terminology, have no need to do this, since they normally consult a glossary simply to find the meaning of a term and not to reconstruct its entire semantic family.

termino-graphic collections

By contrast, product developers and writers, who work with emerging technologies often requiring them to designate unfamiliar concepts, are *active* users of terminology. They must *select* an appropriate term for a given concept. To do this objectively, they need to see *all* the terms for a given concept, with usage information. From this comprehensive view, they can select the most appropriate term for the context. The terminology resource that these users need potentially contains thousands of terms covering multiple product areas. With a lexicographic structure, it is impossible to display all the synonyms and translations of a concept simultaneously. And, for a terminologist, it is impossible to document the unique usage parameters that distinguish synonyms and translations.

active terminology users

Thus the lexicographic structure is suitable for customers and other general users who want to know the meaning of a term, whereas a terminographic structure is required by writers and terminologists for the knowledge and management of synonyms. Translators have hybrid needs: they are passive users of English terminology but active users of the terminology in their language. They therefore benefit from both approaches.

target audiences

5 TransLexis

To support the requirement of a multidimensional, concept-oriented data structure, and to advance from the earlier mainframe-based IBM Terminology Management System (TMS), IBM developed TransLexis. Originally conceived at the IBM Scientific Center in Heidelberg, Germany, TransLexis is now being developed at the German Software Development Laboratory (GSDL). TransLexis is currently used by IBM TCs across Europe.

TransLexis is a desktop terminology management program that is built on DB2. Relational database technology is used by many existing large terminology databases in government and industry. Sager (1990) states that "Trends in database management generally favor an approach based on the relational database model." Relational databases support many-to-many (N:N) relationships, which are common in terminological data, such as where a given concept

relational databases

can be designated by N source terms and N translations. The Standard Query Language (SQL) provides nearly limitless search and display capabilities. And the sophisticated access control, data validation, and security features make it very appealing for large collections with many users.

TransLexis provides a graphical user interface for manipulating the terminology data, so users do not need to be proficient in DB2. A Web interface is also available for querying the data, eliminating the need, for most users, to install DB2 and TransLexis. TransLexis contains a wide range of data elements, including some that are required by IBM's evolving machine translation system. The terminology for a project dictionary can be selected using a search filter, automatically exported from the database into SGML, and then imported into TranslationManager. Conversely, TM project dictionaries can also be imported into TransLexis. As is so often the case for terminology data exchange, SGML is the exchange format in these operations rather than the native storage format. This ability to interchange terminology between TransLexis and TranslationManager provides an integrated workflow between the terminologist researching the data and the translator who uses it.

SEE
Data Categories 8.1.2:552
Terminology Interchange 8.1.5:613

data display

TransLexis can display data in either a lexicographic view or a terminographic view. The lexicographic view presents all the information about a term, including variants such as acronyms and abbreviations, grammatical information, usage notes, and contexts. The terminographic view displays a given term representing a given concept, all its synonyms and all its translations, and detailed concept information, such as the subject field and related concepts. This flexibility is one of the unique strengths of TransLexis. Writers, who are used to conventional dictionaries, can use the first view to quickly verify a term. If they need to reconstruct the semantic field in order to select a term from among synonyms, they can change views with a simple mouse click. Terminologists use the terminographic view to more effectively manage terms within their semantic family.

customization

TransLexis database administrators can create custom values for some data categories, such as subject field, part of speech or usage register (formal, informal, slang, etc.). The translation centers utilized this customizability to design databases that reflected the unique properties of their language, facilitated the migration of their TMS collections, or responded to some other need. It soon became apparent that the program was being used differently by various TCs. Looking ahead to the possibility of data exchange, or the merging of collections, it became obvious to corporate strategists that some standard controls and procedures were needed. And to improve terminology selection, the concept-orientation feature needed to be more fully exploited.

6 Defining Corporate Standards

harmonization

In response to the need for some orchestrated direction, a terminology work group force was formed, with representation from a number of translation centers and the corporate Globalization division. The primary goals were to harmonize the structure of IBM's terminology databases, establish common terminology management methods, and influence the ongoing development of IBM's terminology management tools. Reviewing IBM's terminology structures provided an additional opportunity to direct them along a path that would approach international standards.

corporate solutions

The work group recognized the following potential benefits of a corporate-wide process for managing terminology, possibly including a single terminology management system:

- It enables consistent source terminology use in all languages by providing a centralized repository for terminology.
- It eliminates the need to repeatedly research and develop product glossaries.
- It enables terms and terminology data to seamlessly pass from the development laboratories to the translation centers.
- It eliminates the duplication of source language terminology research at the translation centers.

balanced solutions

To develop standards in a large corporation, all existing processes and tools should be considered. An effective solution in one department or region may be totally impractical in another. Conflicting factors need to be balanced: theory and practice, source language and target language, writer and translator, producer and consumer.

networked solutions

A networked communication channel was therefore set up in Lotus Notes. Employees across the globe can access all the reports of the work group, view the discussions and decisions, and post comments if desired. This consulting process allowed IBM to compile an inventory of terminology resources, tools, and processes across the corporation. Lotus Notes databases, DB2 databases, and ASCII lists were added to the list of tools already being used to manage terminology. Some TCs have chosen the ASCII format because they can easily develop other automated tools to manipulate data in this format.

multitasking

One translation center, for example, developed a series of impressive tools that support multiple tasks carried out locally: collecting key words from the English source, checking for existing translations in the bilingual terminology database at the TC, searching and outputting project dictionaries based on subject field, distributing the data electronically to off-site translators, verifying terminology consistency in the translations, and importing new terminology from the completed translations back into the database.

Asian needs

The consultation process also allowed IBM to determine that the TCs in Asia and the development labs did not use any terminology data category that was not already in use in the European collections. This was an important observation, for it alleviated concerns that the needs of Asian languages or the development laboratories might not be met by the standard data structure being considered in Europe.

practical issues

Thus, corporate standards need to consider, in addition to terminology theory and standards, many practical issues: the dependencies on existing tools, the conservation of existing terminology assets, the preservation of locally-realized economies, and production output requirements. It is also important to ensure that corporate standards streamline, rather than complicate, the terminology process. The terminology standardization efforts at IBM have resulted in the following:

corporate requirements

- Recommendation to use concept-orientation for managing synonyms, variants, and translations
- Establishment of a corporate terminology data structure, including mandatory and optional data elements and permissible values
- Definition of the roles of the various players in the terminology process
- Identification of the dependencies and requirements for existing tools

6.1 Standard Subject Fields

Subject fields exemplify the challenges of harmonizing data structures. Also known as *domains* or *subject areas*, subject fields are areas of knowledge in which concepts and objects can be logically classified. An example of a subject field is *programming languages*. Subject fields are commonly used in terminology databases to organize terms into logical subsets. Terminology management is based on concepts that refer to real-world objects (either concrete or abstract). Every concept is a member of a class of concepts. Terminologists generally concur with Dubuc's observation that, "in terminology, a term cannot be separated from its subject field" (Dubuc, 1992: 26, 75, 95). Interestingly, every existing terminology collection in IBM used subject fields, even ad hoc collections that were created by individuals who had little knowledge of the role of subject fields in terminology management. The challenge was that they did not use the *same* subject fields.

INFO
Classification
INFB 9:334

Determining a suitable subject field classification that serves all needs is one of the most difficult, yet important, tasks in developing a standard database structure. Working in a real business environment, corporations might find it impractical to apply theoretical methods of ontology, whereby a subject field hierarchy is established through a strict analysis of the semantic field, genus, intension, and extension of each term. While recognizing the value of this approach, IBM focused on the practical use of subject fields rather than their theoretical makeup.

INFO
Ontologies
INFB 31:xxx

Thus, to reduce migration impact, IBM considered the subject fields already in use. For the company's business needs, the key business areas were considered; for process needs, the database output requirements for project dictionaries and product glossaries were defined. In the end, a subset of an existing standard corporate product classification was adopted, the *World Wide Offering Classification*, which was developed by a global team of product experts. Among other purposes, this classification is used by search engines to identify IBM products according to their various properties (see http://www.ibm.com/wwoi). And since IBM terminology describes IBM products, defining subject fields based on a product classification made good business sense. Having a terminology classification that is aligned with an existing standard

product classification

product classification preserves the potential for data exchange with other company resources.

6.2. Other Standard Data Values

data category harmonization

In addition to subject fields, the different database collections have different allowable values for key data categories, such as *term source* or *part of speech*. As IBM moves closer to adopting corporate-wide terminology data values, it needs to assess how much of the existing terminology resource should be preserved, and for those collections, the non-compliant values that need to be converted to compliant ones. Where there is a one-to-one match between the old value and the new one, the change can be automated. But often the match is not so obvious and requires human review of the entire terminological record. There is a fine balance between the value of existing terminological assets and the effort required to migrate them.

structural harmonization

SEE *Entry Structures 8.1.3:584*

Another type of data incompatibility relates to its structure. Structural differences, for example, exist between the English lexicographical glossaries produced at the development labs and the terminology databases in the translation centers. The IDWB permits a simple glossary structure consisting of two SGML tags: term and definition, whereas the databases in the translation centers typically contain additional categories describing grammar, usage, meaning, and relations between words and concepts. In IDWB glossaries, there is no place to specify a subject field, part of speech, or other descriptive information. (This structure is currently being reviewed.) Within these structural confines, some writers include multiple types of information within the definition tag: a "use label" (similar to a subject field), a definition, encyclopedic information, usage information, and references to related concepts. This interim solution allows the writer to record more comprehensive information in the glossary than would otherwise be possible if the SGML structure were strictly adhered to. However, it contradicts the SGML principle of elemental data identification, whereby each distinct data type must be identified with distinct markup to facilitate further processing and data exchange. And once multiple data types are merged into one container, it is nearly impossible to automate their separation into multiple containers. Therefore, migrating these glossaries into a more structured terminology database will require human intervention.

terminology planning

This situation points to the need for companies to identify their terminology needs and plan a course of action to meet those needs early in order to reduce work in the future. The SGML tags in IDWB were intended for producing simply structured, unilingual glossaries for product users. But as glossary production evolved to include cross references and other data categories, the tool's ability to meet those needs fell behind.

7 Working with External Organizations

In preparing for the development of corporate standards and tools, it is important to align them with external standards and practices.

7.1 Governments, Companies, and Academia

product evaluation

By evaluating competitor terminology management products, and obtaining feedback from terminologists in government (Canada and Québec) and in other companies (Nortel Networks, Medtronic, and PowerQuest), the work group was able to recommend that IBM use a relational database system and consider updating TransLexis for wider deployment. Advice was also obtained from terminology experts at the University of Ottawa, York University, Kent State University, Université Laval, and the University of Waterloo.

classification

Through the partnership with the University of Waterloo, IBM supervised a Masters research project which served to validate the proposed subject field classification for the *IBM Dictionary of Computing*. Such partnerships are mutually beneficial because they find solutions to specific corporate problems while providing universities and their students with practical applications to test their theoretical constructs.

7.2 International Organization for Standardization (ISO)

SEE

Terminology Interchange 8.1.5:613

In order to consider international standards for terminology, IBM sent delegates to international conferences, researched ISO standards, and consulted with members of ISO Technical Committee 37, which is responsible for terminology. The key ISO standards, *ISO 12620 – Computer Applications in Terminology – Data Standards*, and *ISO 12200 – MARTIF*, motivated many suggestions for the standard database structure and the modifications of the data categories and data relationships in TransLexis. These standards were shared with local database administrators so that they could align their collections at strategic moments, such as during a migration process from one tool or format to another.

data interchange

Complying with international standards for internal terminology processes is not mandatory. However, by doing so, companies benefit from the experience and research of the ISO experts. Furthermore, terminology collections that comply with standards are optimized for data interchange, since any widely accepted interchange format is likely to be modeled on those same standards. And while a company may not need terminology data interchange today, with the focus on globalization, disappearing trade barriers, and growth in departmental and company mergers, there may be a need in the future. Companies that are strategically positioned to handle that need will be more successful in developing global products. Still, companies must balance the effort of compliance with the foreseeable business benefits, and thus IBM may not adopt the ISO standard for *every* data element and structure in its databases.

8 CONCLUSION

process stakeholders

Successful corporate globalization requires a comprehensive terminology management strategy that spans regional and corporate divisions. This strategy should be defined through a consultative process that includes the following groups:

- Key terminology users in all regions and in all stages of product development (writers, editors, developers, translators, etc.)
- Experts in terminology theory
- Experts in terminology management technology
- Technology specialists for all tools used in text production and asset management

proactive management

IBM shares similar challenges with other corporations who strive to improve communication through proactive terminology management, such as the integration of lexicographic and terminographic collections and the deployment of synchronized tools. A corporate terminology process that reflects the spirit of globalization encourages local innovation while balancing it with large-scale efficiencies that can be gained through adoption of corporate practices.

REFERENCES

Dubuc, Robert. 1992. *Manuel pratique de terminologie*. Linguatech, Quebec, Canada.

IBM Dictionary of Computing on the Internet, available at: http://www.ibm.com/software/dictionary (Valid after 2001-01-01)

ISO 12200: 1999. *Computer applications in terminology –Machine-readable terminology interchange format (MARTIF) –Negotiated Interchange.* Geneva: International Standards Organization.

ISO 12620: 1999. *Computer applications in terminology– Data categories*. Geneva: International Standards Organization.

Linnartz, Maude and Mayer, Rene. 1999. *World Wide Offering Classification Scheme*. IBM Corporation.

McDaniel, George. 1994. *IBM Dictionary of Computing*. International Business Machines Corporation. New York: McGraw-Hill, Inc.

Sager, Juan C. 1990. *A Practical Course in Terminology Processing*. Amsterdam and Philadelphia: John Benjamins Publishing Company.

Section 8.3

Terminology and Machine Translation

MURIEL VASCONCELLOS[1]

1 INTRODUCTION

If the right terminology has been supplied to a machine translation (MT) system, the target-language equivalents are retrieved not only automatically but also in their correct place in the output document. This is one of the advantages of MT: it dispenses with the need to look up terms in hard-copy dictionaries or online.

uniform terminology

With large projects and multi-translator teams, fully utilized MT has proven to be very effective for keeping terminology uniform (e.g., Brace 1993). Indeed, in one group of reports from 36 MT users (Vasconcellos 1993), this was the advantage most often cited. Alternatively, even if a translator prefers to create translation in the traditional way, an MT printout can be a valuable guide—assuming that the system's database is well stocked with the required vocabulary. It is worth the spelling assistance alone. To cite a few well-known bugbears in medical terminology, such a program can make the right choice between *gluco-* and *glyco-*; it saves looking up *erysipelas* and *paracoccidioidomycosis*; and it never gives "*military tuberculosis*" for *tuberculosis militar*—a common translator error. The question is not so much *whether* MT is useful for banking and retrieving terminology, but rather what is involved in building up the terminological reserve so that it can produce needed terms on demand.

computer output

MT will be defined here as *the technology whereby computers attempt to model the human process of translating between natural languages*. The computer, rather than a person, generates the "output"—though it is only a rough draft, not yet fit for most types of consumption. The draft is usually polished into final form by a translator or a bilingual editor, though in some cases it may be used directly by a technical expert who is gathering data for ongoing research.

The counterpart to MT is MAT (machine-*aided*—or *assisted*—translation, in which people do the translation and computers "look up" isolated terms on command. The MT-MAT distinction is not water-tight, however, and recent developments are blurring it even more. Innovative approaches are making MAT more MT-like and vice versa. In the case of MAT, the computer has taken on an increasingly active role: the now common translation memory systems monitor human translation input and identify identical or similar strings of text in a database of previously stored documents, which are then displayed on-screen as candidates to be pasted into the ongoing translation. Meanwhile, in the MT world, interactive systems are finding ways in which to capture needed human reactions in midstream in order to produce output of better quality.

translation environments

SEE *Technical Translation 2.1.1:153*

For present purposes, however, the distinction between MAT and MT is quite useful. It defines the difference between terminology management software (TMS), which is covered elsewhere in this volume, and full-text MT, dealt with here, in which surrounding context interacts with terminology and, as a result, terms cease to be isolated entities.

SEE *Evaluating TMSs 8.1.1:552*

2 Getting the Right Translation

ambiguity

When terms are part of a larger text, even in the most limited of subject areas, their translation is often affected by how they are being used. Take for instance, the wholly ambiguous sentence:

(1) Because the clock did not work properly, the conductor failed to make the connection and the bus had to operate with just a driver.

multiple domains

This text could have two entirely different readings, one in the general sense referring to public transportation, and the other a technical description of the failure of a piece of electrical equipment, depending on the meanings assigned to *clock, conductor, connection, bus,* and *driver* (Wheeler 1983).

single domain

This problem is relatively easy to solve, however, compared with the difficulty of getting the right translation when a given word has more than one meaning within the same domain. For example, in a Spanish text on atomic energy the word *núcleo* could be translated as *nucleus* or *core*. In a legal text, *derecho* could be *law* or *right*—and the words that would make the meaning clear are not always nearby.

contextual sensitivity

MT systems use various techniques to generate the right translation for the context—in other words, a *context-sensitive* translation. Some of them have a broad range of strategies and can process many kinds of linguistic information. Linguistic detail helps a system to make subtle distinctions based on context and to produce high-resolution translations, especially in limited subject areas. A number of systems can cope with complex syntactic formulations, and some respond to pragmatic criteria and make use of world knowledge drawn from elaborate databases.

dictionary building

While such capabilities make for powerful MT systems, the truth is that building in the required knowledge is not an easy task. The terminologist or other person working on the system's "dictionary" has to have a fair amount of specific training as well as some appreciation for the principles of general linguistics. Most of the commercial systems have developed user-friendly software for updating their dictionaries (Logos Corporation was a pioneer in this area), but even so, the price of such refinement can be quite high, and some users do not have the time or the human resources that are needed for elaborate customization. Then there are systems that are not so linguistically complex, and they offer a tradeoff that may be worthwhile for certain applications: easy dictionary-building in exchange for less linguistic specificity. This chapter will describe the range of MT systems that were available at the time this study was compiled and give examples of the kinds of problems they are equipped to handle.

3 Types of MT Systems

MT systems are usually classified as *direct*, *transfer*, or *interlingual*. We now also have *example-based* MT. This classification is useful when it comes to understanding what is involved in entering terminology in the dictionaries.

3.1 Direct Systems

phrasal translation

One of the main characteristics of a "direct" MT system is that it does not analyze full sentences. Rather, it builds the sentence phrase by phrase using local routines. Because it doesn't parse, it can make only limited use of syntactic, semantic, and pragmatic information in the translation process. Even if the dictionary allows for such codes and the system has rules that act on them, the lack of a full analysis of the sentence will limit the ways in which the information can be used.

hardwiring vs. parsing

To take a very simple example, if we wanted to say *female rats* in Spanish, in order to get the desired translation *ratas hembras* instead of the default translation **ratas femeninas*, in the simplest of the direct systems we would have to enter a hard-wired phrase on each side of the equation and to repeat the exercise for each animal name that we expected to encounter. If, on the other hand, the direct system features at least minimal parsing capability, we could enter a code "minus human" [-HUM] on animal names. This would make it possible to trigger *hembras* for all the animal names modified by *female*—though without the capacity to parse, the usefulness of this code would be quite localized. In other words, the system would probably *not* be able to handle a sentence like:

(2) The technician separated the rats, placing the females in a cage by themselves.

In such a sentence, a direct system would also miss the fact that the translation of *themselves* requires a feminine ending in Spanish.

form generation

For many translation tasks, it may never be necessary to generate *hembras*. It simply doesn't come up. The possibility of the term occurring is predictable when we are dealing with a particular subject field, such as computer manuals or weather reports. With general translation, on the other hand, there is no way of anticipating when the need will arise.

post editing

Of course, it may not matter whether the correct translation or the feminine ending are generated. For a scientist scanning a database to find material that might be of interest, the sense is clear. The translation has served its purpose. On the other hand, it may be feared that the wrong translation will be offensive to the reader or undermine the client's image. In such cases, there is usually a posteditor who mediates by correcting the machine's output with a few simple keystrokes. Here the operative words are *few* and *simple*. Too many required changes, or operations that are difficult to execute, may ultimately outweigh the advantages of MT, including its potential to produce uniform terminology. (Some maneuvers may be difficult because of the operator's level of word-processing skills. Other difficulties may be inherent in the word-processing software, in which case one solution might be to customize macros that perform repetitive operations more efficiently). The user may decide that MT is not worth the trouble.

platforms

Most commercial systems running on personal computers in he early 1990s were direct systems, but that picture has been changing. Even systems that have traditionally run on mainframes and high-end workstations have been reconfigured to run on PCs, and the earlier PC systems are maturing. Programs like Systran's Babelfish World Wide Web service are bringing indicative translation to Internet users, and Logos' Java-based PC interface will allow PC users to access UNIX power online from a PC workstation. It is therefore unsafe to generalize about the linguistic approach that a system uses based on its platform or its price alone.

public awareness

While it may be difficult to judge the linguistic prowess of an MT system by its platform and its price, one thing is certain: with increased public awareness of MT, people are buying large volumes of software packages off the shelf and through the mail. The heavy demand for multilingual access to Web pages has fueled the funding pipeline to trigger a new spate of projects designed to deliver at least rudimentary translation of online resources.

3.2 Transfer Systems

While direct systems exceed any other kind of MT in terms of number of packages sold and installed, it is safe to say that the bulk of practical machine translation is performed by *transfer systems*, which have been in everyday use since the 1970s and continue to be the mainstay of large, heavy-duty translation operations.

parsing

Transfer systems have three basic components: *analysis*, *transfer*, and *synthesis*. After analyzing the source text, they convert the result of the analysis, or *parse*, into an intermediate structure, called the *transfer component*, which interprets the linguistic information at a more abstract level. They then take the output of the transfer component and feed it into the *synthesis component*, which stitches together the final translation.

modularized approach

This modularized approach emerged in response to several different pressures, mainly generated by the demand for translation of English into foreign languages. First, the need to parse the source text was becoming increasingly evident. Direct systems were challenged by the frequent part-of-speech homographs and syntactic ambiguities of English. Frequently several of the words in an English sentence can function as more than one part of speech, and decisions have to be made in order to work through the labyrinth of possibilities. For example, in the sentence:

(3) Slow sand filters produce more even results.

Every word has at least two potential functions. The possible combinations are astronomical. In addition, words of the same part of speech can have multiple equivalents, as terminologists well know. A parser sets a series of conditions that must be fulfilled in order for a given structure or word choice to be valid. In this way, illegal readings of the sentence are ruled out.

multiple equivalents

The second pressure for developing this kind of system was the demand for MT from a single source language into multiple targets. Manufacturers in the electronics industry and software publishers in particular began to look to MT as a means of speeding up the process of localization, i.e., translating customer support manuals and product-related material into many languages simultaneously for introduction into foreign markets. The transfer component provides an efficient starting point for translation into more than one target language: once the analysis has been performed, the results are captured in a concise and systematic representation to which MT developers can readily attach target modules in a number of different languages. While it is true that direct systems allow developers to add multiple targets to a single source language and to update them simultaneously, the main point about the transfer system is that it analyzes the complete sentence and, in order to do so, organizes massive amounts of in-depth linguistic information in a more efficient way.

SEE

Localization 7.2.5:516

INFO

Localization Defined INFB 27:877

To summarize, then, transfer systems make it possible not only to translate between texts that have very different structures but also to use a single analysis component to capture sophisticated linguistic data for multiple target languages. In addition, they have the advantage that they are open-ended in terms of room for linguistic growth. They can accommodate a wealth of semantic criteria. They are also well-suited to storing information from a pre-existing terminological entry—although what they actually retrieve, and how they are made to do this, is another story. Aspects of this challenge will be explored in Section 5.

single analysis component

3.3 Interlingual Systems

interlingua

Despite all the capabilities of transfer systems, there are some kinds of ambiguity that they do not handle. They also fail to capture many of the discourse-related features of language that are important for maximum communication. From the developer's standpoint, another disadvantage of the transfer system is that it is top-heavy in the analysis module and, in ways, inefficient. These problems have been addressed with *interlingual systems*, in which the transfer component is replaced by an *interlingua*.

generalization inferencing

INFO *Ontologies INFB 32:888*

An interlingua is a comprehensive intermediate representation containing universal linguistic, ontological, logical, and pragmatic knowledge that (ideally) applies to all languages—or at least to all the languages being translated. The ontological knowledge base can be used for generalization and inferencing, two functions that cannot be performed by a transfer system. Since the target translation is rebuilt "from scratch" on the basis of deep-level information returned by the interlingua, the final phase is called *generation*. The interlingual component serves as a pivot from which translations can be produced in several or in many directions with much less investment in language-specific modules at both the input and the output ends. The fact of having an ontological knowledge base also means that it should be possible to build up the lexicon dynamically—and to some extent automatically.

multi-directional pairs

While interlingual systems are costly to build, the economies to be gained from reduced effort to develop specific languages make this approach attractive if there is a demand for translation in a number of directions—especially both from and into one or several pairs. They are particularly interesting in limited subject areas, where carefully constructed domain models reliably produce richly detailed knowledge about the world to which the translation refers.

SEE *Concept Analysis 1.4.2:98*

Some very interesting work has been done on the use of domain models for the development of terminology, (Meyer/ Eck/Skuce in this handbook). Knowledge-based terminology systems—or more precisely, terminological knowledge bases (TKBs)—dovetail nicely with interlingual MT systems. In principle, a TKB could serve as one of the knowledge sources accessed in the course of an automatic translation process. Section 6 below looks at an effort to marry one such TKB to MT, as well as at terminology development for a particular interlingual system.

3.4 Corpus-based Systems

pre-existing translations

A summary of MT types would not be complete without some reference to efforts to develop corpus-based systems. The idea behind these systems is to dispense with labor-intensive rules and deeply coded knowledge sources and instead produce MT by automatically extracting data from corpora of existing translations. In their purest form, there are two basic types of corpus-based systems: (1) *statistical*, in which the translation is derived directly from corpora of matched pre-existing translations and decisions are based on statistical probabilities, and (2) *example-based*, in which the corpora serve as resources for the automatic compilation of dictionaries and other knowledge sources, which are then stored and later invoked by a translation program.

statistical systems

A statistical system—e.g., CANDIDE at IBM (Brown et al. 1990)—builds up its knowledge sources by "training" on immense corpora of existing aligned human translations in a particular subject area. The dictionaries are developed automatically, based solely on statistics, and the "terminology" is merely the reflection of the parallel source and target texts that happened to be available in the domain in question. These systems have at least two advantages: the output tends to sound like natural text, and they do away with manual linguistic coding, thus saving thousands of person-hours. Nevertheless, in their strict form they are inherently incompatible with terminology activities and the incorporation of existing glossaries, knowledge bases, or other external collections of terms.

example-based strategies

There is talk of finding ways in which the advantages of the statistical design can be blended with more classic, controllable approaches. The example-based strategy—which relies on a bilingual database of example phrases extracted from corpora of existing human translations—has been around somewhat longer (e.g., the DLT project in Utrecht and work at Kyoto University, Nagao 1984). The concept is predicated on the belief that translation consists largely of recalling or finding analogous examples (Hutchins 1993:17). Like statistical systems, example-based systems tend to produce natural-sounding translation. Their strength is their greater flexibility: they may be combined with knowledge sources, grammar rules, concept definitions, mapping rules, etc. There is no reason why terminology databases could not be incorporated into such systems as well. Used by itself, of course, the principle of sentence-level equivalents has become widely popular in the guise of translation memory programs, which are more and more being integrated with both terminology management and machine translation systems.

4 TERMINOLOGY MANAGEMENT WITH DIRECT SYSTEMS

protective undercoding

It will be recalled that the example of *hembra* was used to illustrate some of the characteristics of a direct system. It might be possible and worthwhile to tailor such a system to elicit *hembra* in a limited domain, especially if the number of animals to which it applies was relatively small. There are several ways of handling this. If the more usual translation *femenina* will never be needed, the most rudimentary solution is to assume that *hembra* is the dictionary's default translation for female. If it is the *only* option in the dictionary, this is known in the trade as "protective undercoding"—the dictionary simply does not offer the usual choices. But that solution would be quite risky. Many systems offer a safer alternative; the user can enter a customer-specific translation for *hembra* in a dictionary that overrides the general-purpose lexicon supplied by the vendor. In this way the user has flexibility and the basic translation is preserved.

hard-wired translations

Yet another approach, if there are only a few cases in which *hembra* is needed, is to provide hard-wired translations for such fixed strings as *female rats, female guinea pigs,* and *female hamsters.* These are all common strategies that are easy for the user to implement. However, they quickly break down if there is any variation at all in the word string, as, e.g., in the phrase *female rats, guinea pigs, and hamsters* or the sentence *The rats, guinea pigs, and hamsters were all females.*

An example of hard-wired phrases can be seen in selections from a list of English-Spanish dictionary entries prepared for PC-TRANSLATOR by the vendor, Linguistic Products of The Woodlands, Texas. The list of examples in Table 1 shows that it can be helpful to have terminology stored in an MT database. (In these examples, the symbol "X" at the end of each line means that the phrase is entered in the dictionary as an uninflected unit.)

Table 1: *'Hard-wired' Noun Phrases 1*

FLOW CELL ALIGNMENT..........	ALINEACIÓN DE LA CÉLULA DE FLUJO	X
FLOW CYTOMETRY...................	CITOMETRÍA DE FLUJOS	X
LASER LIGHT SCATTER............	DISPERSIÓN DE LUZ LASER	X
LEUKOCYTE DIFFERENTIAL...... COUNTER	CONTADOR DIFERENCIAL DE LEUCOCITOS	X
LIGHT SCATTER ANALYSIS.......	ANÁLISIS POR LA DISPERSIÓN DE LA LUZ	X
LYTIC REAGENT........................	AGENTE REACTIVO, LITICO [NB: should be *lítico*]	X
MEAN CORPUSCULAR HEMO-.. GLOBIN CONCENTRATION	CONCENTRACIÓN MEDIANO DE HEMAGLOBINA CORPUSCULAR	X

automatic parsing

Other examples, however, give an idea how labor-intensive the task really is, as shown in Table 2. Apparently separate entries were required in some cases in order to get *celular* as opposed to *de las células*. However, if the system had been able to parse, it could have generated these distinctions automatically.

linguistic capability

Other entries (Table 3) hinted at the system's strategy for triggering context-sensitive translations (the questions in square brackets suggest why the string was entered as a phrase). The listing from a batch dictionary update tells quite a bit about a system's linguistic capability, but it is not easy to access this information. Another way to find out how linguistically sensitive a system will be in translating user-supplied terminology is to look at the set of dictionary codes that the user is allowed to define— bearing in mind that even if a system is capable of analysis at the sentence level, it cannot act on codes that are not attached to specific entries. A system that expects only a small amount of linguistic coding is likely to perform in a direct manner even though it might have the potential of performing more sophisticated feats.

In the fall of 1992, the American Translators Association conducted a survey of all the PC-based systems then on the market in the United States (Miller, for ATA 1992). As part of this survey, the consultant, L. Chris Miller, tabulated for each system the types of grammatical information that users are allowed to supply for their dictionary entries.

Table 2: *'Hard-wired' Noun Phrases 2*

BLOOD CELL COUNT..................	CONTEO DE CÉLULAS DE SANGRE	X
BLOOD CELLS..............................	CÉLULAS DE SANGRE	X
CELL ANALYSIS...........................	ANÁLISIS CELULAR	X
CELL COMPONENTS..................	LOS COMPONENTES DE LAS CÉLULAS	X
CELL SIZE DISTRIBUTION..........	LA DISTRIBUCIÓN DEL TAMAÑO DE LAS CÉLULAS	X
CELL WALLS................................	LOS PAREDES DE LAS CÉLULAS [NB: should be *las paredes*]	X
DIFFERENTIAL WHITE CELL COUNT......................................	CONTEO DIFERENCIAL DE CÉLULAS BLANCAS	X
RED BLOOD CELL........................	CÉLULA DE SANGRE ROJA	X
RED BLOOD CELL COUNT..........	CONTEO DE CÉLULAS DE SANGRE ROJAS	X
RED BLOOD CELLS.....................	CÉLULAS DE SANGRE ROJAS	X
RED CELL DISTRIBUTION.......... WIDTH	ANCHO DE DISTRIBUCIÓN DE CÉLULAS ROJAS	X
SIZE OF THE CELL.......................	TAMAÑO DE LA CÉLULA	X

Table 3: *Strategies for Context-Sensitive Translation*

A SMALL ORIFICE.......................	UN ORIFICIO PEQUEÑO [To get *pequeño* instead of *chico*?]	X
DISCUSSED THE POSSIBILITY..	DISCUTIERON LA POSIBILIDAD [To get a context-sensitive translation for *discuss*?]	X
DAILY INSTRUMENT CHECKS...	CHEQUEOS DIARIOS DEL INSTRUMENTO [To block the default translation *cheque*?]	X
IT HAS A.....................................	TIENE UNA [To block the generation of a verb phrase with *has*?]	X
ENGLAND ET AL.........................	ENGLAND ET AL [To block the default translation *Inglaterra*?]	X

Most of these systems were capable of making certain localized decisions based on the part of speech that has been specified. They all allow for the identification of nouns, verbs, and adjectives. However, the part-of-speech heading is not necessarily limited to traditional grammatical categories. As with any MT system, this is the first cut in syntactic analysis—the branching point after which separate sets of rules are going to apply throughout the translation process, and it is a useful point at which to make some other basic assumptions as well.

syntactic analysis

For example, in the following list of categories used by MicroTac, the first nine of them are traditional and the last three are system-specific: *Noun, Pronoun, Personal Pronoun, Verb, Adjective, Adverb, Preposition, Article*, and *Conjunction*, on the one hand, and, on the other, *Phrase, Miscellaneous*, and *Unknown*. The *Phrase* category sets aside a "part of speech" for the purpose of triggering multi-word phrases. Many other systems do this as well. Indeed, using one tactic or another, they all offer the user the possibility of forming hard-wired phrases. This capability is sometimes enhanced by the possibility of using a *wild card*—a free variable that functions in a multi-word phrase much the same way as a joker does in a game of cards. In addition, five of the six systems allow the user to specify more than one part of speech for the same word, and some of them also provide for multiple translations within the same part of speech.

INFO

Part of Speech INFB 7:331

Attributes are the other capability of interest for terminology. They are the characteristics of the concept that tie it to the context and help to specify choices within the same part of speech when more than one translation is possible. To some extent they are system-specific. MicroTac, for instance, offered 17 attributes that can be applied to nouns (*Animate, Human, Family, Nationality, Occupation, Place, City, Country, Proper*, etc.), 12 attributes for verbs, six for adjectives, five for adverbs, and two for conjunctions.

attributes

In addition to the foregoing capabilities, all but one of the systems allow the user to specify more than one part of speech for the same word, and some of them also provide for multiple translations within the same part of speech.

Another way of obtaining specialized translations is through subject-specific dictionaries. When such a dictionary is called in at translation time, it is consulted first, and the translation that is retrieved takes precedence over anything in the general dictionary.

subject specific dictionaries

This approach would be useful for translating the sentence cited at the beginning:

(1) Because the clock did not work properly, the conductor failed to make the connection and the bus had to operate with just a driver.

This solution would not work, of course, if the usual meanings of *bus*, *driver*, etc. came up elsewhere in the same translation project.

The premise of this discussion has been that a system which uses more linguistic information has greater flexibility in providing context-sensitive translations. However, as pointed out before, there is a tradeoff. Complex entries tend to be more difficult to add to the existing database—although many systems have user-friendly software that greatly facilitates the task. Still, the direct systems that do not require a great deal of grammatical information are easier to learn and to manage. They may be more labor-intensive and more limited in their capacity to handle technical terminology, but their updatability is an important factor in their appeal to the public.

context sensitivity

5 Terminology Management with Transfer Systems

The resource in a system that parses is *syntax*. Transfer systems contain massive sets of lexical rules that combine syntactic features with other information about a word, often semantic features, to trigger the desired translation. Three examples of such systems viewed in this survey, all of which have extensive resources for eliciting context-sensitive translations, are SYSTRAN, LOGOS, and ENGSPAN.

syntax

5.1 Systran[2]

Of all the transfer systems, SYSTRAN probably reflects the longest and most extensive work on lexical development. The company, based on La Jolla, California, began the arduous process of building its mammoth dictionaries in 1964. At the time of the survey, it had dictionaries in 27 language combinations.

For handling words with multiple meanings or words that can have both general and technical meanings, SYSTRAN relies on a

polysemy

combination of syntactic and semantic analysis plus topical glossaries and dictionaries of customer-specific translations. One amusing review of its capabilities was presented by Wheeler in his article "The Errant Avocado" (1983), in which he guides the reader through a series of exercises including the correct coding of the French phrase *poste d'avocat general* so that it translates as *post of advocate-general*, rather than SYSTRAN'S initial choice—namely, *general avocado station!*

semantic classification

The semantic classification system includes over 500 categories organized in a hierarchical structure. Each word in any of the dictionaries can be encoded with the categories that apply. The system uses this information on both the program and dictionary levels to select the right translation in the target language. An example is the following semantic rule for the English word *frozen*:

> If word is "frozen" functioning as a modifier, check noun modified for the semantic property [FOOD PRODUCT].

This rule would distinguish the sense of *a frozen food* from that of *an immobilized mechanism*.

semantic ambiguity

For handling more complex or more generalized problems of ambiguity, SYSTRAN has specialized programs at the transfer stage of the translation process. One routine determines whether a word is animate or inanimate. This program is called in whenever the system encounters a word coded in the dictionary [+ANIMATE/INANIMATE AMBIGUOUS]—e.g., *seal, parent, acquaintance*.

topical glossary

In the topical glossaries, each word or expression in the source dictionaries is assigned target translations in as many fields as necessary to cover the range of meanings and uses for that word. For example, *stem* in English has a variety of senses when it is used in different domains. The target translations for this word would likely include a default translation for general use plus ones for the sense of *plant stem* in botany and for *brain stem* in anatomy. Before the translation is run, a user may indicate preference for up to 10 fields. For the words that have meanings encoded for the fields indicated, only those specialized translations will be chosen.

multiword terms

The customer can also choose specific translations that override all the others. Companies with special technical or proprietary terms may have expressions encoded so that their translation for the expression will be selected only when a translation is run with their particular code.

Another resource is the set of conditional and nonconditional string replacement features whereby fixed expressions such as *brake shoes* can be handled as a single item with a unique meaning in order to elicit a technical translation in the target language rather than a mere word-for-word replacement.

fixed expressions

In addition to strategies for assigning appropriate meanings to frequently occurring noun phrases, SYSTRAN is designed to remove ambiguity at a very early stage when possible. The development team forestalls problems by examining large corpora and bringing in new terminology with new rule sets for resolving ambiguity in an ongoing effort to expand dictionary coverage and the system's overall scope. Such an exercise yielded the following readings of the Japanese word *ippou*:

ambiguity rule sets

(4) rajio ya terebi no *ippou* tsuukou no jiyouhou wo . . .
'the *one-way* information of radio and television . . .'

(5) *ippou* ni ion bi-mu wo whiyoushiya shite iru . . .
'shining an ion beam on *one side* . . .'

(6) sangyou wo sasaeru *ippou*, . . .
'*While* industry is supported, . . .'

(7) *ippou* LPG ni tsuite ha, . . .
'*On the one hand*, concerning LPG, . . .'

(8) Kyousou ha hageshiku naru *ippou* da.
'Competition will become *more and more* intense'

To capture the differences in meaning in these examples, SYSTRAN encodes not only the knowledge on which they depend, but also rules for selecting one lexical item over another. The expandability of the routines that handle homographs allows for new lexicon growth without an unwieldy proliferation of ambiguities.

lexicon growth

While SYSTRAN has not only addressed the question of ambiguity in depth, it has also dealt with the issues of batch updating and mass encoding of terminology files and integrating terminology so that it produces the desired translations

5.2 LOGOS[3]

LOGOS Corporation efforts to facilitate the development of large dictionaries of context-sensitive technical terminology date back nearly as far as SYSTRAN's. The LOGOS system offers a variety of

terminology differentiation

ways to differentiate terminology depending on the *company* the translator works for, the *country* of destination, the *subject matter* of the translation, and *general vs. technical usage.* (Note: specific values cited below reflect the status of LOGOS software at the time this survey was compiled.)

specialized dictionaries

Clients such as AT&T, Microsoft, and IBM have their own company-coded dictionaries that their terminologists are free to customize. If they need to differentiate, e.g., between terms used by IBM/England and IBM/USA, they can create country-specific dictionaries to accommodate spelling variations and capture such cultural idiosyncrasies as *boot* vs. *trunk* for the German *Kofferraum*. At translation time, LOGOS allows five company codes to be specified, and the system looks the words up in the order in which the user lists the codes.

SEE *Globalization 8.2.3:677*

subfield specification

To deal with semantic ambiguity, LOGOS has up to 500 subdictionaries for different subject areas. The word *Ausleger*, e.g., could have the translation *extension arm* in the general dictionary, *boom* in the automotive dictionary, and *jib* in a specific dictionary for cranes. The user can specify up to five domain-specific subdictionaries when submitting a translation. These subdictionaries are logically linked through generic codes, which means that pointers in the translation program map to several related fields—e.g., electrical engineering, electricity, and electromagnetism. If a word is not found in the specified subdictionary, the system automatically looks for it in one of the related ones.

syntactic & semantic context

Another way to differentiate the meaning of semantically ambiguous words is on the basis of their syntactic and semantic context. For this purpose, LOGOS has a special database that is consulted at various stages of parsing the input sentence. For example, the German adjective *fest* may have the English translation *fixed* in the general dictionary, but it can also elicit *hard* in the context of *Platte* or *Währing*, and it can be translated as *heavy* in the phrase *heavy blow* and *lasting* in the expression *lasting peace*. The tool that performs this task is SEMANTHA, an easy-to-use interactive program for creating and updating customized context-sensitive rules. In the example of *female* applying to rats and other animals, SEMANTHA would enable the user to write a single rule specifying that the English adjective *female* should be rendered as *hembra* in Spanish whenever it modifies a noun that is coded as a nonhuman animal. This guarantees that LOGOS will consistently translate female animals as *hembra*. (It should be noted that this rule is too powerful, since it would translate *female cat* as **gato hembra* instead of *gata*. Exceptions would have to be handled individually.)

context rules

Context rules in the semantic database apply not only to contiguous strings at the surface level but also to all types of transformations based on a single deeper structure. Thus, e.g., a context rule written for the abstract-level construction *Tafel (case= accusative) löschen ⇒ clean N* will apply to very different surface-structure transformations at different stages of the syntactic parse:

parsing levels

(9) Die Lehrerin löschte die Tafel.
The teacher cleaned the blackboard.

(10) Die Tafel wurde von der Lehrerin gelöscht.
The blackboard was cleaned by the teacher.

(11) Die regelmässig von der Lehrerin gelöschte Tafel ...
The blackboard regularly cleaned by the teacher ...

(12) Mit Löschen der Tafel hat die Lehrerin ...
In cleaning the blackboard, the teacher ...

updating

The user first has to decide whether to update an expression (1) as a context rule, (2) through SEMANTHA, or (3) as an entry in the dictionary. The next decision is which subdictionary to update, i.e., under which company code or subject matter code. The tool ALEX is used to update dictionaries. This interactive program asks the user to give minimum semantic, syntactic, and morphological information about the source and target entries. LOGOS has 130 hierarchically ordered semantic categories for the coding of nouns alone, but the user does not have to learn all these categories. If the basic entry (or head noun in the case of a compound), is already in the dictionary, ALEX offers a choice of categories based on the coding already found in the dictionary. If the entry or the head are entirely new to the system, the user can enter a word that has similar syntactic and semantic properties and ALEX will propose possible semantic categories based on that "synonym."

5.3 ENGSPAN

dictionary enrichment

ENGSPAN is a transfer system developed by the Pan American Health Organization (PAHO) in Washington, D.C., for the translation of English into Spanish. While it has a robust capacity to cope with general text, it also has highly developed lexical resources in fields related to medicine, public health, and agriculture.

paired lists

On more than one occasion the developers of ENGSPAN chose to enrich their system through the infusion of large paired lists—including a list of about 30,000 medical terms. When these entries were being augmented with the information required for ENGSPAN's parser, a program was written to generate semantic codes for English words that ended in typical medical suffixes such as *-itis* "inflammation", *-osis* "condition of", *-otomy* "incision into", etc. In this way, a portion of the semantic coding was added to the system automatically. Similar logic is also used in the regular translation program to make educated guesses about words not found in the course of translation.

leveraging existing terminology

SEE

Thesaurus Management 6.1:381

A more labor-intensive exercise was involved in introducing 8,697 English and matching Spanish descriptors from AGROVOC, a multilingual thesaurus developed by the Food and Agriculture Organization of the United Nations (FAO) and the Commission of the European Communities (CEC). In 1987, the ENGSPAN team undertook a terminology project in the field of agriculture to prepare the system for pilot outplacement at two agricultural research centers. The experience was also intended to yield a methodology that could be applied to the incorporation of other specialized terminology in the future. This exercise was reported in detail in *TermNet News* (León, Santangelo, and Vasconcellos 1987).

indexing terms

It was expected that a sizable proportion of the terms would already be present in ENGSPAN's dictionaries. However, given the difference in purpose between a thesaurus of indexing terms and a dictionary for the translation of running text, it was also recognized that some of the AGROVOC Spanish equivalents might not be appropriate for ENGSPAN. Indeed, it was soon found that a number of considerations prevented any large-scale automatic entry of these terms. The set of pairs shown in Table 4 exemplifies the many problems that called for the discerning judgment of a human linguist.

inconsistences

• AGROVOC uses British spelling, whereas ENGSPAN uses American spelling (e.g., *flavour* vs. *flavor*. While ENGSPAN's lookup procedure does have logic for bypassing British spelling at translation time, entries in the dictionary must be in the American style.

Table 4: *Equivalent Pairs*

FLAVOUR	AROMA
FLAVOUR ENHANCERS	REFORZADORES DE AROMA
FLAVOURED MILKS	LECHE AROMATIZADA
FLAVOURING	AROMATIZACION
FLAVOURING CROPS	PLANTAS DE CONDIMENTO
FLAVOURINGS	AROMATIZANTES

inconsistencies

• Many AGROVOC descriptors are given in the plural form; ENGSPAN's entries, on the other hand, need to be in the singular in order to generate the system's full range of possibilities.
• The AGROVOC descriptors are in full caps, while the distinction between upper and lowercase needs to be made in ENGSPAN's entries so that the system will know how to write abbreviations, proper names, scientific words, the names of proprietary products, etc.
• The Spanish words in AGROVOC do not have orthographic accents (acute accent, diaeresis, tilde), whereas these diacritics are required by ENGSPAN.
• Some of the words that are used as count nouns in AGROVOC (*milks* and *flavourings*) would normally be treated as a mass noun (*milk*) or a verb nominalization (*flavouring*) in ENGSPAN.
• Many terms used only as nouns in AGROVOC (*flavour* and *milk*) would have to be coded for other parts of speech (in this case, verb and noun) in ENGSPAN.
• Many of the AGROVOC terms include inflected verb forms (*flavoured*, *flavouring*) that would be unnecessary additions to the ENGSPAN dictionary.
• Some of AGROVOC's Spanish equivalents (e.g, *aroma*) conflicted with existing ENGSPAN translations (in this case, *sabor*).
• Within AGROVOC, the same word is sometimes translated differently in different contexts. In these six entries alone, *flavouring* has been rendered as *aromatizacion*, *de condimento*, and *aromatizante*, while *crops* had been translated as *plantas* and *cultivos*.

encoding duplicates

These problems were solved on a case-by-case basis, beginning with the single words. In the event of conflicts between AGROVOC and an existing ENGSPAN entry, the linguist had three basic options: the PAHO term could be retained, the existing gloss could be replaced, or a second term could be added. Most often both options were needed, and special coding was introduced to trigger different choices on the basis of context. For example, the distinction between *drinker* [+HUM] and *drinker* [+DEVICE] is an analysis problem that has to be handled in the source entry, whereas the choice between *cuello* [+HUM][+BODY PART] and *pescuezo* [-HUM][+BODY PART] gets handled on the target (synthesis) side. When the context had to be specified, the single words were treated as expressions.

multiword terms

The next step was to review the multiword expressions and decide whether they should be entered as substitution units (SU), analysis units (AU), or transfer units (TU). The criteria for choosing between these three types of units are as follows (Leon/Schwarz 1986):

SU There is a high degree of certainty that the particular string will always have the same translation. It is unlikely that other words will interrupt the phrase or that one of its elements will form part of another syntactic structure.

AU The words are likely to occur together but may belong to another syntactic structure.

TU The words may not be contiguous, so the input string must be analyzed syntactically in order to decide whether or not the phrase exists. A rule can be formulated that will apply to a class of lexical items.

human intervention

The list in Table 5 shows cases in which these various approaches would be used. The AGROVOC exercise demonstrated that the incorporation of existing specialized vocabularies into the ENGSPAN dictionaries could only be automated to a certain extent and that many decisions had to be made. While the MT program itself, its supporting software, and word-processing capabilities can be used to speed up some parts of the work, human decisions have to be made at almost every turn. Each set of terms will present problems of its own. Of course, the result of all this painstaking work is a better MT system with increased subject-area coverage and overall improved quality.

Table 5: *Encoding Duplicate Entries*

English term	Type of unit	Spanish gloss
FISH	(Main entry)	PESCAR (verb), PEZ (noun)
FISH	TU	PESCADO
FISH CULTURE	SU	PISCICULTURA
FISH DETECTION	None	DETECCIÓN DE PECES
FISH EXTRACTS	TU	EXTRACTOS DE PESCADO
FISH FEEDING	None	ALIMENTACIÓN DE PECES
FISH LARVAE	SU	ALEVINES
FISH OILS	Same TU	ACEITES DE PESCADO
FISH POISONING	SU	PESCA CON VENENO
FISH PONDS	AU	ESTANQUES PISCÍCOLAS
FISH WASTES		DESECHOS DE(L) PESCADO

6 Terminology Management with Interlingual Systems[4]

knowledge base information

As noted above, interlingual MT relies at least in part on concept systems with elaborate networks expressing relationships between the nodes—relationships such as *is a*, *is a property of*, *composed of*, *precedes*, *follows*, *for the purpose of*, *subtype of*, *supertype of*, *sufficient condition for*, *necessary input for*, *opposite of*, *also known as*, etc. Although online terminological knowledge bases (TKBs) have been developed that exhibit conceptual links of this kind—e.g., COGNITERM, until recently the work has been carried on independently of MT. Many of the elements in a traditional terminology record are excess baggage for machine translation, while on the other hand MT requires specific syntactic and semantic features and attributes that are usually not included in a terminological entry (Melby and Wright 1999). In any case, it seems a rather daunting task to map the information in a TKB so that it can be captured usefully in an automatic translation process that must also consider syntax, case frames, schemata, pragmatics, and the overall discourse fabric.

reusable terminological resources

In the early 1990s, the ET10/66 Consortium in Europe embarked on a project, which, among other aims, sought to overcome the presumed incompatibility between terminology banks, knowledge bases, and machine translation systems. The ET10/66 team, which includes investigators in Dublin, Luxembourg, Lisbon, and

Bonn, worked with EIRETERM in Dublin to build a reusable terminological resource that would provide extralinguistic domain knowledge within the advanced linguistic engineering platform (ALEP) designed under auspices of the European Commission. Working in the domain of telecommunications, they developed an ontology by formalizing the definition of each term and establishing the conceptual links between them (Pearson [ed.] 1992). Since the ALEP framework allows for encoding not only domain knowledge but also linguistic knowledge, the research group was also able to incorporate syntactic and semantic information based on the EUROTRA design. All this information, taken together, may be consulted in a parsing process for machine translation (Schütz/Ripplinger 1993). This model has been used to show how ontological knowledge helps to solve ambiguities in the attachment of prepositional phrases, the identification of collocations, and the selection of appropriate target translations from among several possibilities (Ripplinger [ed.] 1993). The importance of this work is that a common formal device is used to express and represent different types of knowledge, thus avoiding the need to build an interface between them (Schütz/Ripplinger 1993).

SEE *Terminology Interchange 8.1.5:613*

corpus-based systems

A more practical example of terminology development for interlingual MT is the work being done with KANT, a system developed by Carnegie Mellon University in Pittsburgh, Pennsylvania, which was then scaled up by Carnegie Group, Inc., to translate heavy-equipment manuals into a number of target languages for Caterpillar, Inc. The developers relied on KANT's rich-knowledge-base capabilities to generate machine translation that will require no postediting. Experiments have shown that systems like KANT are capable of producing good results in technical domains as long as the domain semantics are "tractable" (Mitamura/Nyberg/Carbonell 1993).

knowledge-based MT

In the past, efforts to scale up knowledge-based MT have entailed a large amount of hand coding to build the necessary lexical resources, ontology, grammatical and other rule sets, etc., representing a sizable investment. Accordingly, the developers of KANT's dictionaries for Caterpillar sought to reduce the labor-intensive aspects wherever possible and take the advantage of online textual resources and corpus analysis software to at least partially automate the process, thereby maximizing human productivity.

automatic tagging

The first step was to compile a comprehensive corpus of documents covering the entire domain of Caterpillar heavy equipment. The words in the corpus were "tagged" with their parts of speech by matching it against the Brown Corpus (a widely used linguistic research tool), which yielded the skeleton of a lexicon containing both single words and phrases. Candidate phrases were identified on the basis of part-of-speech patterns. In the next step, sets of corresponding source-target manuals were aligned automatically, following which the actual translations of terms were identified manually—the initial round being limited to French. The result was a bilingual English-French lexicon.

SEE *Text Corpora 8.4.3:788*

limited domains

A special model was then constructed for the strictly limited domain, which enables the system to make more precise choices. It does not have to allow for the many contingencies that would be irrelevant to the subject field. This step involved constructing both *concept frames* and a *concept hierarchy*.

mapping rules

Other online tools were used to fine-tune Kant's grammar and mapping rules for both the source and target languages. On the source side, *lexical mapping rules* establish the link between a given word (tagged for one of its possible parts of speech) and a set of domain concepts. Lexical mappings are pointers to concept nodes, which are created automatically through the application of relevant grammatical rules. *Argument mapping rules* are constructed from semantic roles identified by the lexical mapping rules and the semantic features present in the interlingua. On the generation side, *target language mapping rules* map the interlingua output onto target *f-structures* (a formalism used in lexical-functional grammar). Next the lexical mapping rules, created automatically from the lexicon, generate one-to-many target mappings. These are further refined by a target-language linguist, who then adds contextual patterns for use in selecting the desired target equivalents.

automatic knowledge acquisition

The team found that the use of automatic knowledge acquisition considerably reduced the human effort required to build up the system's knowledge sources for a particular application. This strategy, coupled with similar approaches using other tools, is designed to reduce costs and therefore increase user acceptance of interlingual MT for commercial production purposes.

7 Final Thoughts

historical perspective

In the 1980s, transfer-based systems were far and away the most widely used workhorses of machine translation, although MT on PCs was emerging as a viable option for production translation. Knowledge-based interlingual MT was still in the laboratory. In 1989 FUJITSU debuted its ATLAS-II based on the interlingual approach. In the user survey mentioned at the beginning of this chapter (Vasconcellos 1993), all of the 15 sites in operation prior to 1990 used transfer systems. In the total population of 36 MT users, transfer systems accounted for 14 of the 16 systems employed. Of the other two, one was direct and the other was an entry-level interlingual system.

The news in the 1990s and beyond is that people have been experimenting at both ends of the spectrum. Direct systems are popular because they are affordable and easy to learn to manage, while interlingual systems are at least as costly as transfer systems and at least as difficult to update and maintain, if not more so.

portability

SEE *Terminology Interchange 8.1.5:639*

One point that was not mentioned before is that it has not been easy to port terminology between systems. Total lossless portability between different levels may well be out of the question, but the OLIF (Open Lexicon Interchange) format provides a good deal of portability, while stripping away much of the proprietary coding that makes MT systems so different from one-another (Thurmair et al. 1999). If terminology management is one of the main concerns, as it is for many MT users, the choice of a system, once it has been tailored to meet their needs, has seemed to be nearly irrevocable. Some user sites have switched, but customization of the new system has proven to be a major adjustment not unlike starting afresh. The new exchange environments, such as SALT (Standards-based Access to Multilingual Lexicons and Terminologies), however, may make it possible to at least leverage basic data for integration into different systems or for use in different kinds of applications.

translation volume

For transfer and interlingual systems, multi-word entries are more difficult to encode than single words, as was seen above particularly in the case of ENGSPAN. This becomes a consideration for the management of terminology, since many technical terms have more than one word. At the same time, however, it is probably in the area of multiword terms that MT makes its most

high-volume users

valuable contribution to the maintenance of uniformity in large projects. In the ATA survey of users (Miller 1992), the advantage of MT that was cited most often was consistency of terminology. Without exception, these were high-volume users. There were sites that use MT for 17, 25, 30 and even 45 million words a year, and a number of them averaged around 10 million. In all, the 24 users that reported statistics were producing 170 million words a year. With the exception of MÉTÉO, which does weather bulletins, and the U.S. operation at Wright-Patterson Air Force Base, the high-volume users were all engaged in some aspect of localization.

integrated approaches

These users have been forging the link between MT and terminology, which will clearly be getting stronger in the coming years as more and more systems adopt an integrated approach that effectively links MT, TM, and MAT applications.

ACKNOWLEDGMENTS

1. With contributions by Brian Avey, Claudia Gdaniec, Laurie Gerber, and Marjorie León. This article reflects to some extent the rapid changes that are taking place in the tools industry. Rather than attempt to update the article with respect to vendor references, the author and editors refer the reader to In boxes 24 on exchange formats and 30, which contains an MT update. Nevertheless, the generic treatment of fundamental aspects of machine translation and terminology remains highly valid.
2. Brian Avey and Laurie Gerber of Systran Translation Systems, Inc., contributed to section 5.1
3. Claudia Gdaniec of Logos Corporation contributed to section 5.2
4. Section 6 draws on Pearson (1993), Ripplinger (1993), Schütz/Ripplinger (1993), and Mitamura/Nyberg/Carbonell (1993). The author is grateful to Ingrid Meyer and Taricha Mitamura for supplying advance copies of their work and to Barbel Ripplinger for a full set of reports of the ET10/66 Project.

COMPILERS' NOTE

Despite the fact that the preceding article was originally written in 1994, the core treatment of the relationship between terminology and machine translation remains valid and unique. Any effort to update the article in a timely fashion would be doomed to obsolescence because of the rapidly changing situation in the field.

Yet, in the current Internet/World Wide Web environment interest in MT is burgeoning as more and more people working in a variety of languages all over the world want and need to be able to access a wide range of information in English in particular, but also in other languages as well. The desire for real-time, online automatic translation has lead to a variety of services offered on the Web. The results of these attempts at instant understanding frequently lack accuracy and elegance, if for no other reason than the fact that the individual requesting the translation has no ability to load specialized terminology in the Web translation engines or to specify the specific domain of the text to be translated. Nevertheless, these sites have a certain value in terms of their gisting capability, i.e., their ability to communicate the *gist* of the message conveyed by the original Web page.

Because of the ever-changing face of the MT scene, rather than try to provide a complete overview, we recommend that interested readers take a tour of some of the most generic Web portals that treat the problems of machine translation and that direct the Web surfer to specific companies and services.

MT Organizations

International Association for Machine Translation (IAMT)
http://www.isi.edu/natural-language/organizations/IAMT-bylaws.html
Association for Machine Translation in the Americas
http://www.isi.edu/natural-language/organizations/AMTA.html
Asia-Pacific Association for Machine Translation
http://www.jeida.or.jp/aamt/index-e.html
European Association for Machine Translation (EAMT)
http://www.eamt.org/

Portals and lists of links:

Rivendel International Communications Limited provides a fairly comprehensive list of commercial MT sites, although, as is the case with most lists, some individual items are sometimes no longer operative.
http://rivendel.com/~ric/resources/mt.html

The *Introduction to Computational Linguistics: Machine Translation* Web page at Georgetown University (Alexandria, VA) also provides a fairly comprehensive list of links:
http://www.georgetown.edu/cball/ling361/ling361_mt_links.html

Machine translation timeline:
In May 2000, *Wired* magazine devoted at least in part to MT, including a partly accurate historical and partly quirky prediction-oriented timeline of MT activity, which can be viewed at:
http://www.wired.com/wired/archive/8.05/timeline.html

REFERENCES

Brace, Colin. 1993. "Making MT work." *Language Industry Monitor*, no. 13 (1-4), Jan-Feb.
Brown, P., et al.1990. "A statistical approach to language translation." *Computational Linguistics*, 16:79-85.
Hutchins, W. John. 1993. "Latest developments in machine translation technology." *Proceedings of the MT Summit IV.* Kobe, 19-22, July 1993, 11-34.

León, Marjorie, and Schwartz, Lee A. 1986. "Integrated development of English-Spanish machine translation: from pilot to full operational capability." Technical Report. Grant DUE-5543-G-SS-3048-00 from the U.S. Agency for International Development. Washington, D.C.: Pan American Health Organization.

León, Marjorie; Santangelo, Susanna; and Vasconcellos, Muriel. 1987. "Terminology work and automatic translation systems: A case study at the Pan American Health Organization." Vienna: TermNet News, 18:21-25.

Melby, Alan K. and Wright, Sue Ellen. 1999. "Leveraging Terminological Data for Use in Conjunction with Lexicographical Resources." *Proceedings of Terminology and Knowledge Engineering '99.* Peter Sandrini, ed. Vienna: TermNet, 544-569.

Miller, L. Chris. 1992. "American Translators Association Report: PC-based MT products." ATA Committee on Machine Translation.

Mitamura, Teruko; Nyberg. Eric H.; Carbonell, Jaime G.1993. "Automated corpus analysis and the acquisition of large, multi-lingual knowledge bases for MT." *Proceedings of the Fifth International Conference on Theoretical and Methodological Issues in Machine Translation*, Kyoto, 14-16 July 1993, 312-328

Nagao, Makoto. 1984. "A framework of a mechanical translation between Japanese and English by analogy principle." *Artificial and Human Intelligence*, Elithorn, A. and Banerji, R., eds. Amsterdam: North-Holland.

Pearson, Jennifer, ed. 1992. "ET10/66, Terminology and extra-linguistic knowledge." Reports 1 and 2 of the ET10/66 Consortium, Dublin: Dublin City University, National Center for Language Technology.

Ripplinger, Bärbel, ed. 1993. "ET10/66, Terminology and extra-linguistic knowledge." Report 3, vers.2, of the ET10/66 Consortium. Luxembourg: Centre Universitaire, Centre de Recherche Public.

Schütz, Jörg, and Ripplinger, Bärbel. 1993. "Machine translation supported by terminological information." *Proceedings of the Fifth International Conference on Theoretical and Methodological Issues in Machine Translation*. Kyoto: 14-16 July, 1993, 102-116.

Thurmair, Gregor; Ritzke, Johannes; and McCormick, Susan. 1999. "The Open Lexicon Interchange Format—OLIF. TAMA *'98—Terminology in Advanced Microcomputer Applications.* Proceedings of the 4th TermNet Symposium: Tools for Multilingual Communication. Vienna: TermNet, 237-262.

Vasconcellos, Muriel. 1993. "The present state of machine translation usage technology, or How do I use thee? Let me count the ways." *Proceedings of MT Summit IV, Kobe*, 19-22 July, 1993

Wheeler, Peter. 1983. "The errant avocado." *Newsletter of the British Computer Society Natural Language Translations Specialist Group*, No. 13, 1-6.

Wheeler, Peter. 1988. "The translator and the dictionary experience." *Technology as Translation Strategy*, Vasconcellos, M., ed. Binghamton (N.Y.): State University Press. American Translators Association Scholarly Monograph Series, 2., 149-158.

8.4 Corpus-Related Applications

8.4.1 Corpus Linguistics and Terminology Extraction

KHURSHID AHMAD AND MARGARET ROGERS

1 TERMINOLOGY AND TEXTS

Human enterprises can be distinguished in the ways in which language is adapted by the community for a specific purpose and the ways in which information is communicated for teaching, learning, research and solving problems. This adaptation of language manifests itself lexically in the coinage of new terms, in changes in the meaning of existing words, or in the conditions of their applicability and use.

term extraction

The language used by members of very different discourse communities—from scientists to hobbyists—is recorded in what might be called the "textual archives" of their common interest, ranging from learned journals and academic textbooks, through professional publications, hobbyist magazines and books, manuals, and even personal letters. Such archives record the emergence, development and demise of terms as they are used. Much like the texts of their parent *general* language, the language of everyday use, special-language texts can be classified into different genres. Structured collections of texts are a valuable source of data on the existence of terms and their behavior in a fast-changing professional world in which knowledge is growing, merging and changing rapidly and where dictionaries provide inadequate cover. The processing of texts using computer programs in order to identify strings that are potential terms is referred to as *automatic term extraction* (sometimes *semi-automatic term extraction* or *term extraction*. Other terms include *excerption,* ISO 1087:1990). Term extraction produces the raw material for terminology databases, material that has to be examined, tested and validated before inclusion in a termbase.

lexical density

Special-language texts are lexically denser than general-language texts, particularly those written by experts for experts. They contain more compound nouns, more plurals and have a higher proportion of nominalized verbs. The systematic variation between special-language and general-language texts, as well as the characteristics of strings that can be identified as terms as opposed to non-terms, can be exploited for purposes of term extraction. Methods include statistically based procedures, linguistically-based procedures and hybrid procedures. Human beings still play a role in all methods proposed to date.

2 Electronic Corpora and Corpus Linguistics

vocabulary variation

The textual archives of discourse communities are now often available electronically. They constitute a valuable source of evidence about the occurrence and use of the specialist vocabularies or "terminologies" of these communities in three principal ways: (1) variation across subject domains and the typicality of terms within their own domain; (2) variation between levels of communication from academic expert to technician or "hobbyist", and their typicality within each level; (3) their variation or constancy across time.

structured corpora

If the interpretation of the data that is gleaned from textual sources is not to be *ad hoc*, then the texts that make up these archives need to be collected together, evaluated, selected and organized in a systematic way according to explicit criteria. The result, a structured body of texts, is known as a "corpus" or text corpus. Today, the importance of using a structured corpus rather than an undifferentiated "collection" of texts is generally accepted in both general-language and special-language work. The concept of a corpus is now also increasingly understood as a computerized or electronic corpus, for which software systems are needed if processing as well as storage is to be automated, or at least, semi-automated. Contemporary corpora tend to consist of whole texts rather than text extracts, and usually comprise written texts, but some, particularly large general-language corpora, may also contain transcriptions of recorded speech.

documentation

Documentation has long been acknowledged in the literature of terminology science as an important source of terminological and conceptual information, not only in the text- and translation-oriented Canadian approach to terminology (Cole 1987) but also in the more normative approach of the Vienna School (Picht & Draskau 1985:164-70). Sager (1990:130) also notes the increased use of "corpora" in the compilation of terminologies. Following evaluation and selection from the available documentation, terminologists are traditionally recommended to "scan" the texts to "extract" terms and terminological data, by means of what Hohnhold has called *termbewußtes Lesen* (reported in Bühler 1987:54). While the 1960s saw the first attempts to automate the storage and retrieval of terminological data in term banks, the first attempts to automate the capture of data from large collections of machine-readable texts, i.e., from electronic corpora, can be dated to the late 1980s. The evaluation of any term-extraction method must go hand in hand with an evaluation of the corpus that is being analyzed, not simply as in traditional terminology management with reference, for instance, to the authority of the authors of texts, but also with regard to the structure and processing of the corpus.

raw & tagged corpora

"Raw" corpora can be used to find the frequency of occurrence of single words or multiword compounds and also for studying the grammatical behavior of these words and phrases through concordance analysis. However, a raw corpus cannot be used effectively where homographs are concerned, i.e., where the same orthographic form occurs in more than one word class (e.g., noun, verb, adjective), leading to further problems in any attempt to group together morphological variants of the same lexeme (e.g., number and case for nouns; person, number and tense for verbs). A tagged corpus, on the other hand, is a linguistically-analyzed or annotated version of a raw corpus that contains not only the constituent words, but typically also part-of-speech information. Given the size of emerging corpora since the 1970s, computer scientists and linguists have developed programs that can automatically tag individual words or phrases with the appropriate parts of speech.

term extraction

The extraction of terms from running text plays a role not only in the compilation of termbases (thematic terminology management), but also in specialist translation (*ad hoc* terminology management) and in information and document retrieval (automatic indexing or automatic keyword extraction). Reference to the use of computer software for "automatic" identification of strings (single words, two or more words) requires comment, since the output of current systems still requires human filtering by, for example, domain experts, to identify string boundaries and to distinguish valid terms from mere term candidates. However, in practice, the concept of automatic extraction can be narrow or broad, depending on the filters employed by the system, such as linguistic filters for statistically based systems (Daille et al. 1996:213). More heuristically based systems tend to overgenerate, requiring more filtering (human or machine), whereas more rule-based systems require less filtering, although they also risk undergeneration. The balance between overgeneration and undergeneration is problematic: systems that produce too much invalid data or "noise" require excessive post-editing by human experts. Systems that miss valid data risk incompleteness.

filtering

SEE *Technical Translation 2.1.1:147*

lexical patterns

The use of large corpora as a source of lexical evidence has a long tradition, although their use in terminology management is still relatively new. Pioneers of "lexical evidence" in corpus linguistics include Quirk, Svartvik, Greenbaum and Sinclair, whose work in collecting and analyzing large corpora of texts for studying a given language has had considerable influence on how general-language dictionaries are compiled. The descriptive linguistic tradition in which the study of "real text", and in particular, the study of lexical patterns in text, plays a central role, owes much to the British linguist J. R. Firth (1890-1960) (Halliday 1966; Sinclair 1966). A corpus-based approach to the compilation of both general-language and special-language lexica reflects trends in linguistic research towards data-driven methods of analysis (Garside, Leech and Sampson 1987; Aarts and Meijs 1990; Aijmer and Altenberg 1991). This trend, which Leech describes as a "resurgence" (1991:9), can be associated with both the increasing power of computers and their greater accessibility, enabling larger and more representative samples of text to be processed by more users.

corpus-driven/ corpus-based

With some justification, corpus linguists argue that texts can be used to support introspective views about language and its use, and, indeed to inform us about the properties of the language. Text corpora have been used in producing grammars for English (Quirk *et al* 1985), and to produce dictionaries of contemporary English, particularly for learners. In this context, a useful distinction can be made between "corpus-driven" and "corpus-based" dictionaries. The *corpus-driven* approach emphasizes the "paramount" role of a corpus of texts in the description of a language (*Collins Cobuild English Language Dictionary*, 1987), whereas the *corpus-based* approach helps to extend and improve the description (*Cambridge International Dictionary of English,* 1995). Indeed, it is common practice these days in many different types of dictionary to use the systematic evidence of corpora rather than the more *ad hoc* selection of citations by readers more traditionally used in lexicography.

standard corpora

Text corpora of British and American English have been developed in the last 30 years or so, beginning from the 1-million word Brown Corpus of American English in the 1960s, shadowed by the Lancaster-Oslo Bergen Corpus of British English (LOB), followed by the 20-million word text corpora of British English by dictionary publishers in the late 1970s and early 1980s including the Longman-Lancaster Corpus and the Birmingham Collection of English. There have been corpora of Indian English, Australian English, Canadian English and so on (Aijmer and Altenberg 1991). The US-based Linguistic Data Consortium (LDC) (Philadelphia) and the International Computer Archive of English at Bergen in Norway now offer various corpora of English. The LDC also offers corpora in French, German, Japanese, Chinese, and so on. General-language corpora for a range of languages, including French, German, Russian and Swedish, have been developed in the recent past (Barlow 1999). The EU-sponsored European Language Resources Association (ELRA) also offers text and speech corpora in a number of EU languages. A recent example of a "national corpus" is the publicly-available British National Corpus (BNC) (Aston and Burnard 1998). While many corpora include some special-language texts, these are often of a more popular kind. Consequently, terminologists usually build their own corpus from scratch.

special language corpora

The use of corpora in terminology management has been accepted both theoretically and practically much more slowly than in lexicography, largely as a result of its generally more prescriptive orientation and onomasiological, i.e., concept-based approach, but also owing to some more practical limitations to which we will return later. Consequently, the use of special-language corpora in terminology management has received much less attention to date than that of general-language corpora in lexicography and language research. However, there are clear signs that corpus-based terminology management, including the identification of terms and translation-oriented terminology, as well as the whole concept of terminology management, is now being discussed (see for instance, Meyer and Mackintosh 1996a; Pearson 1998; Picht 1998).

frequency

The unifying theme of work in corpus linguistics is that the frequency of use of linguistic units, realized at the lexical, syntactic, semantic and pragmatic levels, tells us something about the nature and function of language. Techniques developed by corpus linguists help in pooling evidence about the frequency behavior of the different units, since assessments by native speakers of relative acceptability largely correlate with their assessments of relative frequency (Quirk *et al.* 1985:33). Corpus linguists rely on concordance analyses to assess the morphological, syntactic and semantic behavior of one or more keywords that are selected on the basis of their frequency of occurrence. The relevance of corpus-based studies for terminology can be simply illustrated by reference to neologisms and archaisms (quantifiable through frequency measures of new terms and obsolete terms respectively) and compound terms, the staple of any terminology enterprise (accessible, for instance, through concordance searches, statistical tests of collocational patterns, or the identification of typical patterns characterizing possible forms).

3 Special-Language Texts and Corpora

sense and structure

The word-based or semasiological methods employed by lexicographers have an obvious resonance with a corpus-linguistic approach to the study of language and it is now almost a commonplace that electronic corpora provide both a high quantity and a high quality of evidence to the general-purpose lexicographer in the task of identifying and distinguishing between polysemous words by exploring, for instance, the interaction between sense and structure (see, for instance, Sinclair 1987a; Rundell and Stock 1992a; 1992b; Svartvik 1992). But similar observations can also be made in principle regarding special-language texts. Halliday and Martin, for instance, have remarked that technical terms are used to compact information on the content "plane" and have suggested that such "compacting" can be related to linguistic abstraction (1993:29).

scientific discourse

Specialist knowledge, particularly scientific knowledge, is communicated by specialists through natural-language texts. According to Halliday and Martin (1993:4), scientific English, for example, "is English with special probabilities attached: a form of English in which certain words, and more significantly certain grammatical constructions, stand out as more highly favored, while others correspondingly recede and become less highly favored, than in other varieties of English". Halliday notes that scientists use certain "grammatical resources" in their own language "to create a discourse that moves forward by logical and coherent steps, each building on what has gone before" (Halliday 1993a:64). These grammatical resources include "nominal elements" and "verbal elements". Nominal elements are used in technical taxonomies and for backgrounding and foregrounding information. Verbal elements relate nominalized processes to each other and to the way in which we present nominalized processes and interpret them (Halliday 1993a:54-64). A high regard for the importance of scientific discourse is also shown by Frawley, who makes the strong claim that science *is* primarily discourse: "science is a sign system, a method of creating representations of the world and of institutionalizing these representations into coherent systems of extended talk: science is discourse" (Frawley 1986:68).

special-language restrictions

The special-language texts of particular domains also share certain common features particularly at the level of grammar in terms of a restricted inventory of structures vis-à-vis general language, as well as a relatively high degree of nominalization. The most obvious difference between special languages occurs at the lexical level in the particular realization of aspects of domain knowledge as "terms", which we understand here simply as domain-specific vocabulary. This vocabulary is highly productive, drawing on the inventory of words in general language, as well as on the terminologies of other domains. In addition, the lexical inventories of special languages also contain coinages, lexical items not found in either general language or other special languages. These are the most clearly identifiable as terms.

lexical density

The quantitative and qualitative characteristics of natural language in specialist writing have been known for some time. Svartvik (1966), for instance, noted that the passive voice is used much more frequently in English scientific texts than in a comparable general-language corpus. The analyses of the early corpus linguists carried out on the Brown Corpus of American English (Francis and Kucera 1982), and the British English equivalent of the Brown Corpus, the Lancaster/Oslo-Bergen Corpus (Hofland and Johansson 1982; Johansson and Hofland 1989), show statistically significant variations in the use of certain classes of words between popular science articles (the only scientific texts in these corpora) and other texts including press reportage, editorials, and varieties of fiction. Halliday has noted that lexical density, as measured by the proportion of open-class words, is much lower in general-language texts than in special-language texts: he suggests that lexical density in the former varies between 20%-30%, whereas in the latter it can be as high as 50%-60% (Halliday 1993b:76).

recurrent patterns

The above suggests that not only may potentially useful patterns of various kinds be discerned in special-language text, but that these patterns are characteristic of certain classifications of special-language text in particular. In fact, the study of "recurrent patterns" in text has a tradition going back many decades outside the domain of corpus linguistics, covering both general-language and special-language texts. The exploitation of such patterns of words to examine the properties of text was pioneered by the psycholinguist Zipf (1935) and was later taken up by the mathematical linguist Harris (1982) with the purpose of demonstrating the important contribution of these patterns to efficient knowledge and information processing and management. Modern term-extraction methods exploit these statistical patterns, as well as linguistic patterns of term formation.

special language corpora

As we have seen, machine-readable corpora of special-language texts are not readily available: they usually need to be compiled, whereas general-language corpora are becoming increasingly available off-the-shelf. The purpose of terminology management is usually much more closely defined than that of general-language lexicography, thereby potentially restricting opportunities for re-use relative to general-language corpora. Since terminologies are domain-specific, terminologists will generally need to build a new corpus each time they embark on building a new terminology. By contrast, general-language lexicographers may re-use the same very large corpus: the 100 million-word BNC, for instance, was designed to characterize the state of contemporary British English in its various social and generic uses (Aston and Burnard 1998). The target user group of a general-purpose dictionary is also likely to be much broader and larger than that for a terminology, a fact that is likely to have financial implications for commercial publishers and organizations undertaking terminology projects.

corpus acquisition

In order to build a corpus, selected texts need to be converted into machine-readable form, if not already available in this form. Methods have changed considerably in recent years. Even relatively recently, Burnard (1992:3-5), for instance, described two methods for converting text into machine-readable form: optical scanning and keying. Keying is now no longer used on any large scale and scanning, while feasible in particular languages for which fewer documents are electronically available, is still labor-intensive, since scanned texts need to be proof-read, once any problems with print quality, character sets, layout and graphics have been overcome. Texts may then need to undergo further treatment such as reformatting, encoding or decoding, depending on their original and present purpose. Processing may involve removing or inserting metatextual features such as formatting commands or mark-up codes (e.g., indicating textual divisions or hierarchies) or tagging codes (e.g., indicating word-class membership). The most common method of collecting electronic text is now the re-use of text already in machine-readable form, as organizations are able to exploit electronic archives of material and as text becomes more readily available on the Internet, sometimes free of charge, through, for instance, the publication of electronic journals, traditional journals reproduced in electronic form, academic papers on university websites, company and institutional Web sites, promotional materials, and so on. However, the requirement to get copyright clearance to store and process texts that are not in the public domain and to use the data that is gleaned from those texts in terminologies must be treated seriously. Publishers can also be approached for special-language texts, but they may be reluctant to comply particularly if the intended terminological activity has a commercial purpose.

representativeness

The developers of computer-based corpora of both general-language and special-language corpora have to contend with the still open question of *representativeness*. General-language corpus builders select their texts either deliberately as in the case of the Brown University and Lancaster Oslo/Bergen Corpora of English, or at random from publishers' catalogues like the Birmingham Bank of English and the Lancaster Longman Corpus. The aim in both cases is to obtain samples of exemplar texts from a set that might statistically be referred to as a *universe of texts*. Biber (1993:243) has defined representativeness in corpus linguistics as the extent to which a sample text includes the full range of variability in a population. Leech has a similar view about the notion of a representative corpus: "A corpus is representative to the extent that findings based on its contents can be generalized to a larger hypothetical corpus", but expresses some caution in stating that the "assumption of representativeness must be regarded largely as an act of faith" (1991:27, footnote). For Summers (1993) (Lancaster Longman Corpus), representativeness may be in some sense related to frequency criteria.

text attributes

Representativeness can also be described in terms of a number of so-called text attributes, including how a text was produced and was (intended to be) received, how the authors organized the text in terms of their own goals, when and where, and what language(s) were used in the production of the texts. These four major attributes of a text, i.e., production, reception, author attributes, and language attributes, have been described by Atkins et al. (1992) as the key to the representative design of a corpus in terms of its balance. The balance of a special-language corpus is perhaps more easily judged than that of a general-language corpus, especially where the corpus is being compiled with a particular purpose in mind, such as a bilingual terminology of earth-moving machinery for the translation of marketing materials, or a monolingual terminology of powered industrial trucks for the writers of safety manuals.

unbiased sample

Given that any frequency-based approach relies on the availability of an unbiased sample of a much larger, possibly infinite, population, corpus linguists have debated the issues of optimal size and composition of a general-language corpus that can be used in the compilation of dictionaries. Indeed, the use of corpora for studying language relies on the existence of an unbiased sample. Corpus-based studies also tend to focus on consensual evidence and reference, that is evidence and reference that is used by a large number of the users of a language. The larger the size of a general-language corpus, containing texts of as many different genres as possible, the less biased the corpus is likely to be. General-language usage is highly personalized and it is difficult to argue for or against a standard variety of general language. In addition to the quest for minimal bias, corpus builders at the turn of the 20th century have no longer been constrained by the availability of computer memories and processing power.

corpus size

What is the relevance of the debate on optimal size and composition to corpus-based terminology? It can be argued that, like lexicographers and others interested in language, terminologists should also look for an unbiased sample of texts, size being one indication of this. Empirical studies suggest that the vocabulary used in special-language texts is much smaller than the vocabulary in general-language texts (Alford 1971). The inference that can be drawn is that a smaller corpus (e.g., ca.100,000 words) would be a good starting point for corpus-based terminology management in a highly-specialized discipline. Specialist corpora of 1 million words would be much more challenging to build, since fewer texts are produced relative to the general language. There are a number of arguments for justifying smaller corpora for terminology management:

smaller corpora

- The status of special-language "items" (including terms) may be referred to more or less authoritative sources (e.g., domain experts), whereas in general language the notion of authority is more problematic.
- There tends to be less lexical and grammatical variation in special-language texts, so that patterns of language may emerge more clearly from smaller samples in these semantically-restricted "sublanguages".
- The notion of the "subject field" or domain, while not well defined, is more constrained than the rather ill-defined notion of, say, "modern British English" (Clear 1993:168).

- Special-language texts tend to be lexically denser than general-language texts, as we have seen, and it is the open-class words and phrases that are of interest as potential terms.
- Terminologists are not concerned with all the words in special-language texts, only those that are domain-specific, i.e., the terms of the domain. Other words are considered only in so far as they are illustrative of what we might broadly call phraseological patterns.
- The specific purpose of the terminology that is being compiled may naturally restrict the text type or even the genre selected for the corpus.

As a rule of thumb, special-language corpora already start to become useful for key terms of the domain in the tens of thousands of words, rather than the millions of words required for general-language lexicography.

corpus design

Any corpus that is used for terminological work needs to be organized so that the data gleaned from the texts may be appropriately interpreted by the terminologist, as we have indicated. Information used to structure or organize a corpus should then be reflected in the data recorded in the terminology database being compiled, so that the user is guided to redeploy the terms in an accurate and appropriate way when writing new texts or interpreting existing texts. However, less attention has so far been paid to corpus design in the context of special languages than in that of general language. Design criteria for general-language corpora have been discussed in Renouf (1987) and Quirk (1992), and are in many cases relevant to the design of special-language corpora. Relevant criteria for special-language corpora include: text type (i.e., German *Texttyp* and also *Textsorte*, or genre), domain, subdomain, language, regional variety, original or translated text, spoken or written, date, and so on. Structuring the corpus according to such criteria allows the texts to be "viewed" and processed in a targeted way. For instance, to create a monolingual English terminology of information technology for non-native domain experts aiming to work in the USA, the selection may be: original texts from learned journals and text books on information technology in American English. A bilingual terminology of artificial intelligence in German and English for translators might possibly draw on an overlapping but different set of texts from the same corpus, namely: German and English (American and British English) texts from learned journals, text books, and journalistic sources in original and translated versions including only those information technology texts concerned with artificial intelligence. The selection of a subset of texts from a given corpus leads to a "sub-corpus" that may be organized according to national variant, genre, the publication date of a text, the first language of the author(s), and so on. These possibilities may be regarded as virtual organizations in that a user of a corpus looks at the texts, which actually exist, and then reorganizes the corpus according to his or her criteria. The distinction between physical organization and virtual organization is used to good effect in computing science, and can also be used to good effect in corpus-based terminology. Computer systems for term extraction from text exist to help the user create a corpus organization of his or her choice. Such systems are called *virtual corpus managers* as distinct from conventional text-corpus organization systems that provide the user with pre-determined ways for organizing corpora (see for instance, Holmes-Higgin *et al.* 1994).

subsetting

In addition to selecting a subset of texts from a larger corpus, the terminologist may also wish to view what is already stored, and to add or delete texts. For terminological purposes, the user should be able to view the corpus at a number of different levels, ranging from inspecting lists of text files in a chosen set (e.g., what texts are in the set: German learned journals on four wheel drive systems?), retrieving selected information about a particular text (e.g., author, page numbers, word count, publisher), or inspecting the contents of those files (view the text itself).

text attributes

Adding new texts to the corpus is an important part of corpus management. Texts can be viewed as "objects" that have a certain number of attributes, as we have seen. One possible classification would be: text-oriented attributes, author-oriented attributes, publisher-oriented attributes, domain-oriented attributes, and language-oriented attributes. For terminological purposes, for instance, text type or genre is an important text-oriented attribute, the values of the attribute being defined as, say: advertisement, text book, learned journal article, manual, reportage, official publication such as standard or patent, and so on. Data concerning the values of various attributes can be stored in text headers, which may then be used as the basis of user-driven retrieval for specific terminological purposes. The classification of texts according to these attributes may be straightforward in some cases but problematic in others, e.g., where no indication is given whether a text has been translated or whether the authorship is corporate. The classification of texts according to the attributes proposed can not only facilitate the re-use of the corpus through the definition of sub-corpora, but is also an important means of presenting the terminologist with output data that is systematic and therefore more easily interpretable. Such classifications help to identify variations that are characteristics of different users and situations, as illustrated in Table 1.

Table 1: Examples of terminological variation according to selected text attributes

attribute	**variation**	**example**
text type (genre)	vertical synonyms (according to communicative level)	*catalytic converter – cat*
date	archaisms – current use	*bioscope – cinema*
origin	geographical variants	*silencer – muffler*

lexical ambiguity

Once a corpus has been built, a decision has to be made whether to tag or not. It has been estimated (Lauriston 1994: 150) that 40 percent of English words in running text and between 30-40 percent in French may be "lexically ambiguous", meaning that without tagging to disambiguate homographs into different word classes, the interpretation of output becomes problematic. However, in order to avoid such problems, considerable preparation of the corpus in the form of tagging is needed, i.e., it is necessary to assign grammatical labels to words (part-of-speech or sub-part-of-speech classifications) or, less commonly, syntactic labels (Leech and Fligelstone 1992:124-7). "Automatic taggers" assign morphological classifications to words in running text so that items can be retrieved according *inter alia* to word class. The designation "automatic" refers to the fact that in a restricted way, the taggers learn to tag texts through some form of human instruction, either in the form, for example, of "correctly" tagged text or through a set of rules. For terminological purposes, word tags are particularly useful since the great majority of terms are nominal. Hence, simply looking at all the noun phrases in a lexically-dense special-language text is likely to be a fruitful exercise. However, tagging is a labor-intensive task even at the word level, despite the increasing availability of programs that tag automatically, since each tagged text eventually requires some form of post-editing. Some of these programs are available free of charge on the World Wide Web, e.g., Eric Brill's tagger for English (Brill 1994b). Additionally, further factors need to be taken into account for terminology management such as the availability of taggers for languages other than English. The Xerox Research Centre at Grenoble and Cambridge (UK) now provide a Web-based service for tagging texts in Dutch, French, German, Italian, Portuguese and Spanish, in addition to English (Xerox 1999).

special language structures

Another factor when considering the use of a tagger is the nature of special-language structures, particularly at phrase level, which may cause problems for taggers trained on general-language text. In English, for instance, complex nominals are much more frequent in special-language than in general-language texts, e.g., *ethnic group organization* (sociology), *balance of payments adjustment process* (economics). Furthermore, these nominal expressions may also contain non-linguistic elements including numbers and punctuation or mix full words and abbreviations, e.g., *3-(4-carboxybenzyl)-2-chloro-5-(1,2-dimethoxyethyl)terephthalic acid* (chemistry); *exchange rate OM* (economics). Even with general-language corpora, the error rate for taggers may be as high as 10% (Clear 1993:182). The need to build a new domain-specific corpus for each new terminology also reduces the practicability of using tagged corpora in a working rather than a research environment, as does the need to work in a range of languages where multilingual terminology management is involved. The arguments for or against tagging are therefore balanced between interpretability and accuracy on the one hand (homograph disambiguation and string identification) and practicability on the other (availability of automatic taggers for particular languages, as well as of human and financial resources for training and proofing).

4 Extracting Terms from Text

machine scanning

Within the overall framework of a corpus-based approach to the compilation of terminologies, two considerations stand out when considering the use of a computer to support human activity. Firstly, reading a text to extract terminological data ("manual" scanning) is labor-intensive and potentially repetitive owing to the need to capture different types of terminological data (not only the terms, but also grammatical information, contextual examples, collocational patterns, and so on). By contrast, machine processing is unlimitedly iterative and consistent within its own limits. Secondly, the processing of relatively large quantities of text may allow patterns to emerge that are probabilistic in nature and are unlikely to be detected by manual scanning. It also allows relatively infrequently-occurring lexical units to be included in the analysis. Sceptics may argue that a computer is unable to identify and distinguish relevant data in running text, let alone interpret it. But all term-extraction and retrieval systems are still in some sense support systems that offer proposals to the terminologist who must then decide how to proceed, often in consultation with a domain expert.

terminology management life cycle

The management of terminology spans a range of interdependent phases, including: organizing documentation and capturing data from running text (*acquisition* and *elaboration*); storing, editing, maintaining and updating data using various data structures (*representation*); and finally, "publication" (*dissemination*). All these phases may be carried out electronically, given the appropriate tools to provide the terminologist with various levels of support. We are primarily concerned here with the acquisition and elaboration phases of the "life-cycle", both of which are directly concerned with the use of an electronic corpus during terminology management. The corpus may be used to capture data (e.g., to identify term candidates), to validate data (e.g., to provide further evidence for or about a term candidate), and to elaborate data (e.g., to identify relations of synonymy, or to assist in the compilation of definitions or in the selection of contextual examples).

corpus analysis tools

Tools for analyzing text for lexicographic purposes are now relatively widely reported in the "electronic text" literature (e.g., Burnard 1992; Meijs 1992; Sinclair 1992; Clear 1993; Stigelman

1993). However, since corpus analysis is a relatively new field, the availability and quality of the tools is sometimes variable. Any list of such tools will have to be updated regularly and for this reason alone, the World Wide Web provides the most up-to-date information (Barlow 1999). Barnbrook (1996) and Ooi (1998) have also written practical introductions to corpus-based lexicography and the computer analysis of language, providing information on programming and references to programs used in the analysis of texts. The tagged BNC, for instance, can be analyzed using *SARA*, a low-cost program that is available on the WWW (Aston and Burnard 1998). Given the appropriate tools, a wide range of data can be "captured" from running text that are of relevance to terminology management. We can classify these data broadly into three types, expressed here as questions: What are the potential term "mothers", i.e., those terms that give birth to other terms? Who are their neighbors? And who belongs to their family?

4.1 Identifying Mother Terms

compounding & derivation

Many words have the potential to combine with others to form compounds and to produce derivational forms, but some words are particularly productive in this respect, not only within a given domain but sometimes also across domains. Jumping across domains, these words may retain some of their meaning through metaphorical extension as "simple" terms (i.e., a single word with one base form) or as components of compound terms (i.e., depending on the language, one or more words with two or more base forms) or as derivations (e.g., by the addition of a prefix or a suffix). The term *virus,* for example, has been borrowed from microbiology into information technology, and has its own compounding patterns in each domain: *African Green Monkey Virus, human parvovirus B19, Fowl adenovirus* (microbiology), as well as *virus program, virus protection,* and *anti-virus detection* (IT). Through suffixation the noun *virus* may also become adjectival, *viral.*

stop lists

One of the most straightforward operations in processing a corpus is a simple frequency count, in which each token (word form) in the corpus is listed according to its frequency of occurrence ("noise" from closed-class words can be filtered out using

Results							
particles	171	nitrogen	72	theory	43	velocity	36
atoms	165	mass	67	experiments	43	positive	32
atom	138	scattering	62	elements	41	radioactive	31
nucleus	138	radium	56	scintillations	41	deflexion	31
particle	117	atomic	54	cm	40	electron	29
hydrogen	88	electrons	45	energy	39	nuclei	29
rays	80	field	43	helium	37	collision	29
charge	76						

Figure 1: *Word list from selected texts in nuclear physics sorted according to frequency using an exclusion lexicon of closed-class words*

exclude lists or stop lists). Frequency lists can reveal single-word term candidates and potential mothers, as shown in Figure 1.

compounding patterns

The first 100 most frequent open-class words for a whole range of subjects comprise a large proportion of the "mother" terms of a given specialism or the signature terms of a specialist subject (see also Sager, Dungworth and MacDonald [1980:236-8] for the use of "frequency counts for subject identification"). Such terms nurture more specialized meaning within a specialism whilst retaining their original semantic orientation either in general language or in another specialism. They often carry with them, like mothers, other terms and are used less frequently on their own. Thus we can have *atomic nucleus*, *unstable nucleus*, *compound nucleus* and *halo nucleus* in nuclear physics, but will seldom find *nucleus* on its own. Compounding patterns vary, however, between particular languages: compound terms may be multiword, e.g., *hereditary disorder*, *immunologic disorder*, or single word, e.g., *Erbkrankheit*, *Immunschwächekrankheit*. Frequency distribution patterns of single words will vary accordingly, meaning that mother terms may be harder to isolate in languages like German, but that term boundaries are more easily established than for multiword compounds.

text-analysis software

Often, simple processing techniques are sufficient to indicate possible candidates for mother terms, as we have indicated. Text-analysis software, while exhibiting a range of functionality, can usually carry out at least some of the following basic tasks in order to help to identify candidate terms or components thereof:

- List all the word forms in a text, sometimes with source references, including line numbers, giving frequency distributions.
- List selected items in a concordance showing the immediate context to the left and right of the concorded item and giving frequency distributions, often with an option for changing sorting priorities.
- List lemmatized—or canonical—forms grouping together the morphological variants of a word under a single lemma.
- List tagged forms, giving frequency distributions.

co-occurrence

Each of these outputs may usually be sorted alphabetically or according to frequency with various options for dealing with orthography (hyphenation, punctuation, etc.). Frequency distributions of word forms or larger units help to identify term candidates or parts of potential multiword terms in the first instance. It may also be possible to perform statistical calculations to compute relative frequencies that can be used when comparing statistical outputs from several sources, or even to compute probabilities of co-occurrence for establishing collocational patterns. "Collocations" may be simply understood here as words that regularly occur together in running text. Such collocations often turn out to be compound terms.

lemmatization

The "lemmatization" of word forms, allowing morphologically-variant forms of the same lexeme to be grouped together, can assist in the identification of patterns so that forms showing number and case (nouns, adjectives), gender, number and case (adjectives), and person, number and tense (verbs) are reduced to their canonical form. Since the form that is found as the entry term in a terminology is not a word form, but a lexeme, and it is this element that forms the basis of the whole terminological entry as a kind of abstract label for all possible morphological variants, the terminologist may therefore wish to group together all the data on a particular lexeme through morphological parsing or "stemming" (Lauriston 1994), or in some cases, through simple pattern matching (e.g., grouping *tumor* and *tumors* under *tumor*). Lemmatization helps to establish the frequency (and hence viability) of candidate terms on the system level but it is less useful for term elaboration purposes, where more detailed context-sensitive linguistic information may be needed to provide guidance on the use of the term.

frequency distribution

The definition of terminological patterns according to absolute frequency in running text (even if the lemmatized form is considered) is a useful but unsophisticated technique. Another possible statistically-based technique with untagged corpora is the comparison of the frequency distribution of items in special-language and general-language texts. This technique has the advantage of being language-independent, once the general-language corpus—or even a frequency list—has been obtained. Closed-class words will tend to have low ratios of around 1:1 (represented as "1" in Figure 2 below), whereas terms will have a much higher ratio since their frequency in general-language texts will be low or nil. Computing the "ratio" of word forms in special-language and general-language texts also allows a provisional distinction to be made between general-language open-class words on the one hand, and special-language open-class words on the other, i.e., term candidates. For instance, Figure 2 compares the relative frequency of 11 keywords used in three major 20th century subjects: *agreement, phrase* and *sentence* in linguistics, *nucleus*, *shell* and *interaction* in nuclear physics and *engine*, *function* and *substrate* in automotive engineering. In addition, *structure* and *system* as frequently-used terms in all the three sciences are included. The frequency of the 11 tokens was compared with their frequency in the British National Corpus.

weirdness

The most frequent of the 11 tokens in the BNC are *system*, followed by *structure* and *agreement*. In the automotive engineering corpus, the term *engine* is used nearly 100 times more frequently than in the BNC, and the term *substrate* is used 55 times more frequently, indicating their "weirdness" in relation to general-language distributions and hence their potential as term candidates. Neologisms can be detected by this method, since the weirdness ratio for such terms and variants of terms is infinity, their occurrence in the general language being zero (n/0 = infinity).

term candidates

The particular "ratio" level at which term candidates can be optimally identified needs to be further explored in relation to a number of factors, including the type of general-language corpus used as a comparison and the type of domain covered by the special language. For instance, the use of a contemporaneous

	British National Corpus		Surrey Automotive Engineering (AE)		Weirdness Ratio
Total Tokens (N)	*100106029*		*326621*		
Token	f_{BNC}/N_{BNC}	f_{BNC}	$f_{Auto. Eng}/N_{Auto Eng.}$	$f_{Auto Eng.}$	$f/N_{Auto. Eng}/f/N_{BNC}$
agreement	**0.01327%**	*13287*	0.02113%	69	1.59
engine	0.00501%	*5012*	**0.48221%**	*1575*	**96.31**
function	0.00865%	*8663*	0.04746%	*155*	5.48
interaction	0.00233%	*2334*	0.00429%	*14*	1.84
nucleus	0.00067%	*670*	0.00000%	*0*	0.00
phrase	0.00310%	*3099*	0.00000%	*0*	0.00
sentence	0.00579%	*5796*	0.00000%	*0*	0.00
shell	0.00224%	*2241*	0.00765%	*25*	3.42
structure	**0.01378%**	*13791*	0.01806%	*59*	1.31
substrate	0.00032%	*323*	0.01776%	*58*	**55.04**
system	**0.04463%**	*44674*	**0.39863%**	*1302*	**8.93**
Total	**0.09978%**	***99890***	**0.99718%**	**3257**	

Figure 2: *Comparison of distribution (expressed as relative frequencies) of selected keywords in a general-language corpus (BNC) and a special-language corpus showing "weirdness ratio" (relative frequency of token in special-language corpus/relative frequency of token in general-language corpus)*

corpus may produce less striking distinctions than comparisons with LOB (1961), since some terms tend to filter into general-language use, and a productive metaphorical use of general-language words as the basis for term creation, as in information technology (*mouse*, *window*, *file*, *drag*, *point*, and so on), is likely to reduce the efficacy of the technique. The possibility to select a certain ratio level is an advantage since it enables the system to select automatically those open-class words that are of a certain frequency level in the special-language corpus. In other words, the system selects the items on the basis of a computation, not a tag.

SEE

Term Formation 1.2.1:25

4.2 Identifying Compound Terms

noun phrase structures

There are many different methods for identifying compound terms, which comprise the majority of special language terms in running text. Some linguistically-based—and hence language-dependent—methods of extracting compound terms rely on identifying typical noun phrase structures and morphological patterns, long recognized in traditional terminology management in the context of term formation (e.g., Picht and Draskau 1985: 108-110 for English, German, and Spanish) but only more recently applied for purposes of term extraction (e.g., Drouin 1997:50 for French). Linguistic rules may be supplemented by a statistical measure to assess whether the occurrence of such phrases is statistically significant. Other methods may start from some kind of statistical analysis before filtering the output using linguistic rules or patterns. Some work goes a step further and attempts to conduct a semantic analysis by linking various terms. Other methods look at the typical linguistic/graphetic contexts in which nominal compounds typically occur. Most methods are hybrid, usually involving at least two phases (Drouin 1997).

keyword-in-context

One very simple method is to use mother terms in keyword-in-context (KWIC) searches using as keywords mother terms that have been identified using frequency-based techniques. A KWIC is a concordance search that is focused on particular words or sets of words. In English, sorting the output alphabetically to the left shows up modifier-head patterns where the mother term is the head, e.g., *cardiovascular disease*, *Creutzfeldt-Jacob disease*. Sorting to the right shows up modifier-head patterns where the mother term is the modifier, e.g., *disease profile*, *disease progression* (see Figure 3). The direction of modifier-head relations varies, of course, according to language, and in some languages a large number of compounds may be formed with prepositions, e.g., *base de données*, *table à numériser*, all of which are identifiable using a simple language-independent KWIC tool. However, this method tends to overgenerate, producing "false positives" as well as "true positives" (see Figure 5 below). In a mammography corpus, for instance, true positives would be combinations such as *benign lesion*, *malignant disease*, but false positives such as *screening women* also occur.

cardiovascular	*disease*	
Creutzfeld-Jacob	*disease*	
	disease	*profile*
	disease	*progression*
National Centers for	*disease*	*Control*

Figure 3: *Sample keyword-in-context*

text alignment tools

The work of Church in text alignment is well known and has been recently used for extracting terminology from parallel corpora. Dagan and Church have developed a system called *Termight* for identifying and translating terminology (1994). The authors produce concordances of candidate terms, both multiword noun phrases and single words, from a text corpus. The corpus is already tagged and, hence, the primary input to Termight is tagged text. The authors claim that Termight can help a terminologist to produce 150 to 200 terms per hour, doubling the rate of constructing terminology lists as compared to human efforts that may be supplemented by simpler lexicographical tools. The authors have extended their work to multilingual terminology and have used text-alignment techniques to relate candidate terms in a source text to a target text. Recently, Bourigault and Habert have suggested how automatic text-alignment techniques developed in machine translation and in multilingual corpus linguistics can be used for extracting target-language equivalent terms in parallel texts (see e.g., Bourigault, Gonzalez-Mullier, and Gros 1996 for further references).

rule-based methods

SEE *Information Retrieval 6.3:426*

Various rule-based as opposed to contextual methods to identify compound terms have been developed by computational linguists, especially those interested in complex noun phrases, and those interested in applying information theory, based on the work of Claude Shannon. Voutilainen (1993), for instance, has developed a system called *NPtool* for extracting terms from unrestricted text. This extraction is based on developments in English morphology and theory of constraint grammars for extracting noun phrases with complex structures. These noun phrases are presented as candidate terms, together with evidence of their frequency. Typical statistics used in information retrieval, "recall" and "precision" related to the term candidates, are presented as evidence for the existence of a compound term. "Recall" is the proportion of relevant materials retrieved from a text collection given a set of terms. "Precision" is the proportion of retrieved materials that are relevant. The methodology suggested by Voutilainen has been successfully exploited commercially and NPtool is marketed as a language engineering product (Ahmad *et al.* 1996 contains a description of NPtool, a system developed by LINGSOFT, Finland).

retrieval statistics

The use of information retrieval statistics in identifying compound terms was initially suggested by Justeson and Katz (1995) drawing on their experience in speech processing. More recently, Turney (1997) has produced an empirical evaluation for automatically extracting keywords and keyphrases from documents, again using recall and precision statistics as metrices for evaluating the performance of four text summarization programs: Microsoft Word 97's *AutoSummarize Feature*, a part-of-speech tagger by Eric Brill (Brill 1994b), Verity Inc.'s *Search 97*, and the National Research Council of Canada's *Term Extractor*.

co-occurrence

INFO

Headwords INFB 6:330

Smajda (1994) has proposed a statistically-based method for extracting terms that involves computing the strength of association—or collocation—between words in the same "neighborhood". In this method the neighborhood of a nucleating single word indicates what kinds of other words are "attracted" to it, binding with the nucleate. Collocates of interest in terminology are primarily the contiguous ones, that is the nearest neighbor collocates, like *halo+nucleus* or *neutron+halo* or two+neutron +halo, but there are also instances where the nucleate will co-occur with its potential neighbor in a looser structure: *the **halo** around the **nucleus***. The notion of neighborhood distance is crucial in determining whether the neighbors of a nucleate happen to be neighbors by chance or whether they are collocations, i.e., co-occurring at a probability greater than chance. Smajda's method is based on the computation of the *strength* related to z-score computation, *spread* related to the strengths of collocation between the nucleates and the neighbors, and an idiosyncratic *threshold* computation.

SEE

Collocations/ Sub-Language Texts 8.4.4:788

morpho-syntactic features

Another approach is syntax-based: Heid has focused on morpho-syntactic analysis, which includes the identification of grammatical categories and morpho-syntactic features of single- and multiword terms, particularly in German. Similar to

Bourigault (see 4.3), Heid also uses a part-of-speech tagger for disambiguating morpho-syntactic analysis hypotheses. This disambiguation process, together with the morpho-syntactic analysis, helps Heid *et al.* (1996) to put together a set of candidate terms.

4.3 Elaborating Terms

semantic relations

Since terminology management and terminology research emphasizes the conceptual organization of subject fields, the semantic relations between terms assume considerable importance. According to the traditional terminological view, the knowledge of the domain is represented by concepts and the relations between them. However, these relations cannot be directly accessed but must be conveyed by largely linguistic means. The relations between terms, as labels for concepts are therefore a means of accessing this knowledge (and its structure) through text. Since the interpretation of meaning is still a major problem for machines, attempts to extract lexical meaning relations from running text is largely form-focused. The paradigmatic relationships of terms to other terms include synonymy, antonymy, hyponymy (subordinate), hyperonymy (superordinate), meronymy (part-of).

synonym cues

Dealing as we are here with sense relations, it is not surprising that generally less progress has been made in automating this area than in the former two tasks (i.e., 4.1 and 4.2), where quantitative and formal linguistic measures can be more easily exploited. Nevertheless, there are certain formal clues that may help to identify some of these relations. Synonyms may be marked in text in a number of ways that may be easily detected by a simple pattern-matching routine. For instance, a term followed by an open parenthesis may contain an alternative means of expressing the concept: *one of the assumptions that underlies the proper application of the* ***Pearson product-moment correlation coefficient (r)*** *is that* .... Information in parentheses may also provide support for definitions and explanations: *look at the* ***correlation*** *of scores* ***(how they go together)***. Certain linguistic cues may point to particular relations, also contributing to the definition or explanation (in English: *is a*; *is composed of*; *consists of*; *has a*; *a kind of*). For instance: *An assumption* ***is a*** *precondition that must be met for the particular statistical analysis to be applied.* While such textual cues are language-dependent, the principles are valid across languages.

shared patterns

The use of collocations and compounds to assist in the acquisition of conceptual information from texts has been proposed by Meyer and Mackintosh (1996b). They indicate that shared patterns may, for example, be indicative of synonymy (e.g., in the domain of presentation software, *gallery of charts/library of charts, gallery of clip art/library of clip art*), whereas differing patterns may indicate polysemy (e.g., *processes are born, processes migrate* in computing, but not in general language). Furthermore, compounds may be indicative of generic-specific (hyponymic) relations (e.g., *pre-emptive multitasking operating system* is a kind of *multitasking operating system* is a kind of *operating system*). Meyer and Mackintosh's discussion of a range of ways in which what they somewhat unusually call "phrasemes"—collocations and compounds—may assist in the identification of relations between terms (and hence concepts) takes place in the context of manual scanning, but the animation of such observations is easily conceivable (see for instance, Daille et al. 1996).

candidate terms

Terminology extraction as a basis for conceptual networks has also been discussed in the context of developments in automatic text tagging. Bourigault (1995), for instance, describes how a parser is used to identify complex noun phrases in tagged text, before the head of the phrase and the accompanying "expansion" component are identified. The noun phrases that are extracted from the corpus are considered as candidate terms (e.g., *automatic knowledge acquisition*) but may also contain subgroups that are in themselves candidate terms (e.g., *knowledge acquisition*). Bourigault's parsing module can break up the complex noun phrase into the head (*knowledge acquisition*) and the expansion (*automatic*). The head-position phrase can in this case be further analyzed into its own head (*acquisition*) and expansion (*knowledge*). The terminology can then be structured in a network that relates heads to expansion links. The network actually presents candidate terms that can be viewed and selected for terminology databases and knowledge-based systems. Assadi has extended this work and described not only how terms can be extracted from texts but also how an ontology of the subject domain can be derived using a variant of Bourigault's head/expansion linked network (Assadi 1998).

KWIC concordances

Other aspects of term elaboration include the identification of contextual examples and the construction of definitions. KWIC searches can be used in both the identification of contextual examples and, in some cases, of partial definitions. Figure 4 shows the selected results of a concordance for the keyword *catalytic converter*. The number of words on either side of the concorded word may normally be specified by the user.

4.4 Evaluation

logical possibilities

The evaluation of extraction systems is not yet well developed, although work is beginning to emerge (e.g., Lauriston 1994; L'Homme *et al.* 1996). A simple way of viewing the effectiveness of an extraction system is to measure its performance against the four logical possibilities set out in Figure 4 showing the correct identification of terms (cell A), the correct omission of non-terms (B), the incorrect omission of a term (C), and the incorrect identification of a non-term as a term (D). Cell D also covers cases of partial hits such as *acquired genetic* for *acquired genetic disease*.

of ***noble metals*** in a	**catalytic converter**	is approximately 2 to 3
a	**catalytic converter**	***consists*** of a ceramic honeycomb
a	**catalytic converter**	***is not*** a filter but
a	**catalytic converter**	***looks like*** a cylindrical bulge
combustion of fuel in the	**catalytic converter,**	both substrate and soldered
thermal shock load , the	**catalytic converter**	was ***sprayed*** with water in
heat in the	**catalytic converter**	is *preserved* over a longer

Figure 4: *A concordance (KWIC) of* catalytic converter *indicating composition and function extracted from Surrey's Automotive Engineering Corpus (132 texts, comprising 300,000 words of British and American English)*

	correct	incorrect
identified	term A	non-term (noise) D
omitted	B non-term	C term (silence)

Figure 5: *A schematic view of the four logical possibilities for "hits" and "misses" in term extraction*

false results

"False positives" (D) produce noise but can be filtered out by human intervention or possibly by linguistic rules, depending on the extraction system and the structure of the false positive. A potentially greater problem is posed by "false negatives" (C), i.e, failure to identify terms.

5 Some Known Problems of Identifying Terms

term boundaries

Some of the known problems in term extraction are shared with manual scanning, in which the human terminologist must also, for instance, make decisions concerning term boundaries and relevance to the domain (see Kageura and Umino 1996 for a discussion of what they call "unithood" and "termhood"). In moving from running text, i.e., word forms with graphical form, to terms that may be entered in a termbase as lexemes at the system level, a number of problems arise.

5.1 Polysemy and Terminologization

metaphorical extension

Many words do double duty as words in general language and terms in special languages, e.g., *platform* and *bus,* which are familiar terms in information technology, and as terms in more than one domain, e.g., *morphology* (geology and linguistics). These are both instances of polysemy, but in the former case, also of terminologization or the process of metaphorical extension of a general-language notion to a more precise concept within a special-language domain. The problem for term extraction is to distinguish these different instances. Some indication of the status of a word form may be given by certain formal indicators such as pluralization in special languages, e.g., *cancers*, *fuels*, as well as collocational patterns of various kinds (Meyer and Mackintosh 1996b), and in tagged corpora by typical term structures, although these do not filter out all instances of non-terms.

word class polysemy

As we have seen, many linguistic forms, particularly in a language such as English, may be partially polysemous, belonging to more than one word class, e.g., *test* (noun), *test* (verb); *exhaust* (noun), *exhaust* (verb). In untagged corpora, statistical data on frequency distributions may be skewed by the failure to distinguish between word classes for homographs.

5.2 Ellipsis and Variants

equivalence relations

Human terminologists develop heuristics for interpreting different forms as related to the same concepts, but the problems for machine processing are clearly of a different nature, since term-extraction systems are weak on knowledge (e.g., linguistic and domain knowledge) and strong on counting (e.g., statistical patterns). There are, for instance, a number of areas in which human beings may successfully identify different forms as variants of one another that are in some kind of equivalence or partial equivalence relationship, as well as "disaggregating" contiguous forms that condense two formally-related terms.

omissions

It is well known, for instance, that in using natural language we can omit certain elements that can be recovered from the context, often according to rules linked to language type. Terms are no exception to the rules of ellipsis. So, for instance, instead of a reference to *heart disease and lung disease*, the reference in running text may be to *heart and lung disease*, although *heart and lung disease* is not a term. The problem for both human and automatic identification of terms is to retrieve the deleted element, if appropriate, and to eliminate the conjunction in order to correctly identify the two terms (conjunctions have been identified as a source of problems for some term-extraction software systems, [Lauriston 1994:164]). Distinguishing those cases that represent elided terms from others, e.g., *Problems with the heart and lung disease are known to cause* ... is problematic for a machine, but easier for a human being, who can exploit the surrounding text.

rule-based typology

As we have seen, inflectional variations, e.g., of number and case, may be dealt with by lemmatization, but what has been called "syntactic variation" is more problematic. Involving relations between terms and a range of phrase structures including what could be described as "pre-terms" (*association with chromosome—chromosome association*), as well as various patterns of ellipsis (e.g., *cable or microwave links*), it has been suggested that a detailed rule-based typology of "variations" may form the basis of an enriched terminology through the generation of "base terms" from such "variants" (Daille *et al.* 1996).

orthographic variation

While there are a number of different understandings of the term *variant*, the most common one has to do with orthographic variation, including capitalization (*Automotive Engineering – automotive engineering*), hyphenation (*three-way catalytic converter – three way catalytic converter*), the use of numbers (*3-way catalytic converter*), and abbreviations (*TWC*). Such variations are simpler to deal with (since they largely involve changes in form) than morphological or syntactic variants, and those which concern the order in which elements of a compound appear viz. *candidate term* or *term candidate* (Bowker 1998).

5.3 Term Boundaries

term elements

Whilst the majority of terms are nominal compounds, the number of elements in such compounds is sometimes hard to determine, particularly in languages in which compounds tend to be multiword rather than single-word. The problem is a conceptual one as much as a formal one since domain experts may also disagree about the location of term boundaries. For instance, while a *single-bed oxidation catalyst* is a kind of catalyst in the automotive engineering sense of a device to reduce toxic emissions, is a *closed-loop three way catalyst system* also a kind of catalyst or is it a kind of system? Consistency and concision would indicate that the term *closed-loop three way catalyst* is better motivated than *closed-loop three way catalyst system*. The left-hand boundary may also be problematic: ADJ + N is a common term structure in English, e.g., *harmful emissions* (automotive engineering), but the same structure and even the same adjective may be part of a much looser collocation, e.g., *harmful pollutants*, showing semantic redundancy when viewed in a strict terminological light, but arguably well-motivated in texts directed at non-experts, as the modifier serves as a kind of explanatory descriptor.

terms in text

It would be misleading to suggest that the problems of term extraction as outlined here are easily solved, since the subject matter dealt with, i.e., the behavior of terms in running text of a specialist nature, is complex. In attempting to distinguish between terms and non-terms and to identify the boundaries of terms, we are dealing with continua rather than dichotomies, and so there are always bound to remain areas of uncertainty, even in the human interpretation of such data.

6 Outlook

Our discussion has ranged from the special nature of scientific writing through the design and representativeness of special-language corpora to the frequency of classes of words in both general-language and special-language texts. We have described a range of term-extraction methods, including statistically-, morphologically- and syntactically-based systems, as well as various combinations of these. To date, it would be more accurate to talk of "candidate term extraction" rather than "term extraction", since all known systems require human support. Terminology management faces substantial problems in the decades to come: the proliferation of scientific disciplines, the explosion in the volume of specialist texts, the diversification in the genres of scientific and technical communication. The techniques and methods for the identification and elaboration of terminology will have to match the speed at which such communication is growing. Methods and techniques *based* on corpora—rather than *driven* exclusively by corpora—and informed by corpus linguistics as well as information retrieval techniques, appear particularly relevant to terminology development, especially in view of the availability of text in electronic form, and the growing availability of suitable tools, which are now beginning to move from the research lab into the market place.

REFERENCES

Aarts, Jan and Meijs, Willem, eds. 1990. *Theory and Practice in Corpus Linguistics*. Amsterdam: Rodopi.

Ahmad, Khurshid. 1995. "Pragmatics of Specialist Terms: The Acquisition and Representation of Terminology." *Machine Translation and the Lexicon. Proceedings of the 3rd International EAMT Workshop,* Heidelberg, Germany, (April 1993). Petra Steffens, ed. Berlin: Springer-Verlag: 51-76.

Ahmad, Khurshid; Ogonowski, Antoine; Dauphin, Eva; Sta, Jean-David; and Arppe, Antti. 1996. "Engineering Terminology—A case for a Linguistically-Informed Terminology Database." In: Galinski, Christian; and Schmitz, Klaus-Dirk, eds.: 166-178.

Aijmer, Karin and Altenberg, Bengt, eds. 1991. *English Corpus Linguistics*. London and New York: Longman.

Alford, Mark. 1971. *Computer Assistance in Learning to Read Foreign Languages: An Account of the Work of the Scientific Language Project*. Cambridge, UK: Literary and Linguistic Computing Centre.

Aston, Guy and Burnard, Lou. 1998. *The BNC Handbook: Exploring the British National Corpus*. Edinburgh: Edinburgh University Press.

Atkins, B.T. S.; Clear, Jeremy; and Ostler, Nicholas. 1992. "Corpus Design Criteria." *Linguistic and Literary Computing*. Vol.7/1: 1-16.

Assadi, H. 1998. "Construction of a Regional Ontology from Text and Its Use within a Documentary System." *Frontiers in Artificial Intelligence and Applications*. Vol. 46. Nicola Guarino, ed. Amsterdam: IOS Press.

Barlow, Michael. 1999. "Corpus Linguistics Bibliography." http://www.ruf.rice.edu/~barlow /corpus.html (Accessed 1999-06-16).

Barnbrook, Geoff. 1996. *Language and Computers: A Practical Introduction to the Computer Analysis of Language*. Edinburgh: Edinburgh University Press.

Bazell, Charles E.; Catford, John C.; Halliday, Michael A.K.; and Robins, Robert H., eds. 1966. *In Memory of J. R. Firth*. London: Longman.

Biber, Douglas. 1993. "Representativeness in Corpus Design." *Literary and Linguistic Computing*. Vol. 8/4: 243-257.

Bourigault, Didier. 1995. "LEXTER, a Terminology Extraction Software for Knowledge Acquisition from Texts." *Proceedings of the 9th Banff Knowledge Acquisition for Knowledge-Based Systems,* February 1995. Banff, Canada: University of Calgary.

Bourigault, Didier; Gonzalez-Mullier, Isabelle; and Gros, Cécile. 1996. "LEXTER, a Natural Language Processing Tool for Terminology Extraction." *Proceedings of the 7th Euralex Congress on Lexicography*. Gellerstam, Martin *et al.*, eds. Gothenburg: Goteborg University, Swedish Department, 771-780.

Bowker, Lynn. 1998. "Variant Terminology: Frivolity or Necessity?" *Euralex '98 Proceedings*. Thierry Fontenelle et al., eds. University of Liège: English and Dutch Departments.

Brill, Eric. 1994a. "Some Advances in Transformation-Based Part of Speech Tagging." *American Association of Artificial Intelligence Conference, 1994*. Vol. 1, 722-727.

Brill, Eric. 1994b. *Eric Brill's Code Page: Natural Language Processing Systems*. http://www.cs.jhu.edu/~brill/code.html (Accessed 1999-06-16).

Bühler, Hildegund. 1987. "Terminologieseminar für zukünftige Sprachmittler—das Wiener Modell." *Lebende Sprachen*, Vol. 2: 52-6.

Burnard, Lou. 1992. "Tools and Techniques for Computer-assisted Text Processing." In Christopher S. Butler, ed.: 1-28.

Butler, Christopher S., ed. 1992. *Computers and Written Text*. Oxford and Cambridge, MA: Blackwell.

Cambridge International Dictionary of English. 1995. Cambridge, UK: Cambridge University Press.

Clear, Jeremy H. 1993. "The British National Corpus." In: George P. Landow and Paul Delany, eds., 163-187.

Cole, W. 1987. "Terminology: Principles and Methods," *Computers and Translation*, Vol.2: 77-87.

Collins Cobuild English Language Dictionary. 1987. London and Glasgow: Collins.

Dagan, Ido and Church, Kenneth W. 1994. "Termight—Identifying and Translating Technical Terminology." *Fourth Conference On Applied Language Processing*. Oct 13-15, 1994. Stuttgart, Germany: Association for Computational Linguistics, 34-40.

Daille, Bétrice; Habert, Benoit; Jacquemin, Christian; and Royauté, Jean. 1996. "Empirical Observation of Term Variations and Principles for Their Description," *Terminology*. Vol. 3/2: 197-258.

Drouin, Patrick. 1997. "Une méthodologie d'identification automatique des syntagmes terminologiques: l'apport de la description du non-terme," *META*, Vol.42/1: 45-54.

Francis, W. Nelson and Kucera, Henry. 1982. *Frequency Analysis of English Usage: Lexicon and Grammar*. Boston: Houghton Mifflin.

Frawley, William. 1986. "Science, Discourse and Knowledge Representation: Towards a Computational Model of Science and Scientific Innovation." *The Language of Creativity: Models, Problem-Solving, Discourse Studies in Science and Culture,* Vol. 2, Mark Amsler, ed. Newark: University of Delaware Press: 68-91.

Garside, Roger; Leech, Geoffrey; and Sampson, Geoffrey, eds. 1987. *The Computational Analysis of English: A Corpus-based Approach*. London: Longman.

Galinski, Christian; and Schmitz, Klaus-Dirk, eds. 1996. *TKE'96—Terminology and Knowledge Engineering—Proceedings of the. 4th International Conference On Terminology and Knowledge Engineering*. Frankfurt: Indeks Verlag.

Halliday, Michael A.K. 1966. "Lexis as a Linguistic Level." In: Charles Earnest Bazell *et al.*, eds., 148-162.

Halliday, Michael A.K. 1993a. "On the Language of Physical Science." In: Michael A.K. Halliday and J.R. Martin, eds., 54-68.

Halliday, Michael A.K. 1993b. "Some Grammatical Problems in Scientific English." In: Michael A.K. Halliday and J.R. Martin, eds., 69-85.

Halliday, Michael A.K. and Martin, J.R., eds. 1993. *Writing Science: Literary and Discursive Power*. London and Washington, DC: The Falmer Press.

Harris, Zellig. 1982. "Discourse and Sublanguage." *Sublanguage: Studies of Language in Restricted Semantic Domains*. Richard Kittredge and John Lehrberger, eds. Berlin and New York: de Gruyter: 231-236.

Heid, Ulrich; Jauß, Susanne; Krüger, Katja; and Hohmann, Andrea. 1996. "Term Extraction with Standard Tools for Corpus Extraction—Experiences from Germany." In: Christian Galinski and Klaus-Dirk Schmitz, eds., 139-150.

Hofland, Knut and Johansson, Stig. 1982. *Word Frequencies in British and American English*. Bergen: Norwegian Computing Centre for the Humanities.

Holmes-Higgin, Paul R.; Ahmad, Khurshid; and Abidi, S.S.R. 1994. "A Description of Texts in a Corpus: 'Virtual' and 'Real' Corpora." *EURALEX'94 (Proc. of the 6th EURALEX International Congress on Lexicography)*. Willy Martin, *et al.*, eds. Amsterdam: The Free University of Amsterdam, 390-402.

ISO 1087:1990. *Terminology—Vocabulary.*

Johansson, Stig and Hofland, Knut. 1989. *Frequency Analysis of English Vocabulary and Grammar*. Oxford: Clarendon Press.

Justeson, John and Katz, S. 1995. "Technical Terminology: Some Linguistic Properties and an Algorithm for Identification in Text," *Natural Language Engineering* Vol. 1/1: 9-27.

Kageura, Kyo and Umino, Bin. 1996. "Methods of Automatic Term Recognition: a Review," *Terminology*. Vol. 3/2: 259-290.

Landow, George P. and Delany, Paul, eds. 1993. *The Digital Word: Text-Based Computing in the Humanities*. Cambridge, MA/London, UK: MIT Press.

Lauriston, Andy. 1994. "Automatic Recognition of Complex Terms: Problems and the TERMINO solution." *Terminology*. Vol.1/1: 147-170.

Leech, Geoffrey. 1991. "The State of the Art in Corpus Linguistics." In: Karin Aijmer and Bengt Altenberg, eds., 8-29.

Leech, Geoffrey and Fligelstone, Steve. 1992. "Computers and Corpus Analysis." In: Christopher S. Butler, ed.: 115-140.

L'Homme, Marie-Claude; Benali, Loubna; Bertrand, Claudine; and Laudique, Patricia. 1996. "Definition of an Evaluation Grid for Term-extraction Software." *Terminology*. Vol. 3/2: 291-312.

Meijs, Willem. 1992. "Computers and Dictionaries." In: Christopher S. Butler, ed.: 141-165.

Meyer, Ingrid and Mackintosh, Kristen. 1996a. "The Corpus from a Terminographer's Viewpoint." *Corpus Linguistics*. Vol. 1/2: 257-286.

Meyer, Ingrid and Mackintosh, Kristen. 1996b. "Refining the Terminographer's Concept-analysis Methods: How Can Phraseology Help?" *Terminology*. Vol. 3/1: 1-26.

Ooi, Vincent B. Y. 1998. *Computer Corpus Lexicography*. Edinburgh: Edinburgh University Press.

Pearson, Jennifer. 1998. *Terms in Context*. Amsterdam/Philadelphia: John Benjamins.

Picht, Heribert. 1998. "Terminology and Specialised Communication." *Eugen Wüster: His Life and Work—An Austrian Pioneer of the Information Society*. E. Oeser and Christian Galinski, eds. Vienna: Termnet: 117-131.

Picht, Heribert and Draskau, Jennifer. 1985. *Terminology: An Introduction*. Guildford: University of Surrey.

Quirk, Randolph. 1992. "On Corpus Principles and Design." In: Jan Svartvik, ed.: 457-469.

Quirk, Randolph; Greenbaum, Sidney; Leech, Geoffrey; and Svartvik, Jan. 1985. *A Comprehensive Grammar of the English Language*. London: Longman.

Renouf, Antoinette. 1987. "Corpus Development." In: John McH. Sinclair, ed. 1987b: 1-40.

Rundell, Michael and Stock, Patricia. 1992a. "The Corpus Revolution." *English Today*, Vol.8/2: 9-14.

Rundell, Michael and Stock, Patricia. 1992b. "The Corpus Revolution." *English Today*, Vol.8/3: 21-31.

Sager, Juan C. 1990. *A Practical Course in Terminology Processing*. Amsterdam/ Philadelphia: John Benjamins.

Sager, Juan C.; Dungworth, David; and MacDonald, Peter F. 1980. *English Special Languages: Principles and Practices in Science and Technology*. Wiesbaden: Oscar Brandstetter Verlag.

Sinclair, John McH. 1966. "Beginning the Study of Lexis." In: Charles E. Bazell *et al.*, eds., 410-30.

Sinclair, John McH. 1987a. "The Nature of the Evidence." In: John McH. Sinclair 1987b: 150-9.

Sinclair, John McH, ed. 1987b. *Looking Up*. London and Glasgow: Collins.

Sinclair, John McH. 1992. "The Automatic Analysis of Corpora." In: Jan Svartvik, ed.: 379-97.

Smajda, Frank. 1994. "Retrieving Collocations from Text: Xtract." *Using Large Corpora.* Susan Armstrong(-Warwick), Ed. Cambridge, MA/London/England: MIT Press: 143-177.

Stigelman, Sue. 1993. "Text-Management Software: A Taxonomy." In: George P. Landow and Paul Delaney, eds., 69-83.

Summers, Della. 1993. "Longman/Lancaster English Language Corpus—Criteria and Design." *International Journal of Lexicography*. Vol.6/3: 181-208.

Svartvik, Jan. 1966. *On Voice in the English Verb*. The Hague/Paris: Mouton.

Svartvik, Jan, ed. 1992. *Directions in Corpus Linguistics.* Berlin and New York: Mouton de Gruyter.

Turney, Peter. 1997. *Extraction of Keyphrases from Text: Evaluation of Four Algorithms* (Report No. ERB1051). National Research Council of Canada.

Trimble, Louis. 1985. *English for Science and Technology: A Discourse Approach*. Cambridge: Cambridge University Press.

Voutilainen, Atro. 1993. "Nptool, a Detector of English Noun Phrases." *Proceedings of Workshop on Very Large Corpora*. June 22, 1993. Columbus: Ohio State University.

Xerox. 1999. http://www.xrce.xerox.com/research/mltt/Tools/pos.html. (Accessed 2000-07-15).

Zipf, George Kingsley. 1935. *The Psycho-biology of Language: An Introduction to Dynamic Philology*. Boston: Houghton Mifflin.

8.4.2 Terminology Compilation: Consequences and Aspects of Automation

JUAN C. SAGER

1 INTRODUCTION

new approaches

The introduction of information technology into the process of terminology compilation has given this activity a new orientation that differs in almost every respect from the principles established in pre-automation days. Initially "automated terminology compilation" simply meant either making existing term collections machine-readable or inputting conventionally compiled terminology into a computer, reproducing in the computer the datastructure devised for traditional (paper slip) compilation. An exploitation of the potential of the computer for handling natural language, coupled with developments in Information Processing, has, however, resulted in a number of new approaches to compilation.

INFO

Lexicography vs. Terminology INFB 3:328

The possibilities offered by automatic analysis of large quantities of text have modified the very basis of terminology compilation, as well as of our views on the relevance of certain kinds of data and the degree of human intervention required for the process. The principles and methods of terminology compilation now also have a greater commonality with lexicography than ever before.

terms in context

Terminology is now adopting a corpus-based approach to lexical data collection. By being studied in the context of communicative situations, terms are no longer seen as separate items in dictionaries or part of a semi-artificial language deliberately devoid of any of the functions of other lexical items. The increasing tendency to analyze terminology in its communicative, i.e., linguistic, context leads to a number of new theoretical assumptions and also to new methods of compilation and representation.

traditional approaches

Traditionally there are two separate approaches to terminology compilation according to the motivation of this work:

- Specific cases of terminological information, which may be linked to primary or secondary term formation, usually in response to a particular request for a terminologist's services

- The systematic compilation of terminology, such as for creating a term bank or a technical dictionary

Both systematic and ad hoc term processing are now solidly corpus-based. When terminologists, in the absence of the services of a term bank or other automated collections, advise on single terms, the procedure for obtaining an answer to a specific query must, of course, continue to rely on human judgment in conjunction with searching the existing literature and consulting specialists. But when terminologists have access to textual corpora, this judgment can be supported by evidence based on various types of automatic text analysis and can provide more trustworthy and fully documented information. The production of glossaries or terminological dictionaries is now quite unthinkable without the basis of textual corpora that circumscribe the range and scope of the material on which the collection is based.

SEE
Technical Translation 2.1.1:148

2 The Function and Nature of Corpora

Good practice in terminology management requires that terminological information no longer be extracted from previous lists or by individual searches, but rather from a body of relevant textual material called a corpus. A terminologically relevant textual corpus is a representative body of texts covering all the information and documentation produced and required by a group of subject specialists either in contact with each other or in their communications with non-specialists. A new sub-discipline of general linguistics, namely corpus linguistics, has developed in response to the increasing emphasis in linguistic studies on text grammars, sublanguages, specialized languages, and other pragmatic aspects of language study. This new branch of linguistics, with its antecedents in statistical surveys of texts, is concerned with the identification and description of textual corpora. It can provide terminologists with reliable criteria for many aspects of their work that hitherto were largely a matter of educated guesses and traditional patterns of scientific method. Corpus linguistics can assist in the following steps of terminology compilation and thus provide solid material evidence on which terminological decisions can be based:

SEE
Terminology Extraction 8.4.1:725

corpus linguistics

terminological procedures

- Determining what is a sufficiently large body for reliable term extraction (Statistical means can be used to decide when the addition of more text does not produce any new terms.)
- Determining the boundaries of subject field and subfields (Statistical measures provide frequency patterns and so permit the delimitation of subject fields based on textual evidence.)
- Selecting cohesive and representative bodies of terms and phrases (The delimitation of terms and their distinction from collocations can now be fully justified.)
- Identifying terminological synonyms and describing the conditions of usage for terminological variants
- Supplying documentary evidence for the terminological information provided in an entry or record
- Providing genuine contexts and examples of usage

subject field delineation

INFO

Classification INFB 9:334

Corpus linguistics is in a state of evolution and research. Consequently there are as yet no conclusive guidelines as to what quantity of text constitutes a representative corpus, nor is there any common agreement on the scope of any one subject field. Corpus linguistics can, however, provide more relevant indications of the scope of a subject field by the analysis of documents than the quite different subject field divisions proposed by documentalists that are rarely useful to terminologists. Decisions on the scope of subject fields are changing with our perception of what is a discipline and what is an interdisciplinary field of study or activity as represented in documents and speech situations. Subject field delineation is likely to remain an unresolved issue in the dynamic fields of knowledge representation and terminology. Besides, there will always be an overlap of terms extracted from contiguous corpora, as well as a certain body of terminology that is common to a number of subject-specific corpora. These common terms may respectively represent the interdisciplinary vocabulary linking one subject to another, and the general scientific or technological vocabulary, both of which have been postulated theoretically, but which have not been textually identified so far.

3 Corpus-based Term Extraction

Because text corpora are time- and place-conditioned, they reinforce the principle that terminology compilation is an ongoing and repeated activity. A certain item of information extracted from a particular corpus at a particular time is valid as long as the textual basis maintains its currency. As soon as this is in question, the information has to be revised. There are no hard and fast rules for the life-span of a term. Some basic terms rarely change. Subjects go through periods of relative stability with a steady set of terms and stages of rapid evolution with a considerable growth and turnover of terminology. Textual information is the most reliable indicator of change and consequently it provides the only reliable data for the revision of terminological records.

currency

The increased number of databases containing running specialized texts in machine-readable form has significantly affected the motivation and indeed aided the development of new methods of terminology compilation. Since many technical texts are now stored on computers, they can be preserved in or converted into a suitable format for terminological analysis. This has had several significant consequences, as outlined below.

machine-readable text

Texts that are to be processed by translators or other terminology users can, in the first instance, be analyzed and compared with current machine-readable terminology holdings and a machine-readable general dictionary in order to produce a listing of items not contained in either. On the one hand, such lists will undoubtedly contain many spurious items, proper nouns and names of units of measurement, for example, which are of no interest to terminologists. On the other hand, they will also contain a significant number of new terms, which can be researched by the terminologist before the translator or other end-user edits, translates, or otherwise processes the text in question. This technique could be adapted to suit other requirements. Appropriate text analysis software can be used, for example, to identify terms existing in the term bank but that lack equivalents in a foreign- language or to produce corpus-oriented supplements to machine- translation lexicons and glossaries, as well as indices to documents. The concept of user-oriented terminology compilation is thus taken a stage further with the terminologist anticipating the users' requirements and ensuring that the majority of queries concerning a particular text are satisfied.

user assistance

Processing running text for terminological purposes has a number of advantages. Terminology extracted from running text or discourse offers a greater guarantee of thematic completeness and

data verifiability

coherence. All relevant textual variants will be covered and suitable contexts that demonstrate the linguistic behavior of terms can be selected. Being able to trace a term to its usage in a particular document is very important for such areas of usage as patents, legal contracts, and detailed instruction manuals. The occurrence of a term in a type of document is the greatest reassurance that can be offered a user and is, in fact, a more useful indicator of appropriateness of an expression to a particular context than so-called "quality indicators" that can be found in a number of term banks. Running text is also the most reliable indicator for the actual existence and currency of terms. In the quickly changing terminology of technology and engineering in particular, being able to date a term or a group of terms initially and then to have the means of verifying the continued existence of terms by regular revision and contrastive analyses with new corpora offers the best guarantees for the validity of terminology. Terminology compilation is therefore becoming increasingly text-oriented and less governed by the desire to construct separate conceptual systems.

database checking & updating

Running text can also assist the terminologist in maintaining the correctness of previously entered terminology. Statistical analyses of large volumes of text can be used to pinpoint changes in the frequency or usage of a term. Such analyses may indicate a movement in the linguistic environment in which a term occurs, or more importantly, it may reveal a change in the meaning of the lexical item in question that cannot be detected otherwise. Information about frequency of use in a particular text type corroborates or can even replace usage notes. Similarly, text analysis of large corpora can be used to isolate new terms, and therefore new concepts, to discover the possible obsolescence of terms and their concepts, and to highlight other changes in conceptual systems. While the use of machine-readable text for semi-automated validation of existing term collections, and indeed for machine-assisted terminology identification and compilation per se, is still in its very early stages, it is a feature of terminology compilation whose importance will increase dramatically in the future.

segmentability of work

Increased computer control now makes it possible to defer the task of constructing a terminological record without any loss of efficiency and even to distribute the work among several terminologists in order to increase the objectivity of the information. The possibility of full reference to a corpus makes every terminological annotation directly traceable to a source. Thus a terminologist can separate work on an entry by concentrating on particular aspects at a time. For example, conceptual information may be given priority over linguistic or pragmatic information, or bilingual information may be considered more important in an initial stage of compilation, to which full conceptual information can be added later without wasting time or reducing the quality of the work.

4 THE DISTRIBUTED STRUCTURE OF TERMINOLOGICAL DATABASES

SEE *Entry Structures 8.1.3:572*

Another major innovation that affects principles of compilation is the greater flexibility available for data structures. Hitherto many processing systems for lexical and terminological data have been implemented using standard data processing methods, but the result was normally far from satisfactory. This realization leads to the recognition that the storage of lexical data cannot simply be regarded as just another application of traditional information processing techniques, but rather that new techniques have to be developed for adequately representing the type of information contained in specialized dictionaries. Much progress is being and has been made, but further research is required in both conceptual modeling and computational lexicography and linguistics before anything approaching a comprehensive lexical or terminological data bank can be constructed.

multi-file source structure

A first step in the direction of flexibility is the notion of a multiple terminological database in which source documentation is separated from edited information, and master files are separated from user-access files. An important side-effect of a multiple database approach is the fact that there is, in principle, no longer a physical restriction on the size of a terminological collection. A term bank can be as large as required. Hence it can contain as much information on any one entry as is desired and have as many entries as necessary for covering the entire range of the vocabulary in a subject field. On the other hand, the considerable volume of information that can be made available makes it necessary to structure this information in a user-friendly format. For the end user, instead of a single large database, there are, in this view, only subsets created for practical purposes of, e.g., faster access, separate maintenance, printing of subsections, etc. With such a subdivision it is then also more readily possible to link parts of databases to parallel parts of other database configurations to form even larger units. The availability of corpora and the need for regular checking and updating of lexicographic and especially terminological databases is a further justification for dividing terminological data holdings into separate but interconnected bodies.

The configuration of the multiple terminological database consists of the following parts:

unedited data

The *Source or Documentation Database* (SDB) of raw, unedited data as they are found in the corpora used for gathering the terminological information, are maintained in completely separate databases, so that reference queries and later checks for currency and accuracy can go back to the original texts. End users may also wish to have information beyond that provided in usage examples or definitions, and this can be accessed if the original data have been processed for direct access. The SDB may contain several defining contexts, examples of usage, and also the contexts of foreign language equivalents, and thus offer a substantial amount of supporting information for the construction of the terminological entry.

edited data

The *Central Terminological Database* (TDB) of edited data constitutes the core of the terminological information service, but it contains only information specially selected and edited from the SDB. Instead of only holding bibliographical references to the information source for each separate item of information, it has pointers or references to the separate parts of the SDB so that every decision arising from selection or editing is fully documented. The sources can in this way be accessed directly. The TDB is the master set for terminologists to work on, but it also serves for the production of user or application-specific subsets.

functional subsets

The last constituent of the multiple database is comprised of a database of functionally-determined subsets (*Functional Subsets Database*, FSDB). This resource is not a single structure, but consists of all the various subsets of terminological information that are created for specific purposes and uses at any one time. It relies for its information on the TDB and occasionally the SDB.

Users will normally choose from among different output formats that have been created in the FSDB in response to user profiles analyses. This might consist, for instance, of the typical information required by a translator, i.e., target language equivalents and contextual examples, with the possibility of a check on definitions. Although translators may find these specifically defined data concentrated into one of the subsets of the database, they may also be given access to the edited information in the TDB and the raw information in the SDB. This three-fold division of information on terminology is represented in diagrammatic form in Figure 1.

SEE
User-specific Data Retrieval 8.1.4:600

5 INTERNAL STRUCTURE OF THE MULTIPLE DATABASE

On the same principle by which the three major databases of the multiple terminological database were conceived, it is desirable to divide the SDB and the TDB themselves according to the type of information they contain. The SDB appropriate for terminological purposes features three separate sets of information that contain:

SEE
Data Categories 8.1.2:652

- The definitions for the concepts it seeks to document
- The most reliable source possible for the actual form of a term or phrase and its variants
- A body of examples of usage of the terms and its variants from which to select, or, if necessary, construct serviceable examples of context for the TDB and to write comprehensive usage notes

This set of information categories can then be repeated for each additional language of the database. Alternatively, it may be decided that the definition files are constant and that only the terms and contexts have to be changed for each language. This fundamental distinction between databases is conditioned by the necessity to consider the identity or diversity between conceptual systems developed by different languages and cultures. Thus the information making up any one terminological record or entry may originate from three different file sources. It is obvious that a source document from which the definition of a concept is extracted is rarely the best source for information on usage and for exemplifying the diverse contextual environments of a term.

SEE
Concept Systems 1.4.1:89

This three-fold division is also desirable for the records of the TDB itself:

SEE
Definitions 1.3.1:63

- The definition source file serves to supply the various information items of the TDB concerned with the conceptual specification, namely the attribution of a concept to a subject

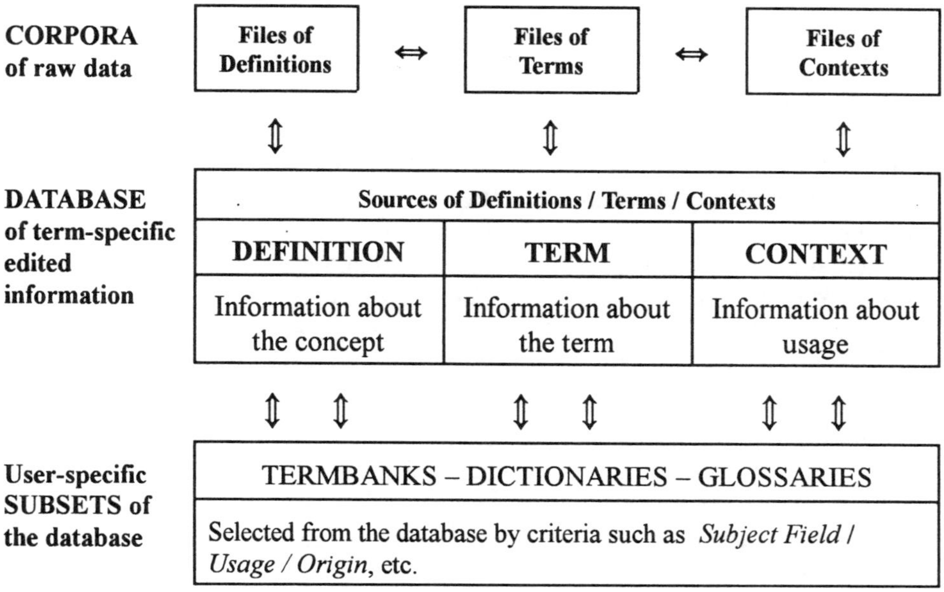

Figure 1: *Multiple Database Structure*

indicating particular subareas of application, its link or relationships to other concepts, and the definition itself.

- The term source files serve to supply the various information items of the TDB concerned with the linguistic specification, namely the language of the term, the term itself and its synonyms, abbreviations and variants, and any grammatical information considered necessary for the database.

linguistic information

- The context source files serve to supply the various information items of the TDB concerned with the pragmatic specification, namely examples for the context and any usage notes or examples considered necessary to illustrate the use of the term and its various alternative linguistic manifestations.

SEE *Contexts 1.3.3:80*

6 OUTLOOK

multi-purpose resources

The natural summary of these observations is that terminological databases can serve multiple purposes if they are appropriately structured. This is not only desirable from the point of view of cost and effort. It is also in the interest of improved communication that there should be only one major database of terminological information for each subject field and language community. A single database, to which all users will refer, exercises a strong unifying influence on terminological usage and thereby facilitates communication across all industrial and institutional barriers. A single database to which all users contribute, even in the minimal form of providing texts for the corpus or by providing specialist advice on the accuracy of the information, would become widely used and thereby act as an informal focus for the unification of terminology. A term bank with which users can identify because they have contributed to it acts as an unofficial standard of usage and is therefore much more effective than officially prescribed forms of language.

unified databases

A single but multiple database for terminology, i.e., one that is based on multiple sources and that is designed for multiple functions, can supply the information needed by publishers for the production of all manner of technical dictionaries; it can produce glossaries with or without definitions for very specialized subject areas; it can provide dictionaries for various forms of natural language processing systems such as machine translation or expert systems; it can supply professional organizations with raw information for future development and completion; it can be developed into various forms of term banks for on-line access or for packaging into stand-alone microprocessor-based systems and, it can, of course, be accessed directly by researchers who want to study the large quantity of data collected on terms, definitions and concepts.

SEE *Terminology Interchange 8.1.5:639*

The corpus and database approach thus can be seen to offer the following direct advantages:

advantages

- It is possible to produce documentary evidence for most information gathered in a terminological record.
- The fullest set of linguistic information is available.
- The various selections and other decisions that a terminologist makes become testable and replicable.
- All information is reliably dated by the date of the corpus and by the date of the terminologist's decision to include the information in the database and can therefore be checked at regular intervals.
- Information on terms is collected with a view to providing as full a record as possible. In this way one terminology collection

can serve a wide range of diverse users and uses, thus removing the costly duplication of effort that application-driven terminology compilation entails.

A multiple terminological database structure can therefore serve such diverse purposes as:

- A tool for translators
- A reference tool for standards experts
- A source for NLP lexicons
- A source for printed glossaries and thesauri
- A reference database for industrial and individual users

8.4.3 Terminological Aspects of Text Production

GREGORY M. SHREVE

1 TEXT AND TERM

terminology & knowledge structure

Technical and scientific terms are always a potential source of trouble for nonspecialist translators, technical writers, and other nonexperts working with specialized texts. For instance, one of the initial phases of the translation process involves scanning the source text and identifying any unknown words. Unfamiliar lexical items are usually part of *special vocabularies* or *terminologies* in specific subject domains and not part of general language. Communications and information specialists, usually fluent in the general language of the source text, may be completely at a loss when terms from an unfamiliar subject area are encountered in a text. Developing competence in scientific and technical texts involves understanding the nature of technical terminologies, their relationship to the knowledge structure of a domain, and the way they are used in special language texts. One of the defining characteristics of these text types is the presence of special vocabularies, thus strategies for creating (technical writing) and for recreating these texts in a second target language (translation) must necessarily include methods for identifying terms, uncovering their meanings, and using their equivalents appropriately in the target text.

knowledge content

It has been said that the terminology of a domain mirrors its knowledge structure. The term is a linguistic label attached to a knowledge content called a *concept*. What is often left unsaid is that terminology almost always occurs embedded in texts. The technical writer, translator, and professional terminologist encounter terms in a textual context: surrounded by other words, embedded in sentences, the sentences embedded in paragraphs. Even the parts list, order form, or catalog is a kind of text. This means that the specific meaning, usage, and appropriateness of a term is always conditioned by its textual environment. While we are used to retrieving the meanings of unknown terms by looking them up in glossaries, technical dictionaries, or reference sources, we should always remember that the target text is the final arbiter of whether the result of our terminological research process "fits" or not. The "fit" between term and text is

terminological fit

- *Semantic* (Is the concept represented by the term appropriate, given the other concepts represented in the text?)
- *Linguistic* (Is the term used "properly" in its linguistic environment?)
- *Pragmatic* (Is the term used the right one, given the intent and situation of the text?)

SEE *Term Extraction 8.4.1:725*

Even more importantly, we should recognize that the text in which the technical term occurs is an important source of information about the term, especially about the relationships it has to other terms. When terms co-occur in a text, they establish conceptual relationships. Indeed, the text-linguist might argue that the larger conceptual structure of a knowledge domain implied by a terminology does not exist except as it has been established by or extracted from a corpus of texts in the domain. The broader information the text provides about the term is not usually adequately captured in glossaries, and perhaps it is unreasonable to expect that it could be.

textuality

SEE *Terminology Compilation 8.4.2:761*

Glossaries necessarily extract terms from texts, usually from a variety of texts. While the terminologist, especially the translation-oriented terminologist, tries to capture information from the text and "import" it into the glossary in the form, for instance, of contexts and notes on register and usage, it is almost impossible to avoid stripping the textuality of a term away, producing what Budin (1994: 218) calls *decontextualized text*. This paper explores some of the relationships between term and text, with practical examples, and then makes some suggestions for re-attaching the term to the textual contexts from which it was derived. The concepts of terminology management should be expanded to include managing the corpora from which terms are extracted (source, parallel, and background texts) and in which terms are used (target texts). New mechanisms for linking glossary terms to corpora of translations and parallel texts through tagging and retrieval systems could provide the necessary apparatus (Shreve 1994).

2 TERMINOLOGY AND THE SEVEN STANDARDS OF TEXTUALITY

communicative constraints

A text is a complex linguistic object that reflects certain social and communicative constraints. It is the linguistic manifestation of a conventionalized interaction between communicative partners. Thus, e.g., a patent, or an assembly manual, or a scientific article, is more than just a series of words and sentences on paper: it is a representation of an accepted and conventionalized form of social interaction and is usually culture bound. The social conventions and interactions underlying a patent, for instance, imply particular social roles, legal statuses, communicative needs and intents, specific required information contents, and required formats. Thus, while a text always has a linguistic surface (words, sentences), it also possesses other features which, taken together, make it possible for human beings to recognize it and use it as a device for communication in specific circumstances and for specific purposes.

textual influences

Terms, insofar as they appear in texts, are under textual influence. A great many aspects of this textual influence have practical value for the terminologist, and for the technical writer and translator, who, in particular, must be concerned with how the text and the term interact.

features of textuality

Beaugrande and Dressler (1981) enumerated seven defining features of textuality. Neubert and Shreve (1992) have examined the implications of the seven standards for translation practice. The purpose here is to identify the practical implications of the standards of textuality for the terminologist or the professional translator doing terminology research or resolving terminological problems during the translation process. Before we can proceed to identify the implications, it would be useful to briefly review the seven standards of textuality as defined by Neubert and Shreve:

1. *Intentionality* is the interaction between the communicative purpose or intent of a text and the reader's need for the information in the text, that is, texts are designed to *do* something: they have an interactional aim.
2. *Informativity* comprises the substantive knowledge content of the text, that is, the information units delivered in and through the text to the reader.

features of textuality

3. *Acceptability* is the text's adherence to the conventionalized norms for communication that apply in a particular social circumstance.
4. *Situationality* comprises the set of conditions that describe the social and communicative context for the text, conditions such as time, place, audience.
5. *Coherence* comprises the logical relationships established in the text between information units, i.e., an information structure or global logical or semantic structure.
6. *Cohesion* is the expression of the logical and semantic information structure through linguistic means.
7. *Intertextuality* is the relationship between an instance of a text (that is a particular text), and other instances of texts recognized by readers as being of the same type

terminological implications

The author of a scientific or technical text chooses terminology carefully. A terminology is developed, sometimes even standardized, in a specialized field of knowledge to improve communication. Scientific and technical texts are most often descriptive and attributive texts. They describe processes, objects, phenomena, and their properties, and terminology is developed and applied to improve the precision of the descriptions and attributions. The development of terminologies proceeds parallel with the evolution of scientific and technical texts to improve their clarity, brevity, and organization, and as the texts strive for clearer, more accurate and more economical communication, terminologies are then further developed and honed to achieve this purpose. This means that each of the seven standards of textuality has a terminological implication, or, conversely, that terminology plays a role in establishing the textuality of scientific and technical texts. The practical terminologist, and especially translators, should never lose sight of the fact that the text is both the source and the destination of the term.

2.1 Intentionality and the Term

communicative intent

The author of a scientific or technical text never uses all of the terms available from a domain, for the particular purposes of the communication entail a purposeful selection of terminology from the total set available. The terminology extracted from any single text is a direct reflection of the communicative intent. Thus, for instance, an operator's manual for a machine tool has a primary intent of describing the machine's features, operation, and maintenance. This communicative intent circumscribes the terms that will be found in the text. This also implies that the texts from which a glossary is constructed each contain only a subset of the total potential terminology of the domain. Therefore terminology managers should be careful in their selection of the range of texts from which terms are documented.

SEE
Descriptive Terminology 2.1.4:184

It should also be noted that texts may use terms from a variety of domains. Our glossaries and dictionaries tend to extract terms from texts and by processes of categorization and classification separate them into discrete domains, but the text itself is usually less discrete in its demarcation of the conceptual domains addressed. Many domains will be evoked and a multitude of semantic relationships established between the domains. The terms in a text create a complex associative network that is unique to the text and a reflection of the specific purpose and intentionality of the text. Terms in their textual context are *multiply related* to one another. These complex relationships are most often stripped away in the reductive processes of glossary creation.

SEE
Technical Translation 2.1.1:147

INFO
Classification INFB 9:334

2.2 Terminology and Informativity

orders of informativity

In scientific and technical texts the terms selected by the communicative intent project the primary information content of the text. They are chosen deliberately by the author because of their precision, and a very heavy information weight is assigned to them. Terms have a high information content on purpose. Standardizers and terminologists have designed them to be that way. Beaugrande and Dressler have spoken of the *orders of informativity* of a text, which is a measure of the significance or relevance of an information unit in a text relative to other information units. In scientific and technical texts, terminological units (words and phrases functioning terminologically) have a very high order of informativity. The weight the author ascribes to the term usually matches the *relevance* of the term to the reader. It should be noted that terms in isolation, for instance in a glossary, have no information value. The informativity of a term derives from its placement in a text, in combination with other terms and the concepts they represent.

The information value in a glossary is information latent, i.e., it represents a potential information value that can only be assessed in textual context. This implies that the more information a glossary contains that helps the terminologist or translator determine textual fit (semantic, linguistic and pragmatic) the more information value can be imparted to the glossary item. One might conceive of the glossary item, for instance an entry in a terminology database, as an unopened surprise package, whose real value can only be determined when it is read, i.e., when the package is opened by the reader.

information value

2.3 Acceptability, Term, and Textual Context

Terms do not exist in a vacuum; as we have claimed, terms are almost always experienced in a textual context; they are read and used as part of texts. This means that their usage in texts is textually conditioned. We say that terms appear in *textual context*. Terminologists recognize that the documentation of terminology almost always includes capturing and documenting *contexts*, usually a small adjacent linguistic "chunk" of a text. While these contexts are important, they of course capture only the immediate local environment of the term. The term may have attachments or relations to other terms and to linguistic structures that occur elsewhere in the text. The acceptability of a term is partly a function of its proper usage in a text. In scientific and technical texts, conversely, much of the acceptability of the text, because of the high order of informativity of the term units, is dependent on the correct and appropriate usage of the term. A great portion of the acceptability of scientific and technical texts, therefore, rests in the ability of the writer or translator to choose the term that "fits" in the text.

textual context

As an example of this concept let us examine the following sentence from an English translation of a popular science article:

semantic correctness

Eine Blende mehr: Color Super SR 200.

This sentence was rendered into English as:

One aperture more: Color Super SR 200.

Taken in isolation, nothing appears to be wrong with this rendition. However, appropriateness of a term in a scientific text is heavily dependent on the global semantic correctness of terms. This correctness cannot be judged from the glossary entry alone, or without expanding the linguistic context from the sentence to larger units. The problems with this translation emerges only after the previous subhead and previous paragraph are read:

> *Bewährt: Color Super SR 100*
> *Der SR 100 mit einer Filmempfindlichkeit von ISO 100/21 ist ein Film mit ultra-feinem Korn . . . Ein Allroundfilm für höchste Ansprüche.*

textual-semantic context

Here it is clear that the primary "dictionary equivalent" of the term *Blende* (aperture of a camera) cannot be used. More specifically, the decision to use *light stop* or *film speed* as an equivalent instead of *aperture* could only be made on a textual basis. It is the only acceptable usage in the given textual-semantic context, a context that includes a textual span of some seven sentences. Now, it is clear that a knowledge of the domain would have precluded this error, but even for a translator with domain expertise the decision to use *film speed* instead of *aperture* would have been reached on a textual basis because the semantic constraints raised by the preceding text would have led him or her to that decision.

acceptability constraints

Acceptability also includes conforming the term to its linguistic surroundings. As an example, from the same source text cited above, the term *Filmempfindlichkeit* occurs in the semantic field of *film speed* or *sensitivity*. The term appears in linguistic micro-context with an adjective, *erhöhte Empfindlichkeit*. While this usage matches the English rendering, *higher film speed*, these combinations, in this case a collocational usage, do not always match between languages. Further the "literal translation" of *erhöhte Empfindlichkeit* is not *higher film speed*, but something more like *heightened film sensitivity*. The appropriate translation equivalent is partly established terminologically (the *film speed-Filmempfindlichkeit* relationship) and partially by textual acceptability constraints, with the result that *higher film speed* is what we would write in this kind of text.

term fit

base corpora

The fit of a term within a sentential context to its adjectival and prepositional neighbors is extremely important and all too often insufficiently detailed. While many of these surroundings can be captured by terminologists who collect contexts, it is clear that contexts are not always collected faithfully. Even if the terminologist collects contexts, it is not clear how many and what kind of contexts should be collected. Ideally, the terminologist needs to collect entire context sets, which would include a variety of the extant usages of the term in typical textual contexts and import them into a glossary. But would it not be more useful to leave the contexts *in situ*, i.e., in the text in which they are found, and point the term to those contexts so that the glossary user can examine them? This, of course, implies that terminologists should seriously consider developing base corpora of parallel texts and selected translations to accompany their glossaries. The suggestion also implies that mechanisms such as SGML (Standard General Markup Language) are introduced and developed to attach glossary entries to multiple contexts embedded in texts (Shreve 1994).

INFO
SGML
INFB 21:863

While several terminology data category systems, such as the ISO 12620:1999, list data category sets for terminological acceptability, these categories reflect only one aspect of terminological acceptability, i.e., whether the term has been deprecated, or is obsolete. This definition of acceptability, while it is also ultimately textual—it is the textual usages that are deprecated or obsolete—does not and cannot include the full range of features that determine the acceptable usage of a term in a text.

SEE
Data Categories 8.1.2:552

2.4 Situation and Term

situationality

The acceptability of a text is conditioned on the expectations of the people who read it. The situationality of a text is encoded in presumptions made by the writer, or by the translator, during the process of writing or translation. While the primary selection of terms is made on the basis of the matching of intent (i.e., to describe the operation of a stamping press) with informativity (the selection of the term-concept assignment to match the information requirements of the message), there can be situational effects on the selection of terms.

intended audience

For instance, the writer's projection of the text production situation most often includes a projection of the intended audience, the audience's need for the information in the text, and a presumption of the readers' level of knowledge of the domain. Many terminologists are familiar with the fact that technical concepts sometimes are labeled with different term forms reflecting the different professional communities that have evolved them and used them in discourse. Most terminologists recognize that bench-level terminology, for instance, is a situationally conditioned term subset: terms addressed to or used by a specific audience.

situational appropriateness

As an example of this phenomenon, consider the use of the term *Rohkarosserie* in a German text on the safety-testing of children's car seats. The translator of this text, in a testimony to dedicated research, consulted automotive engineers and uncovered a potentially obscure English terminological equivalent for the German term: *body-in-white*. When I asked the translator who the audience for the text was, it became clear that while the chosen English term was correct for one audience, but it was not appropriate for another. The research correctness of the term was not at issue, but it's situational appropriateness was. The text was from the German automotive magazine *Auto Motor und Sport*, a text whose audience would not be restricted to automotive engineers and for whom a term like *stripped chassis* might have been more appropriate.

full usage range

Many students of terminology are satisfied with documenting in their glossaries only what they perceive of as the correct or most precise, or even most standardized forms of a term, and these forms are extracted most often from other glossaries, from standards, and from bilingual technical dictionaries. There is a danger here in that first and foremost, this practice abdicates to others the responsibility for extracting the term from text. Some care should be taken to ensure that the full range of valid terminological forms be documented in a glossary, as these forms will be selected by the situation type of the text, and some will be more appropriate in one kind of text than in another. Full ranges of usages and of alternative pragmatic forms related to the situation of the text can only be extracted from primary sources, i.e., from real texts in which the terms are used for real communication.

term types

Many data category systems for terminology management recognize the situationality of terms in category classes such as term type (here attributes like *standardized term*, *legal term*, or *international scientific term* make explicit reference to the situationality of the terms) or rather of the texts in which the terms appear.

2.5 Coherence and Cohesion

information content

The terms that appear in a text are selected by the author to fulfill an intent and to present a certain information content. The text serves as the vehicle for delivering the information content. The terms themselves may be perceived of as linguistic pointers to knowledge units, but it is the text that provides linkages between the units to create more complex knowledge structures. For instance, examine the following text segment from a longer article about laser light applications in urology:

> *Was ist das Besondere am Laserlicht?*
> *Das Laserlicht zeichnet sich durch drei Eigenschaften aus, die es von praktisch allen anderen natürlichen und künstlichen Lichtquellen unterscheidet:*
> *1. die* Einfärbigkeit,
> *2. die hohe* Strahlungsdichte *und*
> *3. die* Gleichschwingung.
> *Mit diesen Eigenschaften tritt das Laserlicht mit biologischem Gewebe in Wechselwirkung und erzeugt durch die Parameter* Energiedichte, Einwirkzeit, *sowie* Wellenlänge *eine Bandbreite von Gewebsreaktionen, die von nicht-thermischen über thermische zu ionisierenden Effekten reicht.*

term-concept relationships

Now, the laser light specialist would, of course, understand the relationships between the terms and concepts before reading the text, but most readers depend on the text to establish the important relationships between the terms. The more complex term-term, or rather concept-concept, relationships are established for the reader by the text during the process of reading. Concept structures are built up in the mind of a reader by using linguistic indices (in this case terms) to point to specific areas of knowledge domains (frames). Elements of the general language are used to establish connections between the terms.

terminological linkages

It is particularly important in technical writing, and therefore in good translation, to understand the linguistic devices that allow the reader to build a mental representation of the concept structure of the text. In most cases, terminological linkages are made by the general vocabulary, i.e., the special vocabulary uses the general vocabulary to establish conceptual relationships. Thus, the text above, using linguistic mechanisms such as *zeichnet sich durch drei Eigenschaften aus* [is characterized by three properties], establishes *monochromacity*, *collimation*, and *coherence* as three properties of laser light. The three terms become part of a concept structure composed of three concept relations.

coherence and cohesion

The linguistic mechanism *zeichnet sich durch...* stands for a logical relation—let us call it the PROPERTY OF relation—that links the three properties to their argument. The logical structure of the text is its *coherence*, and the surface linguistic structure of a text is its *cohesion*. Cohesion serves to establish coherence.

term-concept relationships

The relationship between concept structures and linguistic-terminological structures at the textual surface is complex. For the purposes of this paper, we want to define several special usages, terms if you will, for the discussion of term-concept relationships as they occur in texts:

- *term*—a linguistic label or lexical structure occurring within a text that points to a concept in a specialized knowledge domain
- *term cohesion*—a textual phenomenon established by the linguistic co-occurrence of terms in a text
- *concept relation*—a local semantic structure created by the co-occurrence of one or more terms in a text
- *concept coherence*—(also concept structure) a textual phenomenon established by the global set of concept relations in a text
- *domain structure*—a global semantic structure derived from a corpus of texts
- *terminology*—a global set of linguistic labels associated with the concept structure and derivable from the corpus

textual cohesion

There are several terms in the laser light text cited above: *Laserlicht*, *Einfärbigkeit*, *Strahlungsdichte*, and *Gleichschwingung* (*laser light*, *monochromacity*, *collimation*, and *coherence*). Each of these terms is the linguistic representation of a concept. These terms co-occur in the text. The co-occurrence of the terms is part of what we call textual cohesion, but it also establishes terminological cohesion. It is important to understand the distinction between cohesion and coherence in a text. Some analysts do not make the distinction (Halliday and Hasan 1976), but others have argued that the differentiation has benefits for translation practice (Neubert and Shreve 1992).

Cohesion is a linguistic phenomenon that occurs in texts. It is established by devices such as collocation and iteration (Halliday and Hasan 1976). Coherence, on the other hand is a logical framework or conceptual structure that is created in the mind by the co-occurrence of terms (or words) in a text.

terminological cohesion

Without getting into the details of the theory, let us pose a simple explanatory example. In the text above the author has chosen to use the linguistic forms (term forms) *Einfärbigkeit*,

Strahlungsdichte, and *Gleichschwingung*. The Latin-based terms *Monochromasie*, *Kollimation*, and *Kohärenz* could have been substituted, with no special loss of information value and no change in the conceptual relationships or coherence established. Yet the choice of the former terms *is not without textual effect*. The effect is cohesive; it fits with the other terms used in the text (technical but not too technical) and with the popular scientific style of the text. Furthermore, it reflects a conformance to the norm established for this particular audience, whereby the author uses transparent Germanic roots instead of flaunting more erudite, but potentially opaque Latinate forms.

alternate forms

Terminological cohesion would dictate that other terms in the text be selected with attention to style and register and would have a profound effect on the technical writer's or translator's selection of alternative forms for a term. Thus, while a text directed to a small group of advanced laser scientists might always use the most standard (correct) term, a popular scientific text might not. Terminological-textual cohesion might require that we not mix forms of different register in the same text.

terminology standardization

INFO

Controlled Language INFB 25:872

Terminology standardization attempts to reduce the variation in the linguistic forms of terms, rather, variation in the various labels that apply to the same concept. But until standardization is complete and universally accepted (a situation which modern authoring tools and adherence to Controlled Language are actually achieving in some environments), terminology, because it appears in texts, will be affected by a variety of cohesive mechanisms such as the use of synonyms, near synonyms, substitution by hyponym or hyperonym, etc. Translation-oriented terminology management systems need to document these variations by collecting and commenting on their appearance in texts.

knowledge frame

SEE

Artificial Intelligence 8.4.5:809

Terms have fewer alternate forms (synonyms, register variants) and collocational combinations than lexical items in the general language. This is because most of them have been developed to avoid this linguistic variation. Terminology is intended to point *directly* to the core of a knowledge frame and select a concept, avoiding the imprecision of general language. From a textual perspective, terms are useful precisely in pragmatic texts because they so accurately point to knowledge frames. But the devices for connecting the terms to one another to build concept relations and even larger concept structures are not so precise. A persistent problem in cognitive science and artificial intelligence is the representation of knowledge structure, or, in the case of terminology, conceptual dependency. Note in the text above that the concepts of *Gleichschwingung* and *Einwirkzeit* are placed into a conceptual relationship by the text. *Synchronous oscillation* and *duration of irradiation* have no preexisting relationship of any relevance, only a relationship established in the text via mechanisms of textual coherence. The underlying logical-semantic structure of the text places the concepts in association using the relatively imprecise words of the general language.

Some terminology management systems recognize the importance of documenting what are called concept structures. But these are not always extracted from their natural occurrence in current texts. They are most often constructed from background texts, reference works, and by consultation with experts. These concept structures are, in fact, attempts at developing *domain structures*, attempts to understand the knowledge organization of a subject area. The knowledge structures yielded by this research are, of course, indispensable, but they simplify the structural possibilities and often yield only generic or partitive concept hierarchies, which rarely appear in their pure form in natural texts.

SEE *Concept Systems 1.4.1:89*

pragmatic relationships

Further, because of the differences between the precision of the term and the imprecision of the devices used to connect them in sentences, it is not always apparent how concept structures in texts relate to the concept structures detailed in glossaries and thesauri. This implies that it might be useful to document *in situ* concept structures by creating *associative maps* of inter-term relationships in a text so we can see how the general language is used to establish conceptual relations and concept structures, that is, the linguistic devices for establishing partitive, generic or pragmatic relationships (e.g., the relation USED-IN-THE-MAINTENANCE-OF). The intent here is not to underestimate the importance of domain structure documentation, but simply to point out that terms which occur in texts establish much more complicated and much less "pretty" conceptual arrangements that defy reduction into generic or partitive hierarchies. Some concepts may be related in texts by purely pragmatic relations. *Duration of radiation* and *radiation density*, for instance, are not related to one another generically or partitively. Only one is a defining property of laser light. They are related because radiation density is one of three properties of laser light that allow it to interact with biological tissue. The nature of this interaction is determined by three parameters, one of which is the duration of the radiation. This relation is probably not documented in any glossary, for it is purely a creation of the text as it expresses the knowledge of an expert or author.

singular concept structures

It is clearly impossible for a glossary to document all of the kinds of relationships and structures that can be built with terms used in texts. We would not want to document some of these relationships because each text creates a unique set. Indeed, the terminological coherence of a text is not a mirror of a domain structure, but a singular concept structure existing only in the text at hand. Yet, some individual components of these singular structures are stable and re-occurring and should be documented.

3 Intertextuality, Terminology, Glossary and Corpus

terms in textual context

What are the implications of the textual character of the term—the term in its textual context—for terminology management? A dictionary or a glossary can only be an aide or a guide to making a correct decision about the fit of a term in a text. The more information it contains to assist in this process, the more useful it is. The intentionality of the text, its informativity, its situationality, the acceptability conditions for the linguistic usage of the term in the text, and the requirements of cohesion and coherence all condition the appropriateness, information value, and logical function of a term as it appears in a text.

universe of discourse

Terminology management for translators and technical writers is an endeavor with practical goals. It does not exist as an end in itself, but purely to improve the creation of texts and translations. The preceding sections have argued that glossary-oriented terminology management, while indispensable, should be enhanced by the collection and annotation of the texts in which the terms occur. The entire set of texts in which terms occur are a *universe of discourse* or a *text world* that establishes the intertextual character of terms, that is, the range of textual forms the term can take in the texts associated with a speech community.

This implies that translators and technical writers should move toward the collection and annotation of their source and target texts as a parallel to their terminology management activities. If this is not done, then valuable information about the effects that textual context has on the actual usage and appearance of terms in texts will be lost. This information is too complex and varied to be imported into the glossary, rather, the glossary must be linked or associated with a corpus of texts so that for each term entry in a glossary there are a range of textual contexts, both local and global, in which the term appears. This textual information is precisely the *translation-strategic* or *writing-strategic* information that is needed to actually use terms in texts.

source and target texts

How might the construction of such a corpus be accomplished? Ahmad and Rogers writing in this volume (Chapter 8.4.1) introduce several algorithmic approaches to the problem. Regardless of approach, a number of steps will be necessary, and some time and labor will be required. The time and labor could be reduced if certain conditions were met:

encoding standards

- The development of standards for encoding textual and translation-relevant information (fundamental interactional aim, situation, intended audience, text-type) in sets of parallel texts and model translations; this information should be encoded as annotations to the texts.
- The creation of standards for aligning terms in glossaries with groups of machine-readable texts in which they appear. This means that writers or translators could consult both glossary and text. Tthe glossary would serve as a point of access to a set of annotated parallel texts and model translations.
- The establishment of standards for encoding concept structure information *in situ* in the text are developed, so that the coherence and cohesion structure of any text-situated terminology could be displayed and compared to the more formal, but less actual, systems documented in the glossary.
- The development of software modules for processing the textually embedded codes and creating the textual alignments mentioned above to improve computer-assisted-translation or glossary management systems.

Modern software and hardware technologies make this an achievable goal. Hypertext systems with string-searching and linking capabilities can host both a glossary and a set of annotated texts. Coding systems such as the Standard Generalized Markup Language could be easily adapted to the purposes described above and be standardized as an interchange system, which would allow not only the interchange of terminological data (as has already been done in ISO 12200:1999), but also the exchange of term-in-textual-context information for the purposes of translation and technical writing. The evolution of XML and of highly sophisticated, inferential ontological systems both promise significant advances in providing tools for mapping complex relationships between more abstract conceptual systems and the terms in context.

SEE
Terminology Interchange 8.1.5:613

REFERENCES

Beaugrande, Robert-Alain de, and Dressler, Wolfgang U. *Introduction to Text Linguistics*. London: Longman, 1981.

Budin, Gerhard. 1994. "Text Production and Terminology Control." *Actes de langue française et de linguistique*; Symposium on French Language and Linguistics. ALFA. Halifax, Nova Scotia: Universitas Dalhousiana, 217-223.

Halliday, Michael A. K. and Hasan, Rugaiya. *Cohesion in English*. London: Longman, 1976.

ISO 12200:1999. *Computer Applications in Terminology — Machine-readable Terminology Interchange Format (MARTIF) — Negotiated Interchange*. Geneva: International Organization for Standardization.

ISO 12620:1999. *Computer Applications in Terminology — Data Categories*. Geneva: International Organization for Standardization.

Neubert, Albrecht, and Shreve, Gregory M. *Translation as Text*. Translation Studies 1. Kent, OH: Kent State University Press, 1992

Shreve, Gregory M. "SGML Representation of Concept Systems: Identifying and Retrieving Term-Concept Structures in Textual Context." *Harmonizing Terminology for Translation and Information Retrieval: Theory and Practice*, ASTM STP 1223, Sue Ellen Wright and Richard A. Strehlow, eds. Philadelphia: American Society for Testing and Materials, 1994, 157-168.

8.4.4 Collocations in Sublanguage Texts: Extraction from Corpora

ULRICH HEID

1 INTRODUCTION

This chapter deals with collocations in the language of specialist communication and discusses ways of describing collocational phenomena linguistically and of presenting the descriptive results in specialized dictionaries and termbases. We also examine tools and methods for extracting collocations from specialized texts, in particular technical ones.

specialist communication

Collocations are a phenomenon of lexical combinatorics: they involve the lexical, semantic, and syntactic properties of lexical items and their syntagmatic co-occurence. Collocational word combinations are a problem for translation because, although many collocations of a foreign language are transparent so far as understanding is concerned and do not cause trouble in translation into one's mother tongue, it is impossible most of the time to "guess" the right word combinations when translating into a foreign language. This is true in specialized language much the same way as in general language: collocations are to some extent a matter of convention rather than being explicitly rule-governed.

combinatory phenomena

Before we attempt to discuss details of the acquisition and representation of collocations, we will summarize a few basic assumptions about collocations as they have been discussed by lexicographers and terminologists over the last decade (Heid 1994). The following are the basic, mainly uncontroversial principles:

definitional elements

- Collocations involve *two* lexemes (plus possibly determiners, quantifiers, and prepositions). This distinguishes collocations from other multiword items, e.g., idioms, since the latter can have more than two elements. From the point of view of concept-based terminological description, one of the two components of the collocation must be a term for which a

two lexemes

conceptual description is (or at least may be) available. Both elements may have this status.

collocational elements

- The elements of collocations can be collocations themselves: German *unbeschränkte Gültigkeit haben* can be seen as being composed of *unbeschränkte Gültigkeit* (which is a collocation and forms a collocational "mini-paradigm" with <*beschränkte, unbeschränkte, befristete, ... Gültigkeit*>) and *Gültigkeit haben* (which in turn forms a "mini-paradigm" with <*(seine) Gültigkeit verlieren, Gültigkeit erlangen, Gültigkeit haben*>).

collocational categories

- For the European languages, we can syntactically classify collocations into the following groups according to the category of the two lexical items making up the collocation:

  - noun + verb, noun + adjective, noun + noun
  - verb + adverb, adjective + adverb

  The first three types are much more frequent in specialized languages than the two in the second group. Almost all of our examples in this article will belong to the first group.

polarity

- Linguists and lexicographers have described collocations as being "polar" in the sense that their two components are in a specific relation, with one element determined, the other determining it. Proposals for a rough classification of the determination operators have been made. Hausmann (1989) provides a condensed description of collocations, while Benson (1989, 1990) summarizes Hausmann's work in English. Hausmann uses the terms "*Basis*" and "*Kollokator*", i.e., "*base*" and "*collocate*", which are the terms we will use here. His most recent terminology (*Autosemantikon* vs. *Synsemantikon*) covers the same phenomena, with more emphasis on the polarity aspect.

terminological collocations

The definitional elements summarized above are shared by many lexicographers and terminologists. A recent bibliography with emphasis on sublanguage can be found in Pavel (1993). We will later focus on specific aspects of terminologically relevant collocations. Figure 1 shows a few examples in English, French, and German from general language and from sublanguage.

n + n	Aktivierung des Filters, un rideau de fumée, technique de paramétrisation, taux d'inflation
n + a	strong protest, merited praise, ressources renouvelables, waagrechte Stellung, récession profonde, inflation larvée
n + v	pay attention, poser une question, créer un fichier, une dévise se déprécie, consolider une dévise, Parameter festlegen

Figure 1: *Examples of collocations from general language and sublanguage*

2 Linguistic Properties of Collocations

NLP support

We need to elaborate on the list of collocational properties given above to provide a few more details about types of phenomena that need to be tracked in a collocational description of a lexical fragment of a given sublanguage. Such properties include syntactic and semantic/conceptual phenomena. We point them out here because of their relevance for terminology management. In particular, if the contents of a glossary or a termbase is meant to support a Natural Language Processing (NLP) system, it is necessary to give a detailed description of the linguistic properties of collocations.

2.1 Syntactic Properties

We need to account for the following factors in addition to the syntactic properties described above:

- Collocational relations also play a role in describing noun groups and/or noun compounds.
- Only some types of collocations, not all, are chunks or phrasal categories in the sense of formal grammars. Some collocational relationships (such as subject/verb collocations are orthogonal to such structural notions).
- Collocations may also involve adverbials (adjuncts), which occur relatively frequently in specialized language.

noun forms

Noun phrases and noun compounds play an important role in terminology. Many complex concepts are expressed using compounds in languages like German, Dutch, or Finnish, for example, while being rendered by noun groups in the Romance languages:

Kläranlage (de), *station dépuration* (fr), *Geldmenge* (de), *masse monétaire* (fr), *Zahlungsbilanz* (de), *balance des paiements* (fr).

combinatory preferences

The choice of the components in such noun groups, like the choice of the components of compounds, is often collocationally determined: there are clear combinatory preferences, often merely conventional, that in many cases go as far as the complete terminological "fixing" of the compounds and noun groups. One says *Geldmenge* in de, not "*Geldmasse*", for fr *masse monétaire*. The relationship and the borderline between collocations of the noun-noun or adjective-noun type and "multiword terms" is not easy to describe. A linguist may look at the collocational aspect rather than at the terminological one, stating that, for a given term, the choice of adjectives or nouns that can be combined with that term, is not free, but defined as a lexical property of that term. From a terminological point if view, we may be more interested in whether the combination of term and collocate can be seen as the denomination of a new concept in its own right. For example, in maintenance literature for automobiles, we found the following three adjective-noun combinations:

- *pneumatische Leuchtweitenregulierung*
- *elektrische Leuchtweitenregulierung*
- *automatische Leuchtweitenregulierung*

These three collocations will likely need to be documented in a glossary or a termbase. As they denote subtypes of "Leuchtweitenregulierung", they may have associated concepts, and they may be considered as terms themselves, along with the noun *Leuchtweitenregulierung* itself.

multiword terms

We are not aware of any broadly agreed standard for distinguishing noun-noun and adjective-noun collocations from multiword terms. Moreover, it is often not the classification of the phenomena, but the additional lexical and terminological description that really matters. When designing a terminological data collection, one should, however, try to establish some explicit internal guideline, e.g., classifying subtype-denoting noun-adjective collocations as (multiword) terms, and noun-verb collocations as combinatory properties of a (nominal) term.

collocational inheritance

An additional factor should be considered: often, collocational preferences of a term for certain verbs or adjectives are "inherited" by the nominalizations of these verbs or adjectives. Thus: *épuiser les ressouces (de qn.)* vs. *épuisement des ressources*; *les salaires évoluent* vs. *évolution des salaires*, etc. (examples from Cohen 1986). A collocational dictionary should reflect this tendency either through cross-referencing, verbal and nominalized expression, or at least by ensuring that both verbal/adjectival and nominalized variants are listed.

Collocations vs. syntactic structure. Most NLP systems use some kind of phrase structure or chunk grammars (Abney 1991; Rooth and Carroll 1996). Some tools for detecting collocation candidates in texts also rely on such notions, especially tools for extracting terminologically relevant noun phrases. As observed, for example, by Estopà and Vivaldi (1998), most such devices can capture noun-noun-collocations and some noun-adjective-collocations, but rarely can combinations of nouns with verbs be extracted. Especially in a language like German, with (semi-)free word-order, nouns and verbs that together form collocations can be "distributed" over the sentence.

SEE
Term Extraction 8.4.1:725

The example in Figure 2 shows that in collocations constructed from verbs and their nominal complements, the distribution of syntactic heads (i.e., headwords) and their modifiers is complementary with respect to that of bases and collocates, contrary to what happens with noun-adjective collocations. In these instances, the head (noun) is also the collocational *base*, and the modifier (adjective) is the *collocate*. These cases are mostly found adjacent to one another in corpora.

INFO
Headwords INFB 6:330

Collocations can also involve *adjuncts*, which is important for both searching in corpora (e.g., prepositional phrases with certain prepositions) and for NLP treatment. Examples include: *abstrahlen mit Sand* (de), *test under pressure* (en), *unter Vakuum destillieren*, (de), *auf (sanfte) Rotglut erhitzen* (de).

Example	*strong*	*protest*	*attirer de*	*l'attention*
Category	adj	noun	v	noun
Phrase Str.	modifier	HEAD	HEAD	modifier
Collocation	collocate	base	base	collocate

Figure 2: *Phrase structure (HEAD vs. modifier) vs. collocational relations*

1.2 Semantic Properties

1.2.1 *Work on a Lexical Semantic Classification of Collocations*

lexical functions

Mel'cuk and his collaborators have formulated a well-known approach towards a semantic subclassification of collocations in the framework of the Meaning Text Theory (MTT, Mel'cuk et al., 1995). Their claim in this respect (extracted from the overall framework of their descriptive theory) is that collocations in general language serve to express quite general meanings, such as aspect-related or *Aktionsart* variants of verbal meanings, or intensification, phases of a process, etc. They have classified these collocations by using "lexical functions" as a representation of this general, abstract meaning (Mel'cuk and Polguère 1987; Mel'cuk *et al.* 1984). These classificatory devices have inspired sublanguage lexicographers, such as Cohen (1986). She describes terminology from the domain of the economy and the stock market in terms of the phases of processes (start, increase, decrease, end).

A sample from Cohen's dictionary (the entry *s.v.* ***emprunt)***, is illustrated in Figure 3. The use of devices for the structuring of collocations according to semantic criteria (as in MTT and in Cohen's dictionary) is most useful for tasks like text production because it not only helps to organize the material, but also to access collocations by the types of "abstract meanings" they potentially express. This perspective is a frequent factor in technical writing where the task may be, say, to write about exchange rates and to express that they "go up" or "go down". We will come back to the issue of access to collocational information in Section 2.1.

1.2.2 *Conceptual Collocation vs. Lexical Collocation*

collocational arbitrariness

In lexicography, examples of collocations are usually treated in terms of a given collocate with a given base being an arbitrary phenomenon that must be memorized. We say *eine Dusche nehmen* (de), *take a shower* (en), but *fare la doccia* (it), and we cannot exchange the verbal collocates (**prendere la/una doccia, *make a shower*, etc.). The choice of collocates in these lexical combinations is idiosyncratic and not necessarily rule-governed, which means that they must be memorized, for example in foreign language learning, or that they must be specifically cited in translation dictionaries.

emprunt	nouns	(subj. of) verbs	(obj. of) verbs	adjectives
START	émission lancement		émettre lancer	
INCREASE	accroissement augmentation	s'accroître augmenter monter	accroître augmenter	considérable élevé gros
UNDETERMINED				
DECREASE	baisse diminution réduction	baisser diminuer	réduire restreindre	petit
END			clore liquider rembourser restituer	

Figure 3: *The entry s.v.* **emprunt** *in the dictionary by Cohen (1986)*

partial compositionality

In specialized language, the same phenomenon occurs (e.g., *Geld abheben, withdraw money, retirer de l'argent*): the choice of collocates is a matter of convention. Not all of the terminologically relevant lexical combinations are "unpredictable" in this way, however, and, related to this, the aspect of partial compositionality, which is often mentioned as a definitionally relevant criterion for general language collocations, plays a much less important role in specialized language (L'Homme 1998: 517). Martin (1992) identifies factors that have an impact on lexical choice in terminologically relevant word combinations:

lexical choice

- Conceptual factors:
  - Collocations can denote subtypes of the concept denoted by the base term.
  - Collocations can denote typical properties, usages, and actions performed with the object denoted by the base term.
- Factors pertaining to lexical or conceptual classes:
  - Linguistic selectional properties of a whole class of items can appear as base terms.
  - The denominations of typical properties, usage, etc., of a whole class of objects can appear as base terms.

- Lexical factors:
  - Idiomatic or conventional preferences can be related to the base lexeme (typical case of a lexical collocation).

Martin (1992) describes *conceptual collocation* as follows:

conceptual collocations

"By conceptual structure we mean the semantic valency or argument structure of a concept (or the conceptual meaning of a lexeme). This way modifying concepts in "collocations" such as *infectious disease, nervous system, etymological dictionary*, etc., are conditioned or expected from "definitional knowledge". In other words: the conceptual meaning of *disease* contains a slot for CAUSE, that of *system* contains one for FUNCTION, that of *vowel* one for ARTICULATION, that of *dictionary* one for INFORMATION, etc. In this respect the above 'collocations' are concept bound."

lexical collocations

To get a feeling for the difference between lexical and conceptual collocations, consider the following data, extracted from two machine-readable dictionaries, i.e., CED (*Collins English Dictionary*) and the third edition of the *Oxford Advanced Learner's Dictionary* (OALD 3e). For the nouns *respect, smell* and *alcohol,* we extracted the adjectives and verbs shown in Figure 4.

noun	adjectives	verbs
respect	deep, deserved, due, exaggerated, excessive, great, profound	command, express, deserve, have, indicate, pay, receive, win
smell	acrid, appetizing, bad, banana-like characteristic, disagreeable, earthy evil, fishy, fragrant, [...], unpleasant, unusual, vile	unpleasant, unusual, vile emit, get, give, have, produce, resemble, take
alcohol	absolute, cyclic, denatured, diatomic, fat-soluble, industrial, natural, perfumed, simple, undiluted, unsaturated, water-soluble	collect, consume, drink, sell

Figure 4: *A few adjectives and verbs cooccurring with the nouns* respect, smell *and* alcohol *in CED and OALD*

Here is what we observe in these data:

- Some of the adjectives we see with *alcohol* denote subtypes of "alcohols" (e.g., *natural* vs. *industrial, perfumed* vs. *denatured*, etc.). We do not find such "subtype" adjectives with the other two nouns, or only rarely. *subtypes*
- Some of the adjectives that go with *alcohol* are related to characteristic properties of the concept of "alcohol", which would be found in a definition of the term. We have listed them in Figure 5, along with the properties they denote. *charateristic properties*

property	collocating adjectives
chemical structure	cyclic, diatomic, unsaturated
physical solubility	fat-soluble, water-soluble
concentration	absolute, undiluted
source of production	natural, industrial
substances added	denatured, perfumed

Figure 5: *Adjectives collocating with alcohol, from CED and OALD3*

- The adjectives we get for *respect* are more situated on a scale of intensity (*deep, profound, great*) or evaluative (*due, exaggerated*). With *smell*, we have both sorts, evaluative (*bad, evil, vile*) and quality-evoking (*acrid*) and, on the other hand, more conceptually oriented ones, e.g., by reference to the source of the smell (*fishy, earthy, banana-like*). *adjectives*

To sum up: there seems to be a cline from merely lexically or idiomatically determined collocations at one extreme to collocations where the collocate is more content-laden. In the former case, the collocate has no, or minimal, "schematic" meaning, as in support verb constructions: *pay* in *pay attention* is a "light verb" with a schematic meaning of "make, do, ..."). Terminologically relevant collocations are more often than not situated at the content-related end of the spectrum, with the meaning of the collocates fairly often derivable from definitional elements of the base, its properties, typical actions, usages, etc.

Thus, for terminology, the "concept-bound" nature of certain collocations is most important:

predictability

- On the one hand, conceptual collocations are closely related to definitions, insofar as the relevant classes of collocations are "predictable" from the conceptual analysis of a term. On the other hand, collecting adjective, verb, and noun collocates for a noun term (e.g., from a corpus) provides an overview of the definitionally relevant properties (see our example in Figure 5).

conceptual properties

- Certain combinatory phenomena are due to conceptual properties of the items involved, but seem to be applicable to sets of items (L'Homme 1998).

idiomatic aspect

- On the other hand, the idiomatic aspect of the actual choice of a collocate remains clearly visible in specialized language: e.g., *abort a process, activate a procedure, allocate data (to a location),* etc. (L'Homme 1992:499f.), for which we can foresee a general slot in the definitions of the base terms (type of operation to be performed), but we cannot predict the collocate lexeme to be used.

1.2.3 *Semantic Classes and Collocational Classes*

sublanguage collocations

Following on from the above observation that a large portion of terminologically relevant collocations are conceptual collocations (with full meaning in the collocates), an additional important fact about terminologically relevant collocations needs to be noted: there is a marked tendency to form series of sublanguage collocations, with nouns from a given field or conceptual class showing the same collocational preferences. For example, L'Homme (1998: 519) observes such series in the collocation behavior of nouns from the field of computing. Investigation into Cohen's dictionary of economic terms (1986) provides a further example[1]. In Cohen's dictionary (see above, Section 1.2.1 and Figure 3), we find verbs typically used to express the phases of a process (start, increase, decrease, end). The collocations are given for each of the about 100 nouns listed alphabetically in the dictionary.

collocational properties

The dictionary does not group the nouns in any semantically informed way. We have looked up the entries for nouns that share certain collocational properties. One type of question we asked was which subsets of nouns share one or more collocates expressing the INCREASE or DECREASE of the process denoted by the base noun, both with subject- and object- taking verbs. One such group consists of <*hausse, baisse, mouvement, progression, recul, repli, reprise*>: all these nouns share the collocates <*(s')amplifier, (s')accélérer, (s')accentuer*> to express INCREASE (pronominal reflexive verbs taking the nouns as subjects, the other verbs as objects), <*(se) ralentir*> and <*limiter qc., freiner qc.*>, to express the DECREASE . It is not hard to identify conceptual properties of this subset of nouns: they all denote changes in the economic evolution or just an economic evolution itself. The subset is quite homogeneous.

So is another subset: <*action, change, indice, titre, valeur mobilière*> which take <*monter, augmenter*> to express INCREASE and <*baisser*> to express a DECREASE (all verbs with nouns as subject). These denote shares or indices. The illustration in Figure 6 shows a few more such clusters.

collocational selectivity

We observe some verbs used across many groups. Examples are <*s'accroître, augmenter*> denoting the INCREASE of the processes recorded. Other verbal collocates are quite selective: for example, in Figure 6, the groups (4) and (5) are easy to distinguish ("action-like" events vs. "state-like" situations), and although the verbs denoting INCREASE are not selective, those denoting DECREASE are: <*freiner, réduire*> are shared by all nouns in (4), and *diminuer* is shared by all in (5). We have only indicated those verbal collocates that are indeed applicable to all elements of the groups. Cohen lists more that are applicable to subgroups of (4) and (5).

shared collocates

Both for specialized conceptual collocations and for collocations expressing more general concepts (such as increase or decrease), sublanguage terms seem to be considerably inclined to group themselves into subsets that share collocates. It makes sense at least when conducting field-related thematic terminology management to pay attention to this issue. Meyer and Mackintosh (1994) have made similar observations in texts on the use of CD-ROM players, and L'Homme (1998) has found the same phenomenon in computing terminology. Mel'cuk and Wanner (1994) have conducted much more detailed experiments for a subdomain of general language. To a certain extent, they find collocational behavior and semantic classes also to be parallel in general language, but the phenomenon is more widespread in sublanguage and thus more relevant to terminology.

2 Collocations in Dictionaries and Term banks

presentational options

In this section, we describe a few approaches to the presentation of collocational information in specialized dictionaries and in termbases. There are not many collocation dictionaries for sublanguage, and only a few terminological databases have a specific policy for the treatment of collocations.

No.	Clusters/Collocates	INCREASE noun=subj	noun=obj	DECREASE noun=subj	noun=obj
1	achat, concurrence, déficit, dépense, emprunt, épargne, excédent	s'accroître, augmenter	accroître, augmenter	diminuer	restreindre
2	action, change, indice, titre, valeur, mobilière	monter, augmenter		baisser	
3	économie, balance des paiements	s'améliorer		s'affaiblir, se dégrader, se de'teriorer	
4	activité, bénéficier, chômage, coût, commande, consommation, cours, exportation, importation, investissement, marché, marge bénéficiaire, masse monétaire, offre, pouvoir d'achat, production, productivité, profit, taux, vente	s'accroître, augmenter	accroître	baisser	
5	charges, concurrence, déficit, dépenses, déséquilibre, dette, écart, échanges, emprunt, épargne, excédent, impôt, liquidités, perte, plus-value, rendement, réserves, ressources, volume des échanges, volume des transactions	s'accroître, augmenter	accroître	diminuer	diminuer

Figure 6: *Nouns sharing verbal collocates, retrieved from Cohen 1986*

dual presentation

The presentation of collocational information in printed dictionaries is not an easy task for the following reasons: collocations involve two elements, which usually each appear separately in the dictionary. The question is where to sort collocations, and, related to this problem, which element(s) to choose to make the collocation accessible to the user.

Information about collocations that the user wants to look up in a dictionary or termbase includes at least the following:

look-up information

- The sheer fact that the lexical combination exists and has the status of (a frequent, conventional) collocation
- The morpho-syntactic properties of the elements and of the collocation as a whole
- Possibly the semantic properties (or lexical semantic classification) of the collocation or its relationship with a (possibly formalized) conceptual description of the base term
- (Near) synonyms, related expressions, and related terms
- Equivalents in other languages

access paths

Depending on the type of information the user is interested in, and on the situation of use (and type of communicative task the user of the dictionary is supposed to accomplish), there should be different access paths to the collocational information. In technical writing, for example, and likely also in translation into a foreign language, the user enters the dictionary with the meaning or the conceptual description of the base (and with the base lemma or a set of quasi-synonyms). The secondary search criterion is the type of "meaning" that must be asigned to the collocation for target text production. Example: "I want to write on taxes and express the idea that the government 'increases' the taxes. I am looking for the verb that best expresses this concept."

2.1 Dictionaries

data model design

Given the above issues, the following problems occur in the treatment of collocations in dictionaries, and to some extent, in the design of the data model underlying a terminological database designed to contain collocational information:

- Under which element of the collocation (base or collocate?) is the collocation listed and from which entry word is it accessible (in a database)? Are collocations entry words themselves?
- Which "access paths" to collocations are supported (e.g. "base term → semantics of collocation → collocation", in the example above) are supported?
- Which information (syntactic, semantic, pragmatic, lexical relationships, equivalence) is given for a collocation?

syntactic properties

One example of a series of collocational dictionaries of specialized language (coordinated by the Canadian Secretary of State for English/French) is the *Vocabulaire combinatoire de la CFAO mécanique* (Lainé 1993). This dictionary, which covers collocations and noun groups from the domain of CAD/CAM in French and English, is corpus-based, and is intended for translators. We

ordonnancement	**scheduling**
~ V.	*~ connaître les ordres lancés*
V. ~	*choisir ~, définir ~, essayer ~, [règles] gouverner ~*
~ Adj.	*~ assisté par ordinateur, ~ dynamique, ~ informatisé, ~ multiconvergent, ~ optimal*
~ (Prép) (Art) N	~ à buts multiples, *~ par dates croissantes, ~ par valeurs croissantes des marges libres*
N (Prép) (Art) ~	*l'art de l'~, coefficient d'~, fonction~, méthode d'~ (de production), rebouclage sur l'~, technique d'~*

Figure 7: *The entry s.v.* **ordonnancement** *(Laine 1993)*

reproduce the entry s.v. *ordonnancement* in Figure 7. The entry contains a listing of collocations, sorted by syntactic properties: verbs for which the base noun acts as the subject or complement, adjectives typically used to denote subtypes (conceptual collocations!), noun groups with and without prepositions.

entry structure

The overall structure of the entry is symbolized in Figure 8. Access to collocations is provided via the following path: "base term → grammatical shape of collocation → collocation (set)." The assumption in the dictionary seems to be that no further sense distinctions are to be made for the base term: *ordonnancement* in *l'art de l'ordonnanement* and in *rebouclage sur l'ordonnancement* are seen as instances of one reading.

multiple access paths

Another important collocational dictionary is Cohen's *Lexique de cooccurrents: bourse–conjoncture économique* (1986). We have discussed it above and reproduced an entry from this dictionary in Figure 3. This dictionary is interesting in terms of its organization because the information is displayed in tables, abandoning the linearity of traditional dictionaries. Thus the dictionary supports access via more than one path:

- Base term → syntactic behavior of the collocate → semantic description of collocation → collocation
- Base term → semantic description of collocation → syntactic behavior of the collocate → collocation

hypertext access

It seems that in the future such a multidimensional presentation of collocational information will not only become more important, but also easier to realize. Typically, computerized hypertext systems support the sort of multi-access dictionary for which Cohen provides a foretaste on paper.

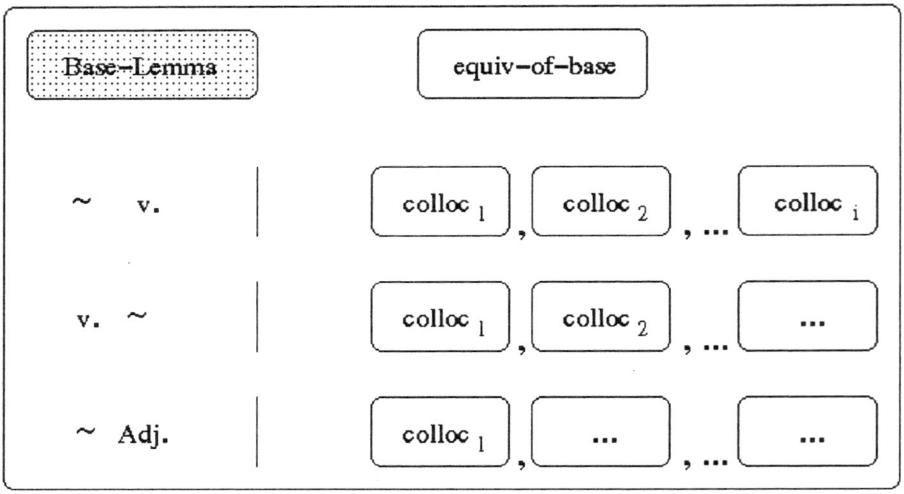

Figure 8: *Schema of the entry structure in Laine 1993*

2.2 Terminological Databases

A very good overview of data models for terminological databases is given by Mayer (1998), with a short summary already in (Mayer 1993). The later treatment provides a very detailed discussion of the problems raised by providing a full account of collocations within a bilingual or multilingual database. *data models*

An early attempt to model collocations in a terminological database is described by Heid and Freibott (1990, 1991). This implementation has shown that at least the problem of access to collocations can be solved within a relational database. The goal in this instance is to provide collocational information for both translation from and to a foreign language. To cater flexibly to this multifunctional aim, the collocations are alternatively made accessible via the base term, the collocate, or the collocation as a whole. Collocations are classified syntactically in terms of the category of the elements (see the introduction to this article). No semantic classification of the collocations themselves is provided, but the collocational description can be integrated with the concept-oriented approach underlying the database. *multifunctionality*

Mayer (1998) shows that this approach, though workable, leads to major overhead in terms of relations and of entries for components of collocations. He proposes a simpler, more efficiently implementable data model that could rely on the assumption that *excess overhead* there has to be at least one term in a collocation for which a full termbase entry exists. Collocations are then linked to these terms, and possibly to their translation equivalents and to quasi-synonymous collocations. All other relationships may need to be expressed by reference to the base term. Access via collocates may be provided through an automatically calculated collocate index (see Groesz et al. 1999).

3 Extracting Collocational Evidence from Texts

collocation extraction tools

There are different types of tools for the extraction of collocation candidates from text based on linguistic or statistical processing, or on combinations of both. We cannot go into much detail here (see Estopà and Vivaldi 1998), but we will try to point to a few of the principles underlying these tools. Collocation extraction tools differ in nature from term extraction tools: term extractors either search for single word or multiword term candidates or for definitionally relevant contexts, such as definition phrases ("an x is a ..."; "we call x a ..."). Pearson provides a detailed overview (1998), and Estopà and Vivaldi offer a comparative assessment of term extractors.

extraction problems

The extraction of terminologically relevant collocations has to deal with two types of problems:

- Criteria for collocation status and discovery procedures for collocations
- Criteria for terminological relevance of single words and of word combinations, such as collocations

orthogonal aspects

These two aspects are to some extent orthogonal: depending on the language under analysis, different linguistic tools are available for identifying collocations of different kinds. These tools tend to extract all collocation candidates contained in a given text, or all that satisfy some sort of frequency criterion. Yet after extraction, the terminological relevance of the extracted word combinations still needs to be established. This may be accomplished by statistical means (including different kinds of cooccurence statistics and/or simpler, e.g., relative frequency-based, indications of the terminological relevance of the base term), or it can be based on lexical lists of candidate base terms.

multiword term recognition

As with multiword term recognition in general language, the ongoing research debate centers on how to combine statistical and symbolic analysis. To the extent that any general comment on these efforts is at all possible, it seems that it is preferable to first run a symbolic extractor to find collocation candidates, and to then experiment with statistical measures to reduce the candidate set to those combinations that are terminologically relevant. Especially when dealing with texts in inflecting languages, or in languages with some degree of freedom in word order, such as the Romance and Slavic languages or German, it makes sense to run a collocation candidate finder based on grammatical knowledge first before applying statistical filtering.

co-occurence finders

Statistical co-occurence finders rely mainly on frequency data. For adjacent collocations (e.g., noun + adjective), bigram frequencies (frequency of any combination of two word forms) can be extracted easily and compared with the frequency of each individual bigram component. There are different measures for co-occurence that are calculated in different ways: Mutual information, the t-test, Z-Score, the log-likelihood measure (see Dunning 1993), etc. Very generally speaking, these calculations are designed to determine frequent and/or typical co-occurence phenomena. Lezius (1999) reports on a comparison of these measures with a view to assessing their usefulness for identifying idiomatic combinations.

log-likelihood measures

Many researchers favor the log-likelihood measure as a collocation finder. In layman's terms, it can be seen as comparing the (potentially significant) co-occurence of two items, say a given noun and a given verb, with three other co-occurence phenomena: the cases where the targeted noun co-occurs with verbs other than the one we are interested in, the cases where the verb occurs with other nouns, and finally the cases where nouns and verbs cooccur, but our candidates do not participate. From these observations, a typicality measure for the candidate collocation is calculated. High scores are likely to indicate collocation status, whereas low scores are interpretable, in most cases, as indicators of a trivial combination. For details (and a statistical account) consult Dunning (1999) or for an implementation, e.g., Daille (1996).

filtering

We chose to use a log-likelihood measure as a filter designed to distinguish trivial combinations from collocations and applied this tool to a database of about 45,000 collocation candidates (noun-verb combinations) extracted from German newspaper texts. For the noun *Frage* (de "question"), the following combinations received high log-likelihood values (in decreasing order): *eine Frage stellen, in Frage kommen, eine Frage stellt sich, eine Frage beantworten, die Frage lautet, eine Frage klären, eine Frage aufwerfen*. At the bottom of the list, we find (in decreasing order, through to the end): *eine Frage ansprechen, in einer Frage [so] stimmen, eine Frage behandeln, eine Frage hören, lösen, die*

Frage heißt..., die Frage bewegt jemanden. From the top-ranked candidates, only *die Frage lautet* is fully compositional, whereas all other combinations are collocational. The combinations with lowest log-likelihood values seem to be all compositional.

Mutual Information

The other measures have also been used, in particular for term extraction from English texts. For unique combinations, the Mutual Information (MI) approach is most useful. The mutual information measure itself does not say anything about the frequency of the combination or of its elements: it can be high for a combination that occurs three times in the corpus, but each time the components occur, they occur together. Some concordancing tools integrate a mutual information calculation for word pairs. We ran MI on a large English corpus targeting the word form *post* to find collocates. The highest figure was obtained for the sequence *post mortem* (since *mortem*, if it shows up, always appears adjacent to *post*). Mutual information is unreliable for small quantities of data and on rare objects (sparse data problem), and thus should only be used together with other statistical devices, such as t-score and frequency data, or the z-score test (Berry-Rogghe 1974). The t-test allows the user to identify frequent typical combinations and thus to correct the results obtained by computing mutual information.

linguistic analysis

The statistical tools themselves are, of course, not linguistically informed. They may be used to filter the results of previous linguistic analyses. In the most elementary case, the statistical filter may be run on lemmatized data, possibly sorted by category (as in our experiment on log-likelihood for noun-verb combinations). To get such sorted candidate pairs, regular grammars "local grammars" or chunking may be used (Abney 1992; Rooth and Carroll 1996). Heid (1996) describes simple local grammars, which have been used to extract noun-verb and noun-adjective combinations from German texts. For filtering, only a relative frequency/based calculation is used, which allows one to judge whether the potential base, the collocate, or both have term status, i.e., are specific to the text analyzed, insofar as they are more frequent in that text than anticipated from general language texts (see Ahmad et al. 1992).

parsing approach

More linguistically oriented exploration tools include devices for the identification of noun phrases, e.g., in terms of part-of-speech shapes, or by means of (partial) parsing. These tools allow users to extract noun-adjective and noun-noun combinations. They can also be used to identify the subject or object noun (phrase)s used with verbs, thereby giving access to selectional information and collocation candidates at the same time. Smajda provides an example of a partial parsing approach (1994).

Current thinking is that combinations of linguistic preprocessing (ideally to the level of chunks, phrases and grammatical functions) and statistical filtering provide the most adequate extraction results. However, research is still ongoing for the extraction of terminological collocations.

combined approaches

4 SUMMARY

This chapter offers an overview of some relevant aspects of collocations, their description in terminology, their presentation in dictionaries and termbases and some of the general principles underlying tools for their discovery in texts.

It seems that awareness of collocational phenomena in terminology is by now mostly standard, but data models that will allow the terminologist to adequately capture collocational knowledge in termbases and tools to feed such termbases are still in the developmental stage and are only being used as prototypes. In these areas, we expect interesting developments over the next few years.

prototype tools

REFERENCES

Abney, Steven. 1991. "Parsing By Chunks." In: *Principle-Based Parsing*, Robert Berwick, Steven Abney and Carol Tenny, eds. Dordrecht: Kluwer Academic Publishers.

Ahmad, Khurshid et al. 1992. "What is a term? The semi-automatic extraction of terms from text." In: *Translation Studies—an Interdiscipline*, Mary Snell-Hornby et al.,eds. Amsterdam/Philadelphia: John Benjamins Publishing Company, 267-278.

Benson, Morton. 1989. "The Structure of the Collocational Dictionary." In: *International Journal of Lexicography*. Oxford: Oxford University Press.

Benson, Morton. 1990. "Collocations and General-purpose Dictionaries." In: *International Journal of Lexicography*. Oxford: Oxford University Press, Spring 1990, Vol. 3, No. 1, 23-34.

Berry-Rogghe, Godelieve L.M. 1974. "Automatic identification of phrasal verbs." In: J.L. Mitchell, ed.. *Computers in the Humanities*, Edinburgh University Press, Edinburgh.

Cohen, Betty. 1986. *Lexique de cooccurrents; Bourse—conjoncture économique*. Montréal: Linguatech.

Cohen Betty. 1992. "Méthodes de repérage et de classement des cooccurrents lexicaux." In: *T & T 1992*, 505-512.

Daille, Béatrice. In press. "Study and implementation of combined techniques for automatic extraction of terminology." In: *The Balancing Act—Combining Symbolic and Statistical Approaches to Language*. Judith L. Klavans, Philip Resnik, eds. Cambridge, MA: MIT.

Dunning, Ted. 1993. "Accurate Methods for the statistics of surprise and coincidence." In: *Computational Linguistics*, 19 (1), 61-74.

Estopà, Rosa and Vivaldi, Jordi. 1998. "Etat de la question des systèmes d'extraction automatique des candidats à terme: vers une proposition intégratrice." In: *Actes des 7èmes Journées ERLA-GLAT: Lexicologie et terminologie, interculturalité et analyse du discours*. Brest.

Fontenelle, Thierry et al., eds. 1998. *EURALEX'98 Proceedings. Papers submitted to the Eighth EURALEX International Congress on Lexicography*. 2 Vols. Liège, Belgium: Université de Liège.

Groesz, Inger; Maks, Isa; and Martin, Willy. 1999. "Data Model DOT." In: *Report No. 2 of the Project 'Database Overheidsterminologie', DOT*. Amsterdam: Vrije Universiteit.

Hausmann, Franz Josef. 1989. "Le dictionnaire de collocations." In: Hausmann et al. 1989, 1010-1019.

Hausmann, Franz Josef; Reichmann, Oskar; Wiegand, Herbert E.; and Zgusta, Ladislav, eds. 1989. *Wörterbücher: ein internationales Handbuch zur Lexikographie. Dictionaries. Dictionnaires. Wörterbücher,* Vol. 1. Berlin, New York: de Gruyter.

Hausmann, Franz Josef. 1997. "Semiotaxis und Wörterbuch." In: Konerding and Lehr 1997, 171-179.

Heid, Ulrich. 1992. "Décrire les collocations—deux approches lexicographiques et leur application dans un outil informatisé." In: *T&T 1992*, no. 2/3, 523-548.

Heid, Ulrich and Freibott, Gerhard. 1990. "Zur Darstellung von Äquivalenten in einer terminologisch-lexikalischen Datenbank für Übersetzer und technische Autoren." In: Burkhard Schaeder and Burkhart, Rieger, eds. *Lexikon und Lexikographie. Vorträge im Rahmen der Jahrestagung 1990 der Gesellschaft für Linguistische Datenverarbeitung (GLDV)* e.V., Siegen, 26. - 28. März 1990. Hildesheim, Zürich, New York:... , 244-252.

Heid, Ulrich and Freibott, Gerhard. 1991. "Collocations dans une base de données terminologique et lexicale". In: *Méta*, Vol. 36, no.1, Montréal.

Heid, Ulrich; Jauss, Susanne; and Krüger, Katja. 1996. "Term extraction with standard tools for corpus exploration—Experience from German." In: *Proceedings of TKE '96*, Frankfurt: Indeks, 139-150.

Hoffmann, Lothar; Kalverkämper, Hartwig; and Wiegand, Ernst Herbert, eds. 1998. *Fachsprachen. Languages for Special Purposes. Ein internationales Handbuch zur Fachsprachenforschung und Terminologiewissenschaft*, Vol. 1. Berlin: Walter de Gruyter.

Konerding, Klaus-Peter and Lehr, Andrea, eds. 1997. *Linguistische Theorie und lexikographische Praxis; Symposiumsvorträge, Heidelberg 1996.* Lexicographica Series Maior 82. Tübingen: Niemeyer.

Lainé, Claude. 1993. *Combinatory Vocabulary of CAD/CAM in Mechanical Engineering*. Ottawa, Canada.

Lezius, Wolfgang. 1999. "Automatische Extrahierung idiomatischer Bigramme aus Textkorpora." In: *Tagungsband des Linguistischen Kolloquiums*. Germersheim.

L'Homme, Marie-Claude. 1992. "Les unités phraséologiques verbales et leur représentation en terminographie." In: *T&T 1992*, no. 2/3, 493 - 505.

L'Homme, Marie-Claude. 1998. "Caractérisation des combinaisons lexicales spécialisées par rapport aux collocations de langue générale." In: Fontenelle *et al.* (Euralex 1998), 513 - 522.

Martin, Willy. 1992. "Remarks on Collocations in Sublanguage", In: *T&T 1992*, no. 2/3 157-164.

Mayer, Felix. 1993. "Entry models in translation-oriented computer-assisted terminography," In: Schmitz, ed. 1993, 205-214.

Mayer, Felix. 1998. *Eintragsmodelle für terminologische Datenbanken. Ein Beitrag zur übersetzungsorientierten Terminographie*. Forum für Fachsprachen-Forschung, 44 Tübingen: Narr.

Mel'cuk, Igor A.; Arbatchewsky-Jumarie, Nadja; Elnitsky, Léo; Iordanskaja, Lidija; and Lessard, Adèle. 1984. *Dictionnaire explicatif et combinatoire du français contemporain. Recherches Lexico-Sémantiques*, I. Montréal: Presses Universitaires de Montréal.

Mel'cuk, Igor A.; Clas, André; and Polguère, Alain. 1995. *Introduction à la lexicologie explicative et combinatoire*. Louvain-la-Neuve: Ducolot.

Mel'cuk, Igor A. and Wanner, Leo. 1994. "Towards an Efficient Representation of Restricted Lexical Co-occurence." In: *Proceedings of EURALEX-94 International Congress*.

Meyer, Ingrid and Mackintosh, Kristen. 1994. "Phraseological Analysis and Conceptual Analysis: Exploring a Symbiotic Relationship in the Specialized Lexicon.". In: *Proceedings of EURALEX-94 International Congress*.

Pavel, Silvia. 1993. *Bibliographie de la phraséologie (1905-1992)*. Montréal: Secrétariat d'Etat du Canada/Rint.

Pearson, Jennifer. 1998. *Terms in Context*. Amsterdam/Philadelphia: John Benjamins.

Rooth, Mats and Carroll, Glenn. 1996. "Valence Induction with a Head-Lexicalized CFG," ms., Stuttgart: IMS.

Schaeder, Burkhard and Bergenholtz, Henning, eds. 1994. *Fachlexikographie. Fachwissen und seine Repräsentation in Wörterbüchern*. Tübingen: Narr.

Schmitz, Klaus-Dirk, ed.. 1993. *TKE '93. Terminology and Knowledge Engineering, Proceedings of the Third International Congress on Terminology and Knowledge Engineering, Cologne 1993*. 2 Vol. Frankfurt: Indeks Verlag.

Schmitz, Klaus-Dirk, and Galinski, Christian, eds. 1996. *TKE '96. Terminology and Knowledge Engineering, Proceedings of the Fourth International Congress on Terminology and Knowledge Engineering, Cologne*. 2 Vol. Frankfurt: Indeks Verlag.

Smajda, Frank. 1994. "Retrieving Collocations from Text: Xtract." *Using Large Corpora*. Susan Armstrong(-Warwick), ed. Cambridge, MA/London/England: MIT Press: 143-177.

Terminologie et Traduction N° 2/3-1992. Luxembourg: Commission des Communautés européennes. 1992. [*T&T 1992*]

1. Special thanks to Regina Steding, who has much contributed to our experiments by encoding the relevant part of Cohen (1986) and by querying our resulting data collection. Without her contribution, we would not have been able to provide the material discussed in Section 1.2.2.

8.4.5 The Role of Specialist Terminology in Artificial Intelligence and Knowledge Acquisition

KHURSHID AHMAD

1 INTRODUCTION

Possession of knowledge and the ability to articulate and use it in a timely and efficient manner is crucial for the well-being of individuals organized into specialist communities. The possession, articulation, and uses of knowledge generally involve the exchange of information to and from individuals and their environment through the agency of cognitive faculties, perceptive capabilities, and motor skills. Language plays a crucial role in this exchange. Timely and efficient information exchange requires intelligence. *Artificial intelligence* (AI) is a branch of computer science that aims to understand the nature of human intelligence by constructing computer programs that exchange information with human users. If the artificially intelligent programs are deemed to exchange with their human users, then it is hoped that eventually such programs will be able to exchange information with each other.

information exchange

An important way in which human intelligence manifests itself is in the solution of problems by experts in highly specialized domains through the application of specialist knowledge. In many domains, much of this knowledge is undocumented, largely experimental and usually qualitative. A *knowledge-based expert system* (KBES) is an artificially intelligent computer system that comprises problem-solving expertise in well-defined areas, such as meningitis diagnosis and therapy or computer configuration, and helps users, often beginners in a specialism, to use the stored knowledge. This enterprise of simulating problem-solving knowledge is referred to as *knowledge engineering* and it is the *knowledge engineers* who design and build KBESs.

expert systems

Given that much problem-solving knowledge is undocumented and experiential, knowledge engineers acquire problem-solving knowledge by interviewing experts and by consulting a range of

knowledge base

texts within a specialist domain. Once the knowledge is acquired, the engineers go on to create a systematically organized repository of this knowledge in a *knowledge base*, which comprises *objects, facts,* and *rules*. Objects are used to denote and represent entities of the specialist domain, for example in the domain of microbial infection therapy, a bacterium or a drug to neutralize it, as well as abstract entities like infection. Terms like *concept* and *frames* are also used to refer to the objects. The term "fact" is used for statements that propose that a certain relationship exists between some or all the objects of the domain. The knowledge base also comprises *rules-of-thumb* or *heuristics* for solving problems.

knowledge acquisition

The term *knowledge acquisition* assumes different shades of meaning as we cross the boundaries of various related disciplines. Computational linguists, corpus linguists and computational lexicographers use the term to refer to methods and techniques that enable humans to acquire knowledge of a given language from archives of language use. Such archives include text corpora and dictionaries, in both paper and electronic versions. This usage of *knowledge acquisition* is in many ways related to the concept itself, and the associated term, *language acquisition*. The important difference here is that linguists and lexicographers now invariably use *knowledge acquisition* in the context of computer-mediated acquisition of linguistic knowledge. Only a decade ago, an undertaking of this kind was costly and time-consuming because computer mediation depends on a ready supply of digitized input data, but the ubiquitous Internet has now made it possible to collect large volumes of digital texts readily at a low cost.

Cognitive psychologists use the term *knowledge acquisition* in the context of developmental behavior, e.g., children who are acquiring and subsequently articulating their knowledge of their immediate environment comprising animate and inanimate things. Cognitivists also use knowledge acquisition in the context of skills acquisition.

terminology in text

In AI, the use of term *knowledge acquisition* embraces the sense in which the term is used in computational linguistics, computational lexicography, and cognitive psychology. One of the active sub-disciplines of AI is called knowledge acquisition. In this area there are workers who attempt to obtain knowledge from subject experts and from textual sources for building expert systems. And there is a group of researchers in knowledge acquisition who are develop programs that attempt to learn from their immediate environment. AI-oriented knowledge acquisition emphasizes familiarity with the terminology of a given specialist domain and involves the elaboration of the terms for problem-solving purposes. There are ways in which current practice in terminology can assist knowledge engineers to acquire knowledge systematically and efficiently. In the subsequent parts of this chapter, the term *knowledge acquisition* will be used only in the context of artificial intelligence. Furthermore, the remainder of the chapter will focus on knowledge acquisition from text and on the role that terminology plays in such an enterprise.

SEE
Term Extraction 8.4.1:725

INFO
Termbase Terminology INFB 28:878

Knowledge acquisition literature is slowly acknowledging the utility of applying the theory and practice of terminology management and text analysis, and terminology literature shows the interest terminologists are taking in this innovative branch of knowledge management. This can only be good news for the terminology community because the development of a range of information dissemination systems, for example, expert systems, information retrieval systems, data mining, and knowledge discovery systems, can be expedited by understanding how (special) language works at the lexical, syntactic and semantic levels, and by using suitably elaborated termbases. This work can be expedited further by adapting methods and techniques of corpus linguistics for organizing full-text databases that can store and retrieve texts of different genres. Researchers in terminology and linguistics have suggested that knowledge acquisition will do well to look at the ways terms can be automatically extracted from texts by either statistical analysis, especially corpus linguistic, or by morpho-syntactic and semantic analysis. In addition, knowledge acquisition can also be facilitated by the use of adapted termbases.

validation

There is a synergy between the enterprise of knowledge engineering and terminology science and practice. We will elaborate more on the synergy between knowledge acquisition and terminology during the rest of this chapter. However, for a synergy to exist, terminology should not only be for the benefit *of* knowledge engineering but should also benefit *from* knowledge engineering. We will briefly outline how work in *knowledge representation* can be used to build termbases, or rather terminology *knowledge bases* of the future. The difference between the two is that a knowledge base can autonomously "reason" over its contents, terms, and the elaborations that are represented in it, as opposed to encoded, encrypted databases, and can thus self-validate its contents.

2 Special Language and Knowledge Transfer

specialist language

Knowledge engineers interview experts in a specialist domain about how they solve a given problem. Before interviewing the experts, the knowledge engineers have to formulate their questions, and after the interview the answers to the questions have to be analyzed. They have to familiarize themselves with the terminology of the specialist domain, consulting technical manuals, and in some cases learned papers, to see how the experts' knowledge is applied, and they sometimes consult textbooks or encyclopedic texts for understanding the conceptual structure of the experts' domain. In many different ways, the knowledge engineer literally has to come to terms with the language used by the expert and that used in the other texts mentioned above, thus becoming conversant in the specialist language of the application domain.

knowledge acquisition tools

In a review of knowledge acquisition, Boose (1992) lists 110 computer-based tools for facilitating knowledge acquisition and classifies them in six major categories reflecting the methods used in acquiring knowledge: *psychology-based and interviewing methods; problem, method and performance exploitation; modeling; knowledge browsers; multiple experts; and, other sources of knowledge*. Boose lists systems for "textual analysis, natural language analysis" under the "other sources" category, and some of these systems are also listed under the "ontological modeling, linguistic modeling" subcategory of modeling-oriented methods. Our intention here is not to criticize the above categorization, especially since the relationship between knowledge (acquisition) and language is a highly complex one, but rather to draw the attention of the reader to work in terminology science and practice where "ontological modeling" and its relationship with "linguistic modeling" has been debated over the years, and where we find systematic ways of dealing with specialist language, for example, in analyzing conceptual organization of specialisms and using such analysis for building terminology databases (termbases).

It is especially unclear whether the knowledge engineering community is aware of the nature and function of the special language texts used and processed. For instance, such texts can be differentiated from general language texts, such as newspapers, at a syntagmatic level, in that special language mainly uses declarative and imperative sentences, whereas general language additionally comprises exclamatory and interrogatory sentences. In special language, for example, one encounters pre- or post-modified nominal groups more frequently than in general language. There is low frequency usage of personal pronouns. Special language texts can be differentiated from general language at the lexical level in that while a general lexicographer faces problems in distinguishing between inherent and collocational meanings of a lexical item, the specialist terminologist has to refer to the knowledge organization within a specialism. Moreover, the specialist terminologist relies on the evidence of a smaller, yet more authoritative, group of people –the specialist experts–than the authorities used by the lexicographer, who has to rely on the evidence of a much larger group.

special language texts

Lexicographers consider three major processes that affect lexical development–*naming, defining*, and *designation*. The terminologist is usually focused on the designation of terms and has to consider various determining mechanisms such as *affixation, prefixation, backformation, compounding, deprecation, borrowing*, and *neologisation* within specialist terms. Special language has its own lexicogrammar characterized by a relatively large vocabulary, restricted syntax, and overt control of semantics. The large and dynamically changing vocabulary, the restricted syntax, and an emphasis on monosemy are all intended to promote rapid and unambiguous dissemination of scientific and technical knowledge. It may be that the artificial intelligence community in general, and knowledge engineers in particular, have paid little attention to specialist text and, unfortunately, have really not exploited the lexicogrammatical idiosyncracies of such texts. However, key researchers in linguistics have looked at knowledge acquisition from texts.

SEE *Term Formation 1.2.1*:25

In a characteristically prescient manner, Terry Winograd and Robert de Beaugrande have respectively predicted that one of the major applications of computers will be in the area of knowledge acquisition, where "the hope is that after reaching a certain level such systems could be given material in the form of text and incorporate the contents without human intervention" (Winograd, 1983:25) and that it may be possible to "design a computer system capable of reading, understanding and writing technical reports"

automatic parsing

(Beaugrande, 1979:239). Beaugrande (1979:240-250) has specified a system that will perform "read-in of natural language texts", that is, will be able to parse a text, identify key topics within the text, create a meaning network of the identified topics, and reason over the represented knowledge in order to infer new facts from old. He stresses the need for an infrastructure for performing the "read-in", which comprises a parser capable not only of syntactic parsing but also of:

a. Identifying the roles of participants [within a text] with respect to actions and events denoted by a sentence
b. Resolving anaphoras, if any, in the text, and
c. Evaluating the focus and newness of the sentences within a text

conceptual glossaries

The parser's outputs are then processed using a *semantic associator* which can create "frames [of reference] for objects and events in the sentences," and can match "incoming words to proper stored concepts, adding new properties or relations asserted by the text" (Beaugrande 1979:250). The parsers and the semantic associators communicate with each other and are supported by the lexicon component comprising a macroglossary and "a set of microglossaries". The former are used for general vocabulary and the latter for "supplying specialized terminology for technical fields" (ibid:243). The author stresses that these glossaries should be conceptually organized. What we learn from Beaugrande's specification of computers that can "read-in" unrestricted (specialist) natural language texts and create a meaning representation of (key) sentences is that a text first has to be parsed syntactically for case information and for anaphora resolution. For us, the most important aspect that facilitates parsing and semantic analysis is the use of conceptually organized macro- and microglossaries.

text analysis

The prescience of Beaugrande and Winograd is reflected by more recent work in knowledge acquisition of the previous decade, for instance in the work of Gaines and Shaw (1992). Their KITTEN system, categorized by Boose as a "psychology-based and interviewing method" (1992:724), has facilities for creating a text base and for analyzing the texts. The authors' stated aim is to detect "clusters of words" for building repertory grids. In the 1990s, however, there has been a welcome reference to the structure and function of specialist texts in the knowledge acquisition literature. For example the long-running *Knowledge Analysis and*

Documentation System Project (Wielinga et al. 1992) comprises tools for "lexicon construction" and "glossary construction" (Anjewierden et al. 1992).

learning vocabulary

Projects at Stanford University, where expert systems were first created, also emphasize that the knowledge engineers must become experts in that they should "learn a new vocabulary" (Musen 1993). Musen's colleagues refer to their interaction with "medical lexicographers" and to their use of "standard vocabularies" for building medical expert systems (Egar et al.1992). More recent writing in the terminology and knowledge engineering literature acknowledges the symbiosis between the two disciplines in the context of computer-aided knowledge acquisition. Bourigault and his colleagues have developed the LEXTER (Logiciel d'EXtraction de TERminologie) system, which provides text-analysis software tools for knowledge acquisition, comprising *transfer tools* for converting text directly into a knowledge base, *hypertext editors* used to visualize the semantic relations between terms, and *scanning tools* for interpreting certain aspects of texts. The LEXTER system performs a morpho-syntactic analysis of a text corpus to identify nominal expressions, particularly compound nouns. Once identified, these expressions are visualized through a hypertext editor (Bourigault et al., 1996). LEXTER appears to be an implementation of de Beaugrande's specification in that the key part of the analysis relies on syntactic parsing and the semantic association is affected through hypertext links.

speed factors

Some regard knowledge acquisition as a major "bottleneck" in the speedy development of artificially intelligent expert systems. We believe that some of the causes of the bottleneck may be related to the rather casual ad hoc way in which the terminology of specialist domains is treated by knowledge acquisition experts and practitioners. This bottleneck is exacerbated by demands made on the knowledge engineer to read and assimilate facts and heuristics about a specialism by studying a range of specialist texts.

3 Notes on Knowledge Engineering

The articulation and use of specialist knowledge depends crucially upon how much one can understand and elaborate the terminology used by the specialists. Terminologists make it their business to understand and elaborate terms, with the help of experts, documenters and others, and to make terminology collections available to translators, technical writers, novices, and experts. Knowledge engineers are engaged in activities that bear remarkable similarities to the work of terminologists in term identification and elaboration with the help of experts.

understanding terminology

We fully realize that there are important differences between the goals of a terminologist and that of a knowledge engineer. Terminologists are motivated to help humans document and translate technical and scientific texts. The knowledge engineer focuses on incorporating human knowledge into a computer system, a knowledge base to be precise. The cognitive orientation of a knowledge engineer is determined by concerns related to problem-solving, whereas the terminologist's orientation is determined by the concerns of language comprehension and production. But both deal with specialist knowledge in a specialist domain. The burden of our argument is that unlike terminologists, knowledge engineers usually rely on their intuition in organizing terminology collections.

different goals

3.1 Engineering Knowledge on a Computer System

Knowledge-based expert systems are developed to store and, on demand, retrieve the knowledge of an expert or experts for solving complex problems. The knowledge in an expert system is almost invariably based on experience, the so-called heuristics, which in some cases is possessed only by a limited number of experts. An expert system is essentially a computer program which, with its associated data, embodies organized knowledge concerning some specific area of human activity. Such a system is expected to perform competently, skillfully, and in a cost-effective manner. It may be thought of as a computer program that mimics the performance of a human expert. The distinguishing feature of an expert system is that it is a repository of "human knowledge" to which authorized users have access at all times. Once the documentation is saved in a machine, the knowledge can be used by two groups: the (end-) user and the subject specialist or expert. The end-user can use this encoded knowledge in solving problems and the expert can inspect the knowledge for the purposes of verification and validation. The user can, in principle, ask the program to "explain" its input requirement and "justify" its output. An expert system can, for

system objectives

instance, give reasons why it requires a certain item of input data, or how it reaches a decision by enumerating the rules used.

Knowledge engineers "engineer" knowledge in an expert system through the execution of a four phase strategy:

engineering knowledge

- Knowledge Acquisition: The knowledge engineer acquires the knowledge of the domain.
- Knowledge Representation: The engineer stores this knowledge inside a computer system using one of the many formalisms available for knowledge representation, such as frames, semantic nets, predicate logic, and so on.
- Reasoning Strategies and Algorithms: In this phase the knowledge engineer works on ways and means for retrieving knowledge for solving specific problems of the domain.
- Knowledge dissemination: Knowledge engineers spend a substantial proportion of their efforts in designing user interfaces which provide easy access to this pre-stored knowledge to novices, technicians and other professionals in the specialist domain. The special-purpose programs that help knowledge engineers in each of these steps are known as knowledge engineering environments or shells.

Knowledge engineering attempts to render the procurement and subsequent use of knowledge in the realm of engineering: the orderly and demand-led production of an artefact through the delineated phases of knowledge acquisition, knowledge representation, reasoning, and dissemination. Knowledge engineers claim that the cost and time of building this artefact can be accurately given. This proposition is very attractive, particularly where the knowledge of experts is concerned, as this knowledge is not well documented and is not easily accessible to non-experts. The migration of experts to other countries or organizations or their death leads to the disappearance of valuable corporate and community resources. Terminologists are well aware of the consequences of the inaccessibility of experts and the effect the lack of documentation has on the creation of termbases.

SEE

Technical Translation 2.1.1:147

3.2 Knowledge-based Expert Systems and Terminology

Expert systems have been developed in a range of human enterprises and pursuits ranging from diagnosing meningitis to

Table 1: *Knowledge-based Expert Systems*

Diagnostic Medicine: MYCIN (an acronym for Microbial Infection Therapy System), a computing system which can help in the diagnosis and treatment of meningitis, has outperformed some human experts and many novice physicians of the disease. MYCIN contains about 450 rules related to the diagnosis and therapy of meningitis.
Computer Engineering: The Expert Configurer system (XCON), used regularly by the Digital Electronic Corporation (DEC) (now part of Compaq Inc.), arranges the components which comprise a computer system, such as central processor, memory, disk drives, input/output circuits and so on, in a configuration which satisfies the requirements of DEC's customers. XCON has nearly 10,000 rules relevant to the configuration of DEC's computers and has data about several hundred components, and is routinely used to "configure" orders which involve 100 to 200 components.
"Art Expert" Expert System: The knowledge engineering community, and its sponsors, have recently diversified into aesthetically pleasing areas like fine arts: AARON is an "expert artist" program which is capable of "autonomously generating original works of art based on the knowledge of the outside world and of the behavior of things in it [..] and [..] the knowledge pertaining to the building of a "visual" representation of the world and its objects" (Cohen 1992:52).
"Chess Playing" Programs: Chess has received considerable attention from the AI community because "problem-solving" in chess appears to be knowledge-based in that there is a clear distinction between the performance of an expert and a novice. The United States Chess Federation (USCF) developed a numerical scale – the so-called unit of strength – which originally rated chess master Gary Kasparov's strength at 3900. After years of development and rematches, the IBM RS/6000 high-performance computer christened Deep Blue beat Kasparov, both in February of 1996 and in May of 1997. http://www.research.ibm.com/deepblue/meet/html/d.3.html

configuring computers and from producing "modern art" to playing chess.

user-program interaction

The point that the development of each of the above programs involves an understanding of the terminology of the domain can be demonstrated by looking at an interaction between a (novice) user and either of these systems. We present an excerpt of this interaction between MYCIN and its typical end-user: the end-user replies are prefixed by an asterisk and the terminology of the domain is shown in bold:

1. Patient's name: (first - last)
**FRED BRAUN (Questions 1-3 request background patient data, like age and sex).
....
4. Are there any cultures for Fred Braun which may be related to the present illness, and from which organisms have been grown successfully in the microbiology laboratory?
** Y
--- **CULTURE** - 1 ---
5. From what site was the specimen for **CULTURE** - 1 taken?
****BLOD**
=**BLOOD** (System corrects typing error). ....
The first **organism** isolated from the **blood culture** of 20-JUN-77 **(CULTURE-1)** will be referred to as:
--- **ORGANISM** - 1 ---
7. Enter the laboratory - reported identity of **ORGANISM**-1:
**UNKNOWN (Typically identity is not known yet).
8. The **strain (gram** or **Ziehl-Neelsen acid-test)** of **ORGANISM**-1:
**NEG (Preliminary lab results give some clues).
9. Is **ORGANISM** - 1 **rod** or **coccus** (etc.):
****ROD**
10. What is the form of the individual **organisms** (e.g. **lancet-shaped** for **cocci**, **fusiform** for **rods**, etc)?
**** FUSIFORM**
(... more questions follow in order to gather information to infer the identity and significance of the infecting organisms ...)
....
25.Is Fred Braun a **compromised host** (e.g. alcoholic, **sickle-cell-disease, immuno-suppressed**)?
** N
(And so on for about 50-60 questions before the system provides its therapy)

Figure 1a: *A Typical MYCIN-User Interaction*

After another 50 or 60 questions, MYCIN produces its diagnosis of the potential bacterium that may be responsible for the ailments suffered by "Fred Braun".

The interaction in Figure 1b shows that the knowledge engineers involved in MYCIN must have obtained and elaborated the terminology related to genus and differentia of bacteria that cause, and drugs that can cure, meningitis. That knowledge engineers

My therapy recommendations will be designed to treat for organisms that are either very likely, or, although less likely, would have a significant effect on therapy selection if they were present. It is important to cover for the following probable infection(s) and associated organism(s):

INFECTION-1 is BACTEREMIA
<ITEM-1> E.COLI [ORGANISM-1]
<ITEM-2> KLEBSIELLA [ORGANISM-1]
<ITEM-3> ENTEROBACTER [ORGANISM-1]
<ITEM-4> KLEBSIELLA-PNEUMONIAE [ORGANISM-1]

Figure 1b: *MYCIN's Diagnosis*

Typical MYCIN 'rule'

IF: the **stain** of the **organism** is **gramneg**, &
the **morphology** of the **organism** is **rod**, &
the **aerobicity** of the **organism** is **aerobic**

THEN there is strongly suggestive evidence (0.8) that the **class** of **organism** is **enterobacteriaceae**

Typical PROSPECTOR 'rule'

IF: The **igneous rocks** in the region have a **fine** to **medium grain** size
THEN: they have a **porphyritic texture**

Figure 2: *Typical Rules for Expert Systems (Specialist Terms in Boldface)*

elaborate terms, albeit for storing knowledge of a specialist domain on a computer system, can be more graphically demonstrated by the so-called heuristics, or rules of thumb, that comprise much of the knowledge base of a typical expert system. Figure 2 contains heuristics from MYCIN, XCON and PROSPECTOR, a geochemical prospecting program (see for instance Waterman 1985:58).

4 Knowledge Acquisition and Well-Grounded Terminology

Knowledge engineers have to study and comprehend bodies of material that contain the knowledge of a given domain. This study *comprehension* is essential for asking questions of the expert, for debugging the knowledge base, and for testing the response of an expert system. This requires interaction with experts, consultation of textbooks, research papers, etc. The knowledge engineer, at a macro level, creates and analyzes a text archive or corpus, and at a micro level, focuses on the building blocks of a specialist domain: the terms.

domain entities

A typical knowledge engineers work in many ways like terminologists. For instance, they collect and catalogue the so-called domain entities in a manner that is similar to the building of glossaries or term lists. The domain entities can be names of concrete objects in a specialism, for example in MYCIN this category would include the names of bacteria and of their various attributes, like shape, size, color on staining, names of their genus and *differentia*, diseases they cause (and the signs and symptoms of those diseases), the drugs that can neutralize the bacteria, the instruments that can be used to detect the presence of bacteria and so on.

rules and tasks

Some knowledge engineers draw complex hierarchies in which objects are organized in inheritance networks and/or partonymies, a process that resembles the development of conceptual structures in terminology management. Each domain entity is assigned attributes, which may have one or more value. Assigning attributes and values to the objects can be compared to defining and elaborating a term by a terminologist. The knowledge engineer extracts "rules or tasks" that the experts are supposed to use to solve a well-defined problem. These rules interrelate the objects of the domain and, at one level, one can construe such rules as an elaborate way of cross-referencing the terminology of a given domain.

unequivocal meanings

Despite the terminological import of what the knowledge engineer does and despite the fact that the engineer is required to analyze large tracts of specialist text, there is not much in the knowledge acquisition literature that refers to terminology management or to text or corpus analysis. We have drawn inspiration from Winograd, who has argued that the results of research in the structure and function of (natural) languages may be used to build systems that may incorporate "bodies of material", including journal articles and textbooks, into the knowledge bases of expert systems, without "human intervention". Winograd and Flores have argued that knowledge acquisition is the explicit construction of a systematic domain—a formal description of the expert's task in words that have unequivocal, agreed meanings. There appears to be some progress in the knowledge acquisition literature on the need to treat the domain more systematically than before: the progress is steady but slow, as our brief survey of the relevant work in the last three decades shows below.

4.1 Terminology as "Systems Dictionary" and "Selected Words" (c. 1970-1980)

MYCIN

MYCIN (an acronym for Microbial Infection Therapy System), is an expert system that contains about 450 rules related to the diagnosis and therapy of meningitis. These rules are of a type similar to the IF-THEN constructs shown in Figure 2 above. MYCIN is among the most cited expert systems and was developed by Shortliffe (1981) working closely with and encouraged by domain experts in infectious disease and in pharmacology and with highly-skilled and original AI researchers.

knowledge chunks

MYCIN's knowledge base was essentially organized by using taxonomies of microbes, assertions about microbes, humans and drugs, causal links between microbes and illnesses, body part meronymies (part-whole relations), statistical correlations, and chance associations amongst domain entities. In computational terms, these diverse and interacting "chunks" of knowledge were organized within the knowledge base as attribute-object-value triples and as rules containing premise (or antecedents) and actions (or consequents). The "MYCIN experiment" clearly shows that one of the first tasks of a knowledge engineer is understanding and elaborating the terminology of the application domain.

system dictionary

In order to understand the "user questions" and also to generate the "expert-like" questions, MYCIN relied on what some of the MYCIN knowledge engineers call, a system dictionary: an "approximately 1400 words" [dictionary that contains] commonly used words in the domain of infectious diseases. It includes all words [sic] that are acceptable values for a parameter [for example, culture, organism, therapy], common synonyms of these words, and words used elsewhere by the system in describing the parameter" (Scott et al. 1984:349-350). Scott et al. do not quote the source from which these 1400 "words" were obtained, but one can only surmise from other published accounts of the MYCIN system that the MYCIN knowledge engineers were helped in this enterprise by the infectious disease experts and pharmacology experts at Stanford University Hospital.

4.2 Terminology, Semantic "Relations" and Hand-crafted Encyclopedias (1980-1990)

conflicting terminology

The post-1985 knowledge acquisition literature contains lengthy discussions about problems that have been faced by philosophers, psychologists, linguists, and terminologists much earlier than 1985. Consider the problems caused by the synonymy of terms or related to inadequate hierarchical classification schemes. For instance, the developers of KNACK, a knowledge acquisition system, (Klinker et al 1988) report that because different people use conflicting terminology (1988:196), they have devised 19 synonymy rules to resolve terminology related conflicts. Elsewhere, Bylander and Chandrasekaran (1988) outline a theory of generic tasks for solving a broad class of problems. The class of generic tasks includes hierarchical classification and knowledge-directed information-retrieval, whereby "The generic tasks provide a vocabulary for describing problems..." (1988:69).

knowledge encyclopedia

Motta et al. have discussed the construction of a "knowledge encyclopedia", the isolation of "knowledge chunks through glossaries and free association" in the context of a knowledge acquisition system, KEATS. According to Motta et al., this involves using a graphics-oriented text editor to analyze interview transcripts. It is optimistically suggested that "mousing on the relevant concept name in the scroll menus" (1991:27) is all that is required to construct the encyclopedia.

language issues

In fact, the current knowledge acquisition literature is beginning to show a growing awareness of language-related issues. For instance, the experts are said to use some kind of a "representation language". Some researchers have developed text analysis tools for analyzing domain specific texts, and a number of researchers stress the need for building glossaries and lexica. However, such discussions contain rather confusing, and at times simplistic, notions about what a lexicon or a glossary of a specific domain is and how to build it. We will use two major knowledge engineering projects, the ESPRIT-sponsored Knowledge Analysis and Documentation System (KADS) project and the MCC-based Encyclopedic (CYC) Knowledge project, to demonstrate that there is some acknowledgment of the need to take terminology seriously.

KADS

KADS (1983-93) sought to develop a "structured methodology" for building knowledge base systems (KBS). This methodology was intended as a precursor to European standards for building such systems. KADS project participants, for instance, Anjewierden et al. (1992) developed SHELLEY, an "integrated workbench for knowledge engineering". *SHELLEY*

This workbench has tools for two of the activities in the knowledge identification phase, that is, "lexicon construction" and "glossary construction", leading to the "production" of lexica and glossaries (Anjewierden et al. 1992: 109-125). The knowledge modeling activities include the elicitation of "concepts", their properties, the inter-relationships between concepts and between properties and the "complex objects" (Wielinga et al. 1992:15-17). SHELLEY provides tools for "creating" concepts from text, like interview transcripts, or for just specifying them through SHELLEY's "Concept Editor". *knowledge modeling*

While KADS is expected to become a de facto standard for knowledge engineering, neither Wielinga et al. nor Anjewierden et al. expanded on how one goes about building lexica or elaborating concepts. The task of building lexica is non-trivial (see Boguraev and Briscoe 1988 and Ahmad et al. 1993), as is the elaboration of concepts (see Picht and Draskau 1985, Ahmad and Rogers 1992). *building lexica*

The ambitious CYC project seeks to build a large "encyclopedic" knowledge base containing "tens of millions" of commonsense facts and concepts "about the world that we all know and that we assume everyone else knows". The leading team members, Lenat and Guha (1992), note that a given rule stored in a knowledge base "comprises a set of terms", but the meaning of the terms is said to remain "in the eye of the beholder". This so-called "free lunch", may give the illusion of "depth of understanding" when it works, but the approach seems ad hoc. Having raised the important issue of terminology in the first chapter of their book, terminology as a subject is not discussed later on. Problems related to the intrinsic and extrinsic properties of terms and problems of how to link terms and concepts are presented by Lenat and Guha as if they were essentially problems related to the CYC project. In mid-1989 the CYC knowledge base (CycKB) was said to contain "about a million primitive 'pieces of data' [...] dispersed roughly across 50,000 units". After five years of effort, the CycKB still only "covered" or "touched upon" the topics in Encyclopedia Britannica Propaedia. This amounts to about one percent of the topics in a desk encyclopedia (Lenat and Guha 1990:351). Perhaps the *common-sense facts*

use of large scale termbases like the Canadian TERMIUM might improve the rather low, c. 1989, one percent coverage of the CycKB.

limitations

4.3 Terminology and Standardized "Vocabulary" (1990-95)

mediating representations

Boose has suggested that the more successful approaches in knowledge acquisition "seem to rely on effective mediating representations: problem modeling languages that help bridge the gap between experts and computer implementations" (1992:719). If the mediating representation matches the expert's "representation language", then it would help avoid the time-consuming and complex task of translating the expert's language to a lower level programming language. Boose specifies a number of levels of mediating representations. He regards "glossaries and lexicons" as "simple structure collections" and suggests that they be regarded as intermediate level mediating representations. But in suggesting that "glossaries" and "lexicons" are simple structure collections Boose underestimates the complexity of natural language artefacts. Furthermore, he gives no indication how these are created.

representative models

Musen (1993), a Stanford University researcher working on medical expert systems, in a survey of knowledge acquisition literature, has noted four major "impediments" to knowledge acquisition: the problem of "tacit knowledge", loosely knowing how; the problem of miscommunication, in that the "experts in the application area and the computer scientists building the knowledge base rarely speak the same language"; the problem of using an appropriate knowledge representation formalism; and the problems of creating representative models of the application domain. In the area of "miscommunication", Musen argues that "in some ways knowledge engineers must become experts. They must learn a new vocabulary and, perhaps, a new way of looking at problems. [....] Knowledge engineers must either make an effort to learn the expert's problem area or risk suffering the consequences of misunderstanding what the expert is trying to convey" (1993:4). Musen concludes by arguing that "researchers in the knowledge-acquisition community seek to understand more fully the nature of human expertise and communication ......" (1993:18).

Egar, again at Stanford University and involved in building medical expert systems, regards knowledge acquisition "as modeling from an unordered set of domain concepts". For example, "physicians can apprehend a clinical scenario from a list of signs, symptoms, and events". The author specifically focuses on "medical dilemmas" and suggests that if a computer system is to formulate models from the unordered set of domain concepts, it would require not only a target knowledge representation language, a traditional knowledge acquisition concern, but also "a vocabulary of standard terms" and "a set of operational instructions" for including each term in a model in the target language" (1992:2). He elaborates the notion of terms (or standard terms?) as linguistic symbols that "ought to match the pertinent concepts of the domain and whatever argot experts have found useful": and, subsequently in the paper we are referred to, what are essentially LSP dictionaries in the application domain, i.e., medicine! For Egar, "the careful formalization of the source language (domain terminology), and the target language (computational structures), knowledge engineers may be able to automate significant portions of model formulation of individual domains" (1992:5-6).

model formulation

Egar goes on to discuss how this "standard vocabulary" can be acquired: "Ideally, the vocabulary is already a standard that is used by practitioners, and has well defined referential semantics, which are found in authoritative text". In developing models of medical decisions, for instance, decisions that involve an element of choice or dilemma, like whether or not to conduct a test for a disease, Egar (1992), and Egar, Puerta and Musen (1992) have used "standard vocabularies", including a Systematized Nomenclature of Medicine and a Physician's Current Procedural Terminology, developed by "medical lexicographers" (terminologists?) who have spent many years designing precise definitions for diagnostic and procedural terms to be used in billing and in coding summaries for medical reporting" (Egar 1992:13). The nodes and links of their graphs were labeled by the terms in the nomenclature and in the terminology collection, and comprise "almost all terminal symbols" in their knowledge base.

standard vocabulary

Skuce has developed a knowledge management system, the Conceptually Oriented Design/Description Environment (CODE), and one of the functions of this multi-faceted system is knowledge acquisition. Skuce emphasizes the role of "conceptual and linguistic analysis", particularly the "careful choice of terminology" (1994:3), elaborated elsewhere as "unambiguous, rational

CODE

terminology", in knowledge acquisition. CODE has a sophisticated knowledge representation system and has facilities to manipulate ontologies. One of the "subsystems' of CODE is the "lexicon subsystem" that "keeps tracks of all terminology, a feature which we have found essential in many applications" (1994:15). The "lexicon subsystem" appears to have three capabilities: to identify terms in texts, and to store and to retrieve terms on demand. The "term identification facility in CODE depends, it seems, mainly on the subsystems string-matching capability which can "check" whether or not the knowledge engineer has come across a term before or not. If the knowledge engineer is unaware of the term, then CODE's lexicon subsystem provides an option for entering the term in the lexicon. The MAC-based CODE system has since been superseded by the Web-based Ikarus project and now by DocKMan, a Document-oriented Knowledge Management system (Skuce 1999).

term identification

The University of Ottawa has developed a system, COGNITERM. The COGNITERM project addresses important questions related to terminology representation using a language-informed framework, including substantial input from the work of modern semanticists, from the work of AI researchers, and from their experience of building large multilingual term bases for real-world applications. COGNITERM researchers believe that if terminology data could be represented explicitly by using structures that effect inheritance of properties, then termbases will have a wider appeal (Meyer, Bowker and Eck 1992).

COGNITERM

4.4 Terminology, Language Engineering, Corpus Linguistics and Knowledge Acquisition (1996-1999)

Language engineering is a term introduced during the last decade or so to indicate that if *linguistics*, with its 20th century mathematical underpinnings, logical argumentation and emphasis on structure, is an exact science, then the object observed by this science can be processed, copied and reproduced using engineering principles. Language engineering is an umbrella term that attempts to synthesise key results in computational linguistics, particularly syntactic parsing and morphological analysis, in corpus linguistics, especially text analysis and text categorisation, and in information sciences, including information retrieval and document indexing,

INFO

Human Language Technologies INFB 31:887

to help in the capture, analysis, processing and distribution of language resources.

terminology extraction

Terminology researchers are using language engineering principles and techniques for terminology extraction and terminology elaboration. This automation of terminology management has considerable implications for knowledge acquisition. As discussed above, the knowledge acquisition community has little or no awareness of the work in terminology. Thus an understanding of the concept of termbases will certainly help alleviate one knowledge acquisition bottleneck, namely the use of the data in termbases for building up the fact base of a knowledge base.

LEXTER

The developers of LEXTER have been working on terminology extraction using a parser to identify complex noun phrases, which are curiously termed *maximal length noun phrases* (Bourigault et al. 1996); the authors work essentially on grammatically tagged texts. Once the noun phrases have been extracted and their "head-position" phrase and the accompanying "expansion-position" phrasal component identified, Bourigault and colleagues structure the terminology in a network that relates heads to expansion links. The network actually presents candidate terms that can be viewed and selected for termbases and knowledge-based systems. Assadi has extended this work and not only described how terms can be extracted from texts, but also has made the interesting claim that one can derive an ontology of the subject domain using a variant of Bourigault's head/expansion linked network (Assadi 1998).

corpus analysis

The elicitation and elaboration of terminology appear to be among the key tasks of a knowledge engineer, including the analysis of documents of different types and varying sizes including textbooks, technical manuals and specialist dictionaries. Transcripts of interviews with domain experts may also be included in the documentation for analysis. The knowledge engineer, in effect, creates a text archive, a corpus of texts and analyzes that corpus. This corpus has a variety of document types, ranging from instructive, e.g., textbooks and manuals, to informative texts, like research papers, and interview transcripts.

text & term management

The organization and analysis of genre-varied text corpora has helped significantly in the study of language in general and more recently in the study of special languages (Ahmad et al. 1994, Sinclair 1996 and Pearson 1998). Above we have suggested that the knowledge engineer also creates a corpus of texts, together with a (informal) termbase, to study aspects of knowledge of a specialist domain. The knowledge engineer in effect is looking at the extraction of terms and their elaboration, albeit for problem-solving rather than for language-based communication. We would like to suggest that a suitably embellished text management system and a terminology management system can be used conjointly to build knowledge bases. The additional task a knowledge engineer has to perform, when compared with a lexicographer or a terminologist using a corpus, is the extraction of the so-called heuristics or rules of thumb.

heuristic rules

A text-based and terminology-informed method for acquiring knowledge for building expert systems will require programs for creating frequency lists and producing concordances of keywords in context (KWIC) for terminology extraction and elaboration (see Section 4.4.1). We show how the suggested method of extracting terms can also help in the automatic extraction of the heuristic rules, many of which are essentially cause-effect relationships (Section 4.4.2). An evaluation of the text-based approach shows that considerable amounts of time can be saved by outlasting the approach in building knowledge-based systems (Section 4.4.3).

automatic analysis

The strategy for automatically (or semi-automatically) analyzing a text corpus and extracting problem-solving knowledge from such an analysis is philosophically provocative in that it challenges the basic notion that only humans can process knowledge. In order to avoid such a provocation, we will present a case study based on an expert system development project.

text-based terminology

The case study is based on a project undertaken by the author during 1993-1995. The case study deals with the development of a knowledge-based system for advising users on the use of water resources within the UK; the knowledge base of the system comprised aspects of legal and engineering knowledge of an experienced water resources manager (Ahmad et al.1994). The case study showed that an approach informed by text-based terminology saves time and effort. The results were substantiated when the method was used in a subsequent project dealing with the design of water carrying networks with specific reference to safety and reliability in the design. The knowledge related to water resources and urban network design is based on engineering, geological, and hydrological theories and practice. However, in both the cases this generic knowledge has to be refined and adapted for local changes in demography, geology, and hydrology. In both the cases experts use a special language, that of water engineering, have complex conceptual schemas, and rely on a number of rules of thumb or experiential knowledge (Ahmad 1997)

4.4.1 *Term Extraction and Knowledge Acquisition*

identifying terms

Terms that form the basis of any specialism are discussed under various names within the knowledge acquisition literature. These names include *objects*, *entities*, *instances* (of a superordinate *object/entity*). And a group of related terms are referred to as *classes*, *sets*, and so on. But how do the knowledge engineers know *a priori* that of the hundreds of noun phrases they may hear from the experts or read in the textual archives of the specialism, some of these phrases do refer to concepts, acts, events, and states within a given specialism? The trivial answer to this question will be to look up the term in a verified and validated termbase, and here the word *trivial* is not used in its pejorative sense. If knowledge engineers do not have access to a termbase, or do not even know that one might exist, they may "ask" the experts. In order to facilitate consultation with the experts, they may expedite their work by analyzing the extant texts at hand, interview transcripts, learned papers and so on, and presenting a set of *candidate terms* to the experts. Recent developments in terminology management may be of help here.

There has been discussion in the literature on terminology about heuristics that may facilitate the extraction of candidate terms (see, for instance, Ahmad 1994 and Ahmad and Rogers in this volume for details). Suffice it to say here that specialist terms, which belong to the stock of words in a given natural language, are used much more frequently in specialist texts than may be the case in general language texts. The use of compound terms will suggest that the probability of finding two or more words used contiguously, like *halo nuclei*, *free-standing sculpture*, *three-way honeycombed catalytic converters,* in general language texts is very low, but relatively quite high in texts in nuclear physics, sculpture, and automotive engineering, respectively. There are statistical techniques, based on the frequency of usage of a single word and the probability of words occurring together that can be used in identifying candidate terms.

SEE *Term Extraction 8.4.1*:725

mother corpus

The method followed for acquiring knowledge from experts in water resources management was as follows: First, a text corpus was created and texts were selected from a broader water engineering prospective in that texts on related topics, including sewer management, urban drainage, and aquatic pollution were included in our corpus. This corpus will be referred to as the *water engineering mother corpus* for reasons which will become clear soon.

statistics

The corpus is genre varied in that it has learned texts from journals (21 texts), excerpts of two books, water resources legislation and statutes (2 texts), (public service) advertisements (9 texts), official notices (31 texts). The corpus comprised over 675,000 words. These texts were analyzed for assessing the frequency distribution of words within the corpus. The frequency count included the computation of absolute frequency and relative frequency (the quotient of absolute frequency of a word divided by the total frequency of all the words). The computation of relative frequency is followed by a look-up of the relative frequency of the word in a general language corpus, for example, the British National Corpus (Aston and Burnard 1998).

weirdness coefficient

If the distribution is significantly different, then the word becomes a candidate term whose significance is measured quite simply by computing the ratio of the two relative frequencies—the so-called *weirdness* coefficient. Knowledge engineers using the text-based method have a list of candidate terms which they need to understand and subsequently elaborate such that these terms can be used in the building of the knowledge base of an expert system. The mother corpus thus prepares the knowledge engineer to communicate with the experts. It would be ideal to have a termbase at hand, but the number of such databases is limited and are not readily accessible.

transcript corpus

Knowledge engineers then interview the experts and record the results of this interaction. The topics of interaction are highly specialized and usually a verbatim record of the conversation is made for subsequent analysis. In our project we created a water engineering *transcript corpus* (4 interviews comprising over 34,000 words in total). The records were used for a number of other analysis, but additionally the transcripts were analyzed for extracting rules of thumb and other terms. We found that a very small number of terms, less than 5, were found in the transcript corpus that were not in the mother corpus. This result is very encouraging in that, once the topics were decided upon, the ensuing collection of texts for the mother corpus was on very opportunistic basis: if there was a text broadly relevant to the topics, then it was included in the corpus. The transcript corpus was also used to extract rules of thumbs as well.

nominalization

Table 2a shows a comparison of some key terms which exhibited interesting distributional behavior in our mother corpus and in the British National Corpus. Recall that the expert system project dealt with water resources management and the problem facing the water resource managers was to issue official assent (*licence*) to withdraw or *abstract* water from *rivers* or *boreholes*. Excessive abstraction will adversely affect (or *derogate*) the quality of the aquatic environment within a given *catchment*. The use of water almost invariably involves *drainage* and *sewers*. *Aquifers*, *reservoirs* and *pipelines* are the key concepts of the water resources domain; one curious term used by the water resources manager is the term *undertaker*, anybody who undertakes to abstract water in one sense. All these terms have a rather high weirdness coefficient. (Note that for nouns we have combined the frequency of the singular and the plurals. Morphological variation is another interesting topic here.) The scientists and technologists have a penchant for nominalizing verbs—hence *abstraction*, *derogation*, and *impoundment* make a disproportionate appearance in the mother corpus: The word licence (including its variant *license*) is used over 4,000 times more frequently in the mother corpus than its corresponding frequency in the British National Corpus (BNC). Ssimilarly, *abstract* and *abstraction* are used 277 and 196 more frequently in the mother corpus than in the BNC.

corpus collection

The knowledge engineer familiarized himself with these terms by discussing the key concepts exemplified by the terms with the experts. The corpus was collected over a period of weeks in the early 1990s, now the deluge of text unleashed by the Internet (which involves a large number of journal publications and legislative texts as well as popular science articles and the electronic newspapers) means that such an exercise can be conducted over a matter of days to collect the same number of texts.

Methods described for the extraction of compound terms are discussed in Ahmad and Rogers (Chapter 8.4.1) and such methods were used quite effectively to obtain candidate compound terms

Table 2a: *Identifying Candidate Terms Based on the Weirdness Coefficient*

	Relative Frequency in BNC corpus (102,606,247 tokens) (1)	**Relative Frequency in 'mother' corpus (675797 tokens) (2)**	**Weirdness (2)/(1)**
Key action-oriented words			
abstract	0.0000014 (*2242*)	0.000400 (*270*)	**277.3**
derogate	0.0000002 (*17*)	0.0000252 (*17*)	**152.1**
impound	0.0000001 (*14*)	0.0000163 (*11*)	**14**
licence	0.0000410 (*4208*)	0.002295 (*1551*)	**4208**
Key Nouns			
aquifer	0.0000252 (17)	0.0000017 (173)	**14.92**
borehole	0.0004143 (280)	0.0000023(235)	**180.90**
catchment	0.0003329(225)	0.0000031(316)	**108.11**
pipeline	0.0000577(39)	0.0000069 (708)	**8.36**
reservoir	0.0004720(319)	0.0000107(1101)	**43.99**
sewer	0.0004410(298)	0.0000045(461)	**98.15**
undertaker	0.0009707(656)	0.0000039(401)	**248.38**
water	0.0098077(6628)	0.0003842 (39419)	**25.53**
Key nominalization			
abstraction	0.001249 (*844*)	0.000007 (*651*)	**196.84**
abstractor	0.000022 (*15*)	0.000000	**INFINITY!**
derogation	0.000061 (*41*)	0.000001 (*71*)	**87.68**
impoundment	0.000067 (*45*)	0.00000012 (*12*)	**569.36**

term frequency

The British National Corpus does contain a number of interesting compounds including *waterway* and *watercourse*. Again these key terms are used quite disproportionately in the mother corpus. The frequent use of a term sometimes is an indication that the authors of the text regard the term highly and considers the term crucial for communicating their messages or intentions, etc. (Table 2b shows the relative distribution of the two terms discussed above.)

In the expert system project reported here, the experts were interviewed over a period of three months and four interviews were conducted. A contrastive analysis of the frequency distribution of the terms in the transcript corpus and the mother corpus

Table 2b: *Comparison of Relative Distribution of Compound Terms with Absolute Frequency (Values in Parentheses)*

	Relative Frequency in BNC corpus (102,606,247 tokens) (1)	**Relative Frequency in 'mother' corpus (675797 tokens) (2)**	**Weirdness (2)/(1)**
watercourse	0.0012400 (*838*)	0.0000011 (*117*)	1087.47
waterway	0.0000607 (*41*)	0.0000041(*424*))	14.68

showed that there were only two candidate terms found in the transcripts -*drawoff* and *acidisation*, in addition to the terms within the mother corpus.

term candidates

The distribution of words and terms, as judged by 10 most frequent words, mainly closed class words like determiners, conjunctions and prepositions, and by candidate terms, shows key similarities and differences in the distribution of the words and the terms. The first ten most frequent words, all closed class, in both the corpora comprise just over 26% of the total text and the first ten (candidate) terms cover just under 4% of the total text (see Table 3 for details).

most frequent words

A closer look (see Table 4) at the morphological variance of the key terms used in the two corpora show that the interviews were focused on *abstractions, derogation, impoundment* and *licensing* in that these terms were respectively used 18.95, 7.12, 97.38 and 574 times more in the transcripts than are used in the mother corpus. The lemmas or the roots show up as candidate terms in the analysis of the mother corpus and the morphological variances in the analysis of the interviews. A keyword in context analysis of the candidate terms will help in the elaboration of the terms. Such a method has been discussed in the context of terminology elaboration by Ahmad and Rogers (this volume).

morphological variance

4.4.2 *Extracting Problem-solving Rules and Domain Objects from a Text Corpus*

A number of knowledge-based expert systems use the so-called *production rules* to store and retrieve aspects of problem-solving knowledge, particularly the "rules of thumb" or *heuristics*. A production rule has been defined as a procedural response triggered by a pattern. These rules are commonly structured in an ***if...then....***

production rules

Table 3: *Relative and Absolute Frequencies of the 10 Most Frequent Words in the "Mother" Corpus and the Transcript Corpus Together the Frequencies of First 10 Open Class Words in the Two Corpora.*

"Mother corpus" (675797 words): First 10 most frequent closed class words					
	the	**of**	**to**	**a**	**in**
Relative (%)	7.2	5.06	3.5	2.22	2.19
Absolute	48977	34212	23677	15047	14838
	or	**and**	**any**	**be**	**for**
Relative (%)	2.09	1.914	1.37	1.289	1.22
Absolute	13823	12940	9291	8697	8276
"Mother corpus": First 10 most frequent open class words/candidate terms					
	authority	**water**	**section**	**order**	**land**
Relative (%)	0.9	0.7	0.4	0.37	0.36
Absolute	6191	5129	3151	2545	2467
	person	**drainage**	**state**	**licence**	**application**
Relative (%)	0.33	0.27	0.24	0.23	0.22
Absolute	2236	1830	1650	1551	1531
Transcript corpus (34698 words): First 10 most frequent closed class words					
	the	**to**	**of**	**and**	**a**
Relative (%)	6.49	3.8	3.5	2.6	2.5
Absolute	2255	1334	1226	912	886
	that	**is**	**in**	**be**	**it**
Relative (%)	2.16	2.13	1.91	1.32	1.3
Absolute	750	741	665	460	452
Transcript corpus: First10 most frequent open class words/candidate terms					
	water	**licence**	**abstraction**	**river**	**application**
Relative (%)	0.86	0.81	0.57	0.29	0.28
Absolute	300	282	198	101	99
	section	**act**	**resources**	**abstractions**	**licenses**
Relative (%)	0.25	0.22	0.219	0.214	0.20
Absolute	87	79	76	73	70

Table 4. *Relative Frequency of Keywords and Their Morphological Variants in the Water-engineering "Mother" Corpus and the Water Engineering Transcript Corpus (Absolute frequencies in parentheses).*

	Relative Frequency in 'mother' corpus (675797 tokens) (1)	**Relative Frequency in Transcript corpus (34698 tokens) (2)**	**Weirdness (2)/(1)**
	abstract		
(lemma)	0.000400 (*270*)	0.000317 (*11*)	0.79
abstracted	0.000451 (*304*)	0.000173 (*6*)	0.38
abstracting	0.000105 (*71*)	0.000115 (*4*)	1.10
abstraction	0.001138 (*769*)	0.0057 (*198*)	5.01
abstractions	**0.000111 (*75*)**	**0.00210 (*73*)**	**18.95**
abstractor	0.0000133(*9*)	0.000115 (*4*)	8.60
abstractors	**0.0000088 (*6*)**	**0.000144 (*5*)**	**16.20**
	derogate		
(lemma)	0.0000252 (*17*)	0.0000865 (*3*)	3.44
derogated	0.0000163 (*11*)	0.0000288 (*1*)	1.77
derogating	-	-	-
derogation	**0.0000607 (*41*)**	**0.000432 (*15*)**	**7.12**
	impound		
(lemma)	0.0000518 (*35*)	0.000144 (*5*)	13.911
impounded	0.0000163 (*11*)	0.0001173 (*6*)	10.62
impounding	0.000354 (*239*)	0.001355 (*47*)	3.83
impoundment	0.0000651 (*44*)	0.001989 (*69*)	30.54
impoundments	**0.00000148 (*1*)**	**0.000144 (*5*)**	**97.3827**
	licence (license)		
lemma	**0.002295 (*1551*)**	**0.00812 (*282*)**	3.54
licenced	-	-	-
licencing	**0.00000296 (*2*)**	**0.0017 (*59*)**	**574.5579**
licencee	0.0000237 (*16*)	- (-)	0
licencees	0.00000296 (*2*)	- (-)	0
licenceds	0.000474 (*320*)	0.002017 (*70*)	4.26

format. In Section 3.2 it was shown that a number of well-known expert systems (*MYCIN*, *XCON*) do use this kind of format:

IF a pattern is matched, THEN schedule a procedure for execution

IF...Then patterns

The *IF* part of the clause is referred to in knowledge-based systems literature by various names, including *condition part, antecedent* part, the *left hand side of the rule* and so on. Similarly, the *THEN* part is respectively referred to as the *action part, consequent* part, and the *right hand side of the rule*. In a number of specializms one might find the two-clause *IF ..... THEN* pattern in texts of the specialisms. However, the *antecedent* clause may not appear simply as a clause beginning with the word IF. Furthermore, more frequently, the *consequent* clause may not actually have the word THEN in it.

Consider the following example from the legal domain, for example, from one part of the British legal system, the Scottish legal system. The *Divorce (Scotland) Act 1976* comprises the following statute law formulation:

> **[In an action for divorce] the court may grant decree of divorce if, but only if, it is established [in accordance with the following provisions of the Act] that the marriage has broken down irretrievably.**

The above formulation can be recast as a "statute law-statement that is a type of conditional statement" (Susskind 1987:83):

> **If and only if it is established that the marriage has broken down irretrievably, then the court may grant decree of divorce.**

candidate rules

Note that in the above recasting, the knowledge engineer (Susskind 1987) has (a) ignored items in parentheses and (b) inserted the word THEN himself. The important point here is that the knowledge engineer has to scan the dense legal prose that emanates from the pens (or word processors) of legal experts. This aspect of knowledge acquisition will be facilitated by a computer program that may scan every sentence in the Act, for example, looking for the signal for the antecedent part and also for the consequent part. Once found, these sentences, which may or may not contain a rule or heuristic, may be presented to the knowledge engineer as candidate rules or heuristics.

identifying heuristics

Many of these rules, particularly the heuristics, are not well documented and are literally passed on from experts to novices by word of mouth. When experts are interviewed and asked to explain how to solve a given problem, they often articulate these heuristics. Typically, these interviews are recorded and subsequently transcribed, either *verbatim* or only as selections of the interview. Moreover, such heuristics can also be found in technical manuals, learned papers, text books and other archival material related to a discipline.

The transcripts and the archival material easily can comprise hundreds of thousands of words of text dispersed in a number of documents. A knowledge engineer may have to sift through a large volume of text, containing unfamiliar terminology. A computer program capable of finding such *if* ... *then* patterns will help in finding all such collocation patterns. Such patterns, i.e., the candidate rules, can then be presented by the knowledge engineer to subject experts for verification and validation. Table 5 shows an excerpt of a verbatim transcription of an interview conducted by a knowledge engineer and the interviewee here is a water engineer:

4.4.3 *An Evaluation of the Corpus-based Approach to Knowledge Acquisition*

knowledge corpus

In order to quantify the benefits of a text-based terminology informed approach to knowledge acquisition, consider a knowledge engineering project executed by the author of this article and some colleagues. The objective of the project was build a knowledge-based system which can give advice on the use of water resources in the United Kingdom. The project was sponsored by the UK government's Environmental Agency. The knowledge acquisition involved the use of brainstorming, interviews, interview transcript analysis, term identification and rule extraction. Four experts were interviewed for approximately 60 minutes each, and the transcription resulted in a text corpus of about 35,000 words. The transcript

Table 5: *Deriving Candidate Rules*

Interview Excerpt	**"Candidate" Rule**
The most important thing to remember about a canal is that somewhere a river is suffering which means that every canal gets its water from somewhere which will probably be a river which is being denuded of the flow for the purposes of that navigation and **if** you license more abstraction from that canal it will need more water from the river, so you need to take account of the knock-on impact on the originating watercourse when you consider an Abstraction Licence on a canal.	IF considering an abstraction licence on a canal THEN the knock-on impact on the originating watercourse must be taken into account.

analysis was conducted with the help of System Quirk by a knowledge engineer who identified over 200 topics and propositions comprising problem-solving tasks, rules, and domain objects. We believe that a 25% saving was made in the analysis phase using this approach (Ahmad, Griffin and Shaw 1994).

interview analysis

5 AFTERWORD

term acquisition & elaboration

In this chapter we have noted that in knowledge engineering, term acquisition and elaboration are often cited as an initial problematic stage of knowledge acquisition, but that a systematic approach is rarely adopted. Some solutions are proposed from terminology and corpus linguistics. For instance, a termbase would be one way of storing and retrieving such domain terminology: indeed, it could be argued that the first step in knowledge acquisition should be to determine the existence of termbases in the domain. Termbase building tools, such as text analysis tools, termbase editing and browsing tools, and tools for graphically displaying the interrelationships between terms, like synonymy, hierarchies, antonymy and so on, could be used to capture and store domain terms from relevant texts, thereby helping knowledge engineers to understand the domain by providing them with a list of potential domain terms.

There are three key areas currently under investigation in linguistics, lexicography, and language engineering, that impact on the automatic extraction and validation of terminology data and that have an impact on knowledge engineering as well.

proper nouns

- <u>Unknown Lexicalisation</u>: The use of proper nouns (eponyms) is also used to indicate the existence of terms, for example, a range of units that form the basis of SI units, celebrate the works of famous scientists and engineers, including *Newton*, a unit of force, *coulomb* and *ampere* for electric charge and current, *Pascal* for pressure and so on. Proper nouns are also used in specialisms to refer to innovations of one kind or another. Proper nouns behave in an idiosyncratic manner and are sometimes referred as *unknown lexicalizations* (Boguraev and Pustejovsky 1996:14). The analysis of proper nouns will play a significant role in automatic terminology extraction and validation in that proper nouns comprise semantic information, especially through their usage in expressing units and dimensions. Furthermore, proper noun analysis will help in attributing the creation of neologism and in tracking the use of extant terms by particular authors. Proper noun analysis will certainly help in the validation of termbases and, eventually, knowledge bases.

- Semantic Relational Analysis: Cruse (1986) discusses a wide range of relations, distinguishing between different types of hierarchical relations, such as hyponymy, taxonomy, and meronymy, and considering various kinds of synonymy and antonymy. He stresses that his discussion is *in* meaning rather than *about* meaning. An important contribution Cruse makes is his discussion of what he terms "diagnostic frames" for some of the relations. The frames are essentially what Lyons (1977) refers to as "formulae" for semantic relations, basically comprising phrases to denote a particular relation; for example, the phrase "X is a kind of Y" denotes the relation of hyponymy. Cruse discusses the variety of ways these frames may be expressed in natural language. For instance, the relation of hyponymy may be expressed by such phrases as "X is a type of Y", "X is a species of Y", and so on. Ahmad, Griffin and Shaw (1994) have shown that these diagnostic frames and formulae can be used in extracting knowledge from a text corpus.

diagnostic frames

- Definition Analysis: It has been observed that definitions of general language words in a lexicon (Barnbrook and Sinclair 1995 and Barnbrook 1995), and more recently by Ahmad and Al-Jabir (forthcoming) that terms, are elaborated according to a pattern that indicates the meaning of the defined word. The definitions of terms, largely of noun phrases, comprise the use of optional determiners and usually the super-ordinate term as well interspersed with adjectives. Typical garden path constructs, involving ambiguous prepositional attachments are also avoided by lexicographers and terminologists alike. This means that with a relatively simple parser these definitions can be parsed, and Ahmad and Al-Jabir have shown that the analyzed sentences can be mapped onto a logical proposition. Once mapped these propositions can be readily transformed into knowledge representation schemes, like *frames* and *semantic networks*. These schemes then can passed onto the knowledge base of an expert system. This work shows that contents of a termbase can, in principle, be mapped onto a knowledge base.

SEE *Definitions 1.3.1:64*

elaboration & documentation

In order to exploit the symbiotic link between artificial intelligence and terminology, mainly due to the dependence of both on specialist knowledge, it appears that the use of termbases for automatically acquiring knowledge, for expert systems, from text corpora is a cost efficient proposal In summary, if one has termbases at hand, one can parse definitions and assert the equivalent logical form directly into a knowledge base of an expert system. However, if only glossaries or lists of key terms are available, then the terms have to be elaborated. In the case of unvalidated term lists, one will have to conduct an intermediate process of term validation. In the absence of termbases, glossaries and term lists, one will have to resort to text corpora to extract, validate and elaborate the terms (in close co-operation with the subject experts). Text analysis allows the scanning of literally unlimited material for the description of objects and for the elicitation of rules of thumbs as well. Figure 2 summarizes the methodology outlined in this chapter.

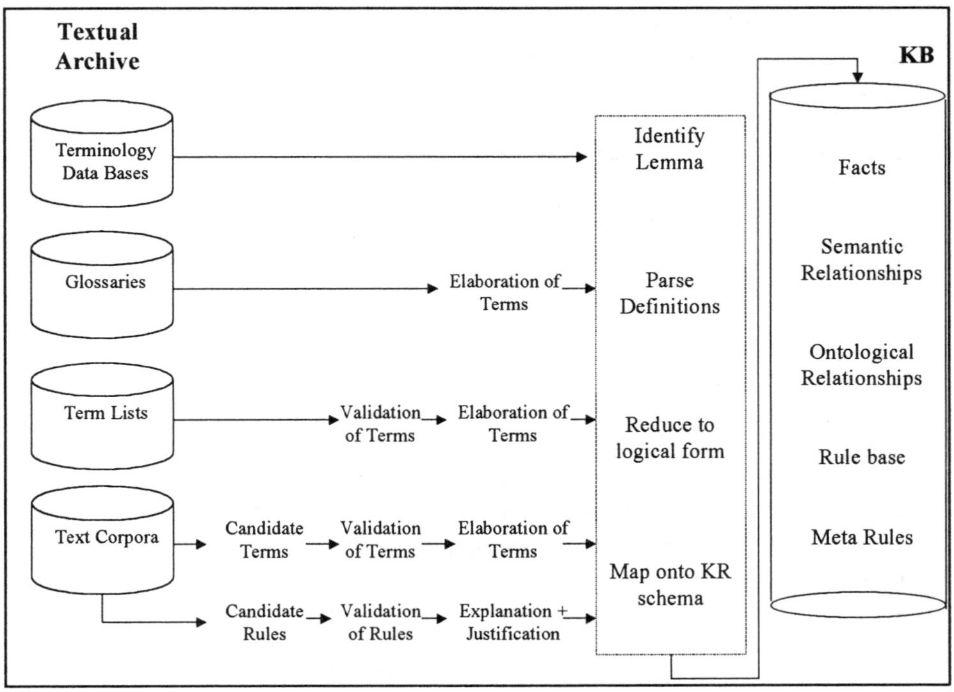

Figure 2: *The Synergy and Symbiosis of Work in Terminology and Knowledge Engineering*

REFERENCES

Ahmad, Khurshid and Al-Jabir, Saika. 2000. "Wüster's Legacy: Automatic Extraction and Representation of Terminology from Termbases." In: Gerhard Budin, ed. *Proceedings of the PROCOM Conference (August 1998)*. Vienna: TermNet (in press).

Ahmad, Khurshid; Griffin, Stephen M.; and Shaw, Paul J. 1994. "A Systematic Knowledge-based Approach to Licensing Water Resources". In: B. Coulbeck, ed., *Integrated Computer Applications for Water Supply and Distribution 1993. Volume 1: Methods and Procedures for System Simulation and Control.*, Taunton (Somerset):Research Studies Press Ltd., 407-424.

Ahmad, K. 1997. "Safety-Critical Systems: Prescient, Presignifying, Public-Good Systems?" In: Felix Redmill and Tom Anderson, eds. "Safer Systems". *Proceedings of the Fifth Safety-Critical Systems Symposium*, Brighton, 4-6 Feb. 1997. London: Springer-Verlag., 1-37.

Anjewierden, Anjo; Wielinga, Robert; and Shadbolt, Nigel. 1992. "Knowledge Acquisition Tools: Supporting Knowledge Acquisition: The ACKnowledge Project." In: Luc Steels, Brice Lepape, eds. *Enhancing the Knowledge Engineering Process: Contributions from Esprit*. Amsterdam, New York: North-Holland, 143-174.

Anjewierden, Anjo; Wielemaker, Jan; and Toussaint, Catherine. "Shelley—Computer Aided Knowledge Engineering." In: Guus Schreiber; Bob Wielinga; and Joost Brueker, eds. *KADS: A Principled Approach to Knowledge-based System Development*. London, San Diego: Academic Press, 247-255.

Assadi, Houssem. 1998. Construction of a Regional Ontology from Text and its Use within a Documentary System. In: N. Guarino, ed., *Frontiers in Artificial Intelligence and Applications, Vol. 46*. Amsterdam: IOS Press, 236-249.

Aston, Guy and Burnard, Lou. 1998. *The BNC Handbook: Exploring the British National Corpus*. Edinburgh: Edinburgh University Press.

Barnbrook, Geoff and Sinclair, John. 1995. "Parsing Cobuild Entries". In: John Sinclair; Martin Hoelter; Carol Peters, eds. *The Languages of Definition: The Formalization of Dictionary Definitions for Natural Language Processing.* Luxembourg: European Commission, 13-58.

Barnbrook, Geoff. 1995. *The Language of Definition: A Cobuild Sublanguage Parser.* Ph.D. Thesis, School of English, University of Birmingham.

Beaugrand, Robert de. 1979. "Theoretical foundations of the automatic production and processing of technical reports." *Journal of Technical Writing and Communication* 9, 1979, 236-269.

Boguraev, Branimir and Briscoe, Ted. 1988. *Computational Lexicography for Natural Language Processing*. Harlow (Essex, UK): Longman Publishers.

Boguraev, Branimir and Pustejovsky, James, eds. 1996. *Corpus Processing for Lexical Acquisition*. London and Cambridge Mass., USA: The MIT Press.

Boguraev, Branimir and Pustejovsky, James. 1996. "Issues in Text-based Lexical Acquisition." In: Branimir Boguraev and James Pustejovsky, eds., 3-17.

Boose, John H. and Gaines, Brian R., eds., 1988 *Knowledge Based Systems Vol. 1 – Knowledge Acquisition Tools for Expert Systems,* Academic Press Ltd.

Bourigault, Didier; Gonzalez-Mullier, Isabelle; and Gros, Cécile. 1996. "LEXTER, a Natural Language Processing Tool for Terminology Extraction". In: Martin Gellerstam, et al., eds. *EURALEX '96: Proc.I-II, Part II – Papers submitted to the Seventh EURALEX International Congress on Lexicography in Göteborg, Sweden.* Göteborg: Göteborg University, Sweden, 771-779.

Bylander, Tom and Chandrasekaran, Balakrishnan. 1988 "Generic tasks for knowledge-based reasoning: the "right" level of abstraction for knowledge acquisition". In: Boose, John H. and Gaines, Brian, eds. 1988, Vol. 1, 65-77.

Cruse, David A. 1986. *Lexical Semantics*. Cambridge: Cambridge University Press.

Egar, John W. 1992. *"Graph-Grammar Assistance for Knowledge Acquisition". Knowledge* Systems Laboratory Report No. KSL-92-37. Stanford, California: Stanford University.

Egar, John W.; Puerta, Angel R.; and Musen, Mark A. 1992. *"Graph-Grammar Assistance for Modelling Decisions"*. Knowledge Systems Laboratory Report No. KSL-92-42 Updated December 1992. Stanford, California : Stanford University.

Klinker, Georg; Bentoilla, Joel; Genetet, Serge; Grimes, Michael; and McDermott, John P., 1988 "KNACK – Report-driven knowledge acquisition". In: Boose, John H. and Gaines, Brian, eds. 1988, Vol. 2, 195-209.

Lenat, Douglas B. and Guha, Ramanathan V. 1990. *Building Large Knowledge-Based Systems: Representation and Inference in the CYC Project.* Reading MA: Addison-Wesley Publishing Company, Inc.

Meyer, Ingrid; Skuce, Douglas; Bowker, Lynn; and Eck, Karen. 1992. "Toward a New Generation of Terminoloigcal Resources: An Experiment in Building a Terminological Knowledge Base." In: *The Proceedings of the 14th International Conference on Computational Linguistics*. Nantes, 956-960.

Motta, Enrico; Rajan, Tim; Domingue, John; and Eisenstadt, Marc. 1991 "Methodological foundations of KEATS, the Knowledge Engineer"s Assistant". *Knowledge Acquisition* Vol. 3, 21-47.

Musen, Mark A. 1993. "An Overview of Knowledge Acquisition". Knowledge Systems Laboratory Report No. KSL-92-58 Updated February 1993. Stanford, California: Stanford University.

Nebel, Bernhard. 1991. "Terminological Cycles: Semantics and Computational Properties". In: John F Sowa, ed.: 331-361.

Pearson, Jennifer. 1998. *Terms in Context.* Amsterdam and Philadelphia: John Benjamins Publishing Company.

Picht, Heribert and Draskau, Jennifer. 1985 *Terminology: An Introduction,* Guildford: University of Surrey.

Scott, A. Carlise; Clancey, William J.; Davis, Randall; and Shortliffe, Edward H. 1984. "Methods for Generating Explanations". In: Bruce G. Buchanan and Edward H. Shortliffe, eds. *Rule-based Expert Systems: The MYCIN Experiments of the Stanford Heuristic Programming Project.* Reading, MA: Addison-Wesley Publishing Company, 338-362.

Shapiro, Stuart M. 1992. *Encyclopaedia of Artificial Intelligence*. Second Edition. New York: John Wiley and Sons, Inc.

Shortliffe, Edward H. 1981. "Consultation Systems for Experts: The Role of Artificial Intelligence Techniques." In: Bonnie L. Webber and Nils J. Nilsson, eds. *Readings in Artificial Intelligence*. Palo Alto, California:Tioga Publishing Company, 323-333.

Skuce, Douglas. 1994. "A Multi-functional Knowledge Management System". *Knowledge Acquisition*. In Press.

Skuce, Douglas. 1999. "A Document-oriented Knowledge Management System." In: Peter Sandrini, ed. *TKE 99: Terminology and Knowledge Engineering: Proceedings of the Fifth International Congress on Terminology and Knowledge Engineering*. Vienna: TermNet, 320-327.

Susskind, Richard. 1987. *Expert Systems in Law: A Jurisprudential Inquiry*. Oxford: Clarendon Press.

Waterman, Donald A. 1985. *A Guide to Expert Systems*. Reading, MA: Addison-Wesley Publishing Company.

Wielinga, Robert; Schreiber, Guus.; and Breuker, Joost A. 1992. "KADS: a Modelling Approach to Knowledge Engineering". *Knowledge Acquisition*. Vol. 4, 5-53.

Winograd, Terry. 1983. *Language as a Cognitive Process - Vol I: Syntax*. Reading, MA: Addison-Wesley Publishing Company.

Winograd, Terry and Flores, Fernando. 1986. *Understanding Computers and Cognition: A New Foundation for Design*. Norwood, New Jersey: Ablex Publishing Company.

ACKNOWLEDGEMENTS

The work was carried under the aegis of the Translator's Workbench Project (1989-92; 1992-94), the MULTILEX Project (1991-94), the TRANSTERM Project (1992-96), the INTERVAL project (1995-97) and the ACE project (1995-98) sponsored by the European Commission's ESPRIT and LRE Programmes. The author is grateful to Shaikha Al-Jabir, Syed Sibte Raza Abidi, Steve Collingham, Lee Gillam, Stephen Griffin, Paul Holmes-Higgin, and Stephen Hook for interesting discussions which they went on to translate into computer programs

Section 8.5

Terminology Resources on the Internet

GERHARD BUDIN AND SUE ELLEN WRIGHT

1 PROSPECTING FOR TERMINOLOGICAL INFORMATION

term mining

The Internet has become THE primary information resource for language professionals. This chapter focuses on language resources found on the World Wide Web. The systematic search for monolingual or multilingual terminology resources on the WWW as well as their evaluation in terms of relevance to one's own information needs, quality, consistency, and reliability has become a valuable skill and activity that we propose to call *term mining*, in analogy to the well known process of *data mining*, bearing in mind that our approach is oriented toward developing human skills rather than modeling computer algorithms for automatic term extraction, which is also sometimes called *terminology mining*.

term reusability

The main tasks of term mining involve:

- Identifying available terminology resources
- Evaluating them for accuracy and credibility, and for their relevance to stated information needs
- Selecting accurate and relevant resources
- Checking the accessibility and applicability of such resources, overcoming, if necessary, technical, legal, and usability barriers

search engines

Methods for term mining are grounded in the efficient use of *search engines*. These Web-based tools reflect different technologies, which is one reason why we obtain different results when we use the same search statement with different search engines. (In point of fact, we can even obtain different results from sequential or even apparently simultaneous, identical search strategies used in the same search engine!) Bearing in mind that the palette of available resources is constantly changing, we can distinguish among:

- ***Generic search engines*** that try to cover the whole WWW, such as *AltaVista* and *HotBot*, or enhanced versions such as Alta Vista's *Raging Search* or Yahoo's *Google*.
These tools create their indexes by sending out computer programs called robots (spiders, webcrawlers, or even ferrets – choose your favorite vermin) to search the Web, collecting index terms from the text found in pages, but particularly from titles and from HTML meta categories such as "Description" and "Keywords". These are lines of information that are invisible to most users and that enable Web masters and Web authors to control to a certain extent the results that the bots report for their sites. The latest trend (*Raging Search* and *Google*) is to eliminate the kind of distracting advertising and portal-style noise that clutters the screen with many engines and to provide a very clean, very efficient introductory page designed to appeal to serious researchers, librarians, and information specialists. Complex analysis and frequency features enable leading-edge engines to provide high-speed results that are both reliable and complete.
- ***Domain-specific search engines*** that focus their search on major disciplines such as medicine
- ***Web-site-specific search engines*** (often called site maps or indexes) that allow visitors to a Web site to search for information within the boundaries of that Web site. In fact, site maps and internal indexes are a good way to search a site for information that is not actually linked to public pages.

site maps

Many people also use what they incorrectly consider to be search engines, for instance:

indexes & directories

- ***Indexes, meta-indexes*** (such as *Metacrawler* or *inFind*), and ***directories*** (such as *Yahoo*). The *inFind* resource, for instance, provides a parallel search capability that combines search results from *WebCrawler, Yahoo, Lycos, Alta Vista, InfoSeek*, and *Excite* and then "clusters" results according to various preset criteria. There has been a great deal of enthusiastic response to this procedure because it facilitates evaluation of resources and eliminates much of the time lost in looking at numerous, only marginally relevant Web sites.
- ***Virtual libraries*** (e.g., the US National Library of Medicine, which is associated with Medline)

- ***Portals***, which act as gateways to certain areas of the Web (*Onelook* and *YourDictionary* and Yahoo's dictionary page are typical portals; see Annex A below.)
- ***Composite technologies***, such as Google, which combine text matching, indexing, and portal methodologies with sophisticated link analysis. Some systems, such as *Northern Light,* search not only the Web, but a collection of other resources as well (major newspaper, magazine, and broadcast archives), offer a wide range of alternate search strategies, and sort the results for more efficient selection of hits.

collegial recommendations

Valuable help is coming from colleagues with recommendations of specific Web sites. One excellent resource that harnesses the collective findings of an extensive number of colleagues is the *GlossPost* listserve (http://www.egroups.com/group/GlossPost), a no-nonsense, no-chat, no-spam place to post or request information on subject-field-specific glossaries. You can also download GlossPost archives (http://www.egroups.com/files/GlossPost/Bookmarks/). Other subject-specific lists may be useful, but this one provides information with a minimum of mailbox clutter.

2 Dictionary and Glossary Resources

online lookup

The first place to search for definitions and even equivalents for a given set of terms can involve searching online terminological resources. There are a number of specialized search engines, indexes, and portals that provide either direct access to online dictionaries and glossaries or bibliographical information on hardcopy lexicographical and terminological collections.

canonical form

SEE

Term Selection 1.1:17

Actually finding terms in online dictionaries involves several caveats, however. Many compilers of online dictionaries are blissfully unaware of some of the cardinal rules of lexicography and terminology research and documentation. For instance, capitalization of terms (either the first letter or maybe all letters) is not uncommon in these resources, and definition form, even in technically accurate definitions, often leaves a great deal to be desired. Consequently, it is very wise to reuse these resources critically and to bring imported information into conformance with whatever criteria have been set as the norm for your particular working environment. Quality Assurance guidelines and checklists can be very useful in this regard because it can be very tempting to just cut and paste information found on the Web without checking for important stylistic features.

corpus trace

Also bear in mind that working translators, technical writers and information specialists place a great deal of value on brevity, so it is inadvisable just to copy encyclopedic chunks of data from the Web into individual terminological entries. Always include the URL for the resource so that users can link to it for further information (provided that the URL remains current, of course), or better yet, if your environment allows you to create a corpus of documents supporting your research, save the entire Web document to that location and provide a path name to that document. This new data category is currently being called a *corpus trace* and allows you to create a hypertext-oriented knowledge base for users of your terminological resource. It is also desirable to document the date you accessed the Web page.

3 AD HOC SEARCH STRATEGIES

content-related sites

Of course, finding a term or set of terms in an existing glossary in any language is wonderful, but the fact remains that most working terminologists are called upon to research leading-edge technologies where no one else has ever conducted terminology research. Furthermore, it is highly desirable to reconfirm information found in glossaries and dictionary resources by finding *in vivo* terms in context in content-oriented sites rather than relying solely on existing dictionary or glossary resources. This is particularly true in multilingual environments, which may not be covered as well on the Web as is English. For example, in researching this article, we found that the term *diaphragm spring* was very nicely and correctly defined and illustrated in dictionary resources on the Web, but we could not find its German equivalent *Tellerfeder* in any glossary. Nevertheless, this term shows up and is fully described and illustrated in the content-oriented Web pages of several major German manufacturers of diaphragm springs.

Web search strategies

Content-related searches of this nature do not need to be haphazard, and success is more a result of careful search strategies than a matter of serendipity. It is not our intent, however, to provide an in-depth treatment of general content-oriented search methodologies. Harris's *Guidebook to the Web* (2000) offers an excellent introduction, as do many Web-based guidelines. For an overview at any given time, we suggest that you simply search the string "Web search strategy" in several good search engines or indexes and follow the many resulting hits. When working with any given search tool, it is smart to read the detailed recommendations for optimizing your search results using that particular tool. Hone your skills for targeting specific topics, and learn to use Boolean logic to pose queries, but bear in mind that not all systems distinguish appropriately between *AND* and *OR* searches. (Some engines, such as Google, do not require Boolean operators.)

4 Evaluating Resources

semantic evaluation

The content-related evaluation of terminology resources found on the WWW is crucial due to the anarchic nature of the Web: anybody can create a Web site and offer terminological and content-related resources, and there is nobody in the world who can prevent or control this. We can distinguish among the following types of evaluation:

Entry level:

entry level

- Semantic evaluation: verifying the correct meanings of words (lexical semantics) or of text segments up to whole texts (text semantics)

verification

- Terminological evaluation: verifying the correctness of semiotic assignments of terms to concepts within one language (including synonyms, abbreviations) and across languages (full or partial equivalences, multiple assignments), verifying the correctness of definitions, checking the plausibility of all information related to a given concept

Resource level:

resource level

- Evaluation for relevance: verifying whether a resource is relevant to a given task from thematic, situational, contextual, and pragmatic perspectives
- Evaluation for coherence and consistency: checking the uniform structure of entries, cross-references among entries
- Formal evaluation (automated or semi-automated procedures)

data level

Data level:

- Syntactic validation: checking the syntactic correctness of sentences
- Grammatical validation: grammar checkers
- Orthographical validation: spell checkers based on standard language varieties (e.g., British or American English)

Meta level: *meta level*

- Tagging: checking the conformance of tagging (resource and entry level) to given standards (e.g., XML)

A comprehensive approach that combines all forms of evaluation is a prerequisite for quality assurance in terminology management.

4.1 Evaluation Criteria

CARS Checklist

Harris (2000:60) has proposed what he calls the CARS Checklist, which is designed to examine Web site information with respect to *Credibility, Accuracy, Reasonableness*, and *Support*. In examining credibility, he focuses on the author's credentials and prefers sites that represent organizations or institutions, on the grounds that materials that have been approved by an expert group are more likely to reflect the values traditionally associated with peer review. Articles that have previously appeared in industry or scholarly publications are also more likely to be reliable. By the same token, companies that specialize in a particular technology may be reliable in their fields, as long as the user watches out for excessive advertising hype. Researching competitors' sites can be very useful, provided you bear in mind that different companies may use different terminology. Governmental agencies and publically funded entitles, for instance, tend to be more reliable as a rule because their goal is to serve the public rather than to sell a product. Despite all these caveats, it can also occur that individual afficionados specializing in a particular interest may very well put together highly accurate and informative pages. If time and resources allow, one of the best ways to check for credibility and accuracy is to try to confirm items of information in multiple sites. If you have some level of understanding in a field, and you find a Web site with parts you know are correct, it is easier to validate parts of the site with which you may be unfamiliar, and, based on this evaluation, to venture a good guess on the reliability of the rest of the site.

4.2 Unreliable Findings

formalistic cues

Anonymity, bad grammar, poor writing style, or an indication that the site is actually a bad translation from another language are cues that the user should be very careful in using material found on that site, particularly terminological information. Misspellings are also a concern: sometimes your own guess at a term will be misspelled, and you will actually have the good (or bad!) fortune to find hits on the Web because other people have misspelled the same term in the same way. This is one of the hardest "errors" to guard against because it can be difficult to check against information that you don't have. Some sites (like Onelist) allow you to do "fuzzy match" searches that may yield variant spellings you can research. If in doubt about the accuracy of alternate spellings, variants, or synonyms, some terminologists rely heavily on frequency of hits. Nevertheless, alas, frequency may not always be the only criterion you should look at because some terms (particularly loan translations, for instance) may enjoy widespread distribution, but experts may still prefer other terms for use in serious writing.

timeliness

Timeliness is important. It pays to check the dates of Web pages that deal with time-sensitive topics. Many sites have old pages sitting around that have not been updated for significant periods. Again, confirmation of search results in more than one site or against reliable hard-copy resources, if they are available, is highly desirable for ensuring reliability.

4.2 Researching Equivalent Terms

unfamiliar equivalents

Probably the most difficult task for the multilingual terminologist is finding unknown equivalents for terms that are known in a source language, particularly if the source language is English. Frequently a search can begin by trying to validate "educated" guesses at those terms that appear to be good candidates for cognates or loan translations in the target language. Searching on known superordinate concepts or closely related terms in the target language can frequently yield informative content-related pages that will provide a wide range of embedded terminological information. Again, keep an eye out for bad translations, especially if the field is very new and the technology originated in the source-language community, with the result that this knowledge is being transferred to the target-language community. In the best of all possible worlds, collaboration with target-language experts provides the ideal validation mechanism. This type of problem underscores the importance of documenting contexts that demonstrate the actual existence of the term in target-language discourse.

5 Types of Web Resources

data availability

We encounter terminological resources on the WWW in basically two presentation forms: HTML text or as databases with Web interfaces. With the increased introduction of XML-based Web resources, it is highly likely that more sites will provide the latter. Due to lack of coordination, there is no systematic coverage of knowledge domains or of languages. Maintenance is another problem: as noted above, there are quite a number of data graveyards on the Web, including those that contain terminology resources that have not been updated for years. Unsolved copyright issues also prevent many (high quality) terminology resources from being made available on the Web. Data owners fear uncontrollable use of their data, and, if they are also "in the business" of selling hardcopy resources, they may be reluctant to "give stuff away for free". We can hope that perhaps in the future methods will be designed to provide reasonably priced search capability in many different kinds of online resources, for instance in the standards arena.

metadata initiatives

Term mining is only in its infancy and prospects for the future are excellent. Several research activities have generated promising results. Metadata initiatives will also enhance the prospects of more multilingual terminology resources being not only made available, but also found by those who look for them.

Annex A: "Currently" Available Resources

In the following we would like to give an overview of major types of terminology resources available on the WWW, as well as provide concrete references to Web sites containing information about how to find information on the Web. Due to the ephemeral nature of information on the Web and the rapid increase in the number of Web sites all over the world, this selection must remain incomplete right from the start. It is rather meant as an appetizer, a starter to inspire you in doing your own searches. We have tried, however, to list useful, reliable and validated URLs and to add brief information on their content.

Disclaimer: *The last visit to all the URLs given in this article was on 2000-07-15. We do not assume any responsibility for changes (disappearance of Web sites, change in URLs, or changes in the information contained in Web sites, etc. The fact that relevant Web sites are not listed does NOT mean that we considered them bad or not worthy of consideration. If you have a site you like, we suggest you post it on the GlossPost list (see above).*

A.1 Keeping Abreast of Search Engines

The *Search Engine Watch* site provides undated reports on how search engines work and rates individual resources. Check out:

How Search Engines Work:
http://searchenginewatch.com/webmasters/work.html
Search Engine Report Newsletter:
http://searchenginewatch.com/sereport/index.html
Web Searching Tips (including tutorials (also on Boolean searching), a search engine glossary, search engine reviews, etc.)
http://searchenginewatch.com/facts/index.html
Domain-specific search engines can be found under:
http://searchenginewatch.com/links/Specialty_Search_Engines/

A.2 Metadata Initiatives, Multilingual Digital Libraries, Cross-Language Information Retrieval

Cross-language information retrieval
http://www.dlib.org/dlib/december97/oard/12oard.html
Dublin core – metadata (including multilingual versions)
http://purl.oclc.org/dc/
Multilingual digital libraries
http://www.dlib.org/dlib/june97/06borgman.html
Multilingual Internet
http://www.ceveil.qc.ca/viag2.html
Metatdata registries
http://www.soi.city.ac.uk/~dy661/metadataregistries.html
Multilingual search engine (experimental)
http://crl.nmsu.edu/Research/Projects/tipster/ursa/Mundial/mundial.html
Multilingual searching:
http://www.dlib.org/dlib/september98/powell/09powell.html

A.3 Multilingualism

European Commission Website on the Multilingual Information Society (MLIS)
http://www.ispo.cec.be/topics/i_multiling.html
http://www.hltcentral.org/MLIS/

A.4 Portals for Terminology Activities, Practical Support

Terminology Forum (the first terminology portal on the web, not regularly updated)
http://www.uwasa.fi/termino/
ETIS European Terminology Information Server (currently under construction, in operation by October 2000)
http://www.unilat.org/dtil/etis/fr/fr.htm
Terminology Pilot at the University of Surrey:
http://www.surrey.ac.uk/LIS/Terminology/surrey_sites.htm
Innsbruck *Terminologie* Web page (useful collection of links)
http://translation.uibk.ac.at/termlogy/index.html

Practical Support in Terminology Management:
Translation, Theory, Technology Group at Brigham Young University:
http://www.ttt.org
including: ISO 12620 Data categories
http://www.ttt.org/clsframe/datcats01.html
Standards and Good Practice for digital resources
http://vads.ahds.ac.uk/guides/creating_guide.html
DINT Project (Development of Innovative Networks for Terminology in Europe):
http://www.qa9000.com/dint/index.html
Terminology training materials, Kent State Univeristy Institute for Applied Linguistics (KSU/IAL):
http://appling.kent.edu/resources/courseware/

A.5 Neology Watch, Term Mining

Balnéo, neology watch
http://rint.org/cadreIndex4.htm
Neologisms in corpora
http://www.compapp.dcu.ie/Projects/1998/CL/sroche.cl4/study.html

A.6 Terminological Resources

Termbases:
Eurodicautom

http://eurodic.ip.lu/cgi-bin/edicbin/EuroDicWWW.pl

Dictionary portals:

http://www.onelook.com provides lookup capability in an increasing number of indexed dictionaries (625 in July of 2000). French, Italian, German Spanish (FIGS) are available, but it is most reliable in English, probably because it has access to a wider range of resources for English.

http://www.yourdictionary.com offers another exhaustive list of dictionaries, but its "quick lookup" feature is limited to English in a single dictionary, the Merriam Webster. Efforts to find technical terms that turned up excellent results with OneLook failed with YourDictionary.

http://dir.yahoo.com/Reference/Dictionaries is frequently the starting point for a search of dictionary-like resources.

http://rivendel.com/~ric/resources/dictionary.html is a commercial site, but it offers an encyclopedic listing of dictionaries in many languages and language combinations.

Representative domain-specific services:

Environment: The U.S. EPA Terminology Reference System:

http:www.epa.gov/trs

Telecommunications: Termite, the terminology database of ITU, the International Telecommunications Union

http://www.itu.int/search/wais/Termite/index.html

Medicine: Medline

http://www.nlm.nih.gov/databases/freemedl.html

Computing:

Whatis.com:

http://www.whatis.com/ is a knowledge exploration tool about information technology, especially about the Internet and computers containing over 2,000 individual encyclopedic definition topics and a number of quick-reference pages.

High Performance Computing and Communications Glossary

http://nhse.npac.syr.edu/hpccgloss/hpccgloss.html

INRIA Institut National de Recherche en Informatique et en Automatique

http://www-rocq.inria.fr/~deschamp/www/CMTI/LAF.html

Délégation generale à la langue française

http://wwwchimie.u-strasbg.fr/membres/GB/Lexique.html

A.7 Institutions, Associations, Networks

European:

EAFT (European Association for Terminology)

http://www.unilat.org/dtil/aet/index1.htm
ELRA European Language Resource Association
http:www.icp.grenet.fr/ELRA/home.html
TDCNet (European Network of Terminology Documentation Centers, with links to them)
http://www.tdcnet.net/

International:
Infoterm (International Information Centre for Terminology)
http://www.infoterm.org
International Network for Terminology (TermNet)
http://www.termnet.at/
International Institute for Terminology Research (IITF),
Current statement on cooperation:
http://www.unilat.org/dtil/etis/resumes/budin.htm
VHG Virtual Hyper Glossary
http://www.vhg.org.uk/

Regional:
Francophonie, Romance languages:
RINT (Le reseau international de néologie et de terminologie)
http://www.rint.org/cadreIndex1.htm
Realiter (Le reseau panlatin de terminologie):
http://www.iula.upf.es/cpt/cptwebes.htm
Nordterm (Nordic countries):
http://terminology.uwasa.fi/nordterm/en.html

National:
Termdat, Berne, Switzerland
http://www.admin.ch/ch/i/bk/termdat/index.htm
TSK, Tepa, Finland
http://www.tsk.fi/index-en.html
Office de la langue française, Canada
http://www.olf.gouv.qc.ca/
Termium:
http://www.translationbureau.gc.ca/pwgsc_internet/english/03_tools/03_termium.htm
German Terminology Documentation Centre, Deuterm, Germany
http://www.fbi.fh-koeln.de/fachbereich/ projekte/Deuterm/Tdcnet_e.htm

A.8 Thesauri

Web Thesaurus Compendium
http://www-cui.darmstadt.gmd.de/~lutes/thesauri.html
WorldWeb English Thesaurus and Dictionary
http://www.x-word.com/thesaurus/
The Art and Architecture Thesaurus Browser
http://shiva.pub.getty.edu/aat_browser/
The Getty Thesaurus of Geographical Names
http://shiva.pub.getty.edu/tgn_browser/

A.9 Projects and Initiatives

Language Engineering, Terminology Engineering, Translation Engineering, Ontology Engineering:
SALT (Standards-based Access to Lexicographical and Terminological multilingual resources)
http://www.loria.fr/projects/SALT/saltsite.html
OLIF (Open Lexicon Interchange Format)
http://www.olif.org (active from August 2000)
ISLE (International Standards in Language Engineering)
http://www.ilc.pi.cnr.it/EAGLES/ISLE_Home_Page.htm
TELRI (Trans European Language Resource Infrastructure)
http://www.telri.de/
DHYDRO (Standardised Multilingual Hydrographic Dictionary)
http:www.loria.fr/projects/MLIS/DHYDRO

Ontology engineering:
General information:
University of Padova bibliography for computer scientists working in areas such as Artificial Intelligence, Databases, or Computational Linguistics, with a specific interest in *Conceptual Analysis* issues
http://www.ladseb.pd.cnr.it/infor/ontology/Papers/Ontobiblio/TOC.html
OASIS, the Organization for the Advancement of Structured Information Standards
http://www.oasis-open.org/
Ontology.Org. an independent industry and research forum focussed upon the application of ontologies in Internet commerce
http://www.ontology.org/
Ontology Projects World-Wide
http://www.kr.org/top/projects.html

Standards:

Simple HTML Ontology Extensions (SHOE)
http://www.cs.umd.edu/projects/plus/SHOE/
http://www.oasis-open.org/cover/shoe.html

XOL - XML-Based Ontology Exchange Language (a format for exchanging ontology definitions among a set of interested parties)
http://www.oasis-open.org/cover/xol.html

Individual ontology projects:

The Beethoven Project, a Java interface to a knowledgebase of metadata generated from US MARC records:
http://www-personal.umich.edu/~peterw/Ontology/Beethoven/demo.html

The Cyc family multi-contextual knowledge base and inference engine
http://www.cyc.com/

Mikrokosmos, the Computing Research Laboratory, New Mexico State University:
http://crl.nmsu.edu/Research/Projects/mikro/

John Sowa's KR Ontology, featuring good explanations of ontologies and how they work:
http://www.bestweb.net/~sowa/ontology/

Stanford's "build-your-own" axiom-based Ontolingua server:
http://www-ksl-svc.stanford.edu:5915/
http://www-ksl.stanford.edu/knowledge-sharing/ontologies/html/index.html

Wordnet: A Lexical Database for English
http://www.cogsci.princeton.edu/~wn/

REFERENCES

Harris, Robert. 2000. *A Guidebook to the Web*. Guildford CT: Dushkin/McGraw-Hill.

For continuous updates of this article, see http://www.benjamins.nl/jbp/htm http://appling.kent.edu/htmweb/htmweb.htm

Information Boxes

INFOBOX 20
UNICODE

The Unicode Standard is a fixed-width (i.e., all double-byte) universal character encoding standard used for the representation of written characters and text for computer processing, i.e., for storage and interchange of textual data. It is universal in that it provides encoding points for the characters used to represent most major modern languages, as well as common technical symbols. It is fully compatible with the International Standard ISO/IEC 10646-1:1993, and contains all the same characters and encoding points, over 33,000 coded entities in all.

Computer programming originated in English-language environments, resulting in the evolution of the original 7-bit ASCII (American Standard Code for Information Interchange), which is limited to the non-accented characters generally used to represent the English language. (There are other encoding systems as well, such as IBM's EBCDIC (Extended Binary Coded Decimal Interchange Code), which will be ignored here for simplicity's sake.) 7-bit ASCII is specified internationally in ISO 646:1991. This set was limited to 128 characters by the simple mathematics of the binary system used to encode computer data, so the inclusion of the diacritics used for other standard European languages required the addition of the 8th bit in order to accommodate expansion to 256 characters.

Even with eight bits, it is impossible to provide for the thousands of characters used in syllabary languages such as Chinese, Japanese, and Korean (CJK), nor even to account for the many variations on the Latin alphabet itself that are required for representing Slavic, Finno-Ugric, or non-European languages. As a consequence, ISO developed a suite of standards to define different code pages for various language groups, many of which are grouped together in the ISO 8859:1986 suite of standards. The most common of these is Latin-1, which is used for Western European languages and is frequently referred to as the ANSI set with reference to the parallel American National Standards Institute document.

There are several reasons why the 8859 standards do not satisfy all needs for information handling in all environments. In the World Wide Web, where information has to flow among many different platforms to a variety of different browsers and through "gateways" from one computer in the network to another, special 8-bit characters in the data stream can get lost or corrupted because of differences in representation between systems. As a consequence, SGML and HTML have substituted complex, difficult-to-read character ISO-646 entitites for letters having diacritics, e.g., the simple French phrase *à côté* will be represented in SGML as the unintelligible string: *à côté* . Non-Latin scripts, which must also be represented in similar ways, become totally illegible without conversion to a reader-friendly form. Currently Web browsers must be specifically enabled to display non-Latin languages, especially ideographic or bidirectional languages like Hebrew and Arabic. Concurrent with the evolution of Latin character sets, the CJK and other language groups have evolved their own encoding systems, in some cases developing multiple, incompatible expressions of the same characters or character sets.

Individual applications that do not have to accommodate all languages at once can get by with loading, say, a code page for Cyrillic and the ASCII set, which means that texts can be displayed in Russian with embedded quotations from English, but the representation of Cyrillic, Chinese, Western European, and Arabic in the same text running a single system poses a formidable, if not insurmountable challenge. Although traditionally only a small percentage of documents may use such a variety of character sets, multilingual bibliographical and terminological resources not infrequently demand access to the full range of character sets all within a single system.

Unicode provides a single internationalization process designed to produce code that will handle character set requirements for all world languages at the same time. It is designed to eliminate corruption problems that plague mixed codeset programs, as well as the need to convert characters back and forth between different systems. It is based on a 16-bit representation, which offers the possibility for over 65,000 possible unique code points arranged in uniform double octets. It cannot be simply assumed that the existence of Unicode will solve all character-set issues, but as more programs adopt the standard, there will be fewer difficulties in achieving universal representation without special intervention.

REFERENCES

ISO 646:1991. Information technology—*ISO 7-bit coded character set for information interchange.* Geneva: International Organization for Standardization.

ISO 8879:1986. *Information processing—8-bit single-byte coded graphic character sets.* Geneva: International Organization for Standardization.

ISO/IEC 10646-1:1993. *Information Technology—Universal Multiple-Octet Coded Character Set (UCS)—Part 1: Architecture and Basic Multilingual Plane.* Geneva: International Organization for Standardization.

Unicode, Inc. 1999. *The Unicode® Standard: A Technical Introduction.* http://www.unicode.org/unicode/standard/principles.html. Accessed 1999-11-21.

The Unicode Consortium. 1997. *The Unicode Standard, Version 2.0.* Reading, MA: Addison Wesley.

Unicode Consortium. 2000. http://www.unicode.org/. Accessed 2000-07-19. (See Web site for updates and amendments to the standard.)

INFOBOX 21
Standard Generalized Markup Language

Sperberg-McQueen (1990 and current) describes Standard Generalized Markup Language (ISO 8879) as an "international standard for the definition of device-independent, system-independent methods of representing texts in electronic form". In a more "gentle" vein, he goes on to simplify the definition by stating that "SGML is an international standard for the description of marked-up electronic text. More exactly, SGML is a metalanguage, that is, a means of formally describing a language, in this case, a markup language." Originally, markup was the annotations or hand-written text that a proofreader or copy editor would add to print galleys in order to instruct the typesetter how to print or lay out the text in question. In this respect, markup is *text about text*, which is what is meant by a *metalanguage* or *metatext*.

Although most people who use word-processing programs never see it, markup is used to encode text, indicating how it will be presented either in a screen display or printed on paper. Traditionally, text is enclosed by paired markers that indicate that everything inside the markers will be treated in a certain way. For instance, the line:

Infobox 1 *Terminology Collections* 35

might be represented internally with the markup something like this:

[FontTimesNewRoman> [FontSize: 11pt>] [Left Tab] [Bold> Infobox 1 **<Bold][Left Tab]**
[Italc>Terminology Collections **<Italc][HdFlushRight]** 35 **<FontSize: 11pt] <FontTimesNewRoman][HRt]**

Here the bracket convention **[...>** (Text goes here) **<...]** is used to represent paired encoding elements that indicate that the text appearing inside the pairs of codes (which are called *tags* in SGML and similar markup languages) should be interpreted in the specified way. The [...> tag is commonly called a start tag, and the <...] notation is an end tag. Some of the [...] codes appear only once, but can work together: the [HdFlushRight] command is only valid until the program encounters the [HRt] code. In the example shown here, this markup is used exclusively to instruct screen and printer drivers how to represent the text physically on-screen or on the printed page. In this way, the tags serve as markers that are interpreted according to a set of rules specified in the screen and printer drivers. Based on this sense of markup, Sperberg-McQueen views markup as *encoding* used for "making explicit an interpretation of a text. ... By markup language, we mean a set of markup conventions used together for encoding texts".

Perhaps the most important thing about SGML markup is that, in contrast to the word-processing markup we see above, SGML does not generally provide instructions on how the text is actually supposed to look. Instead of specifying boldface, italics, or a particular font and point size, i.e., *presentational markup*, SGML ideally provides information about the logical structure of the text. The following example provides a contrast between representational and SGML *structural markup*:

HANDBOOK OF TERMINOLOGY MANAGEMENT, Vol. II

can be represented internally in word-processing markup as:

[HdCenter on Marg][Italc> HANDBOOK◊OF◊TERMINOLOGY MANAGEMENT, Vol. II **<Italc]**

but can be represented in SGML simply as:

<hd1> HANDBOOK OF TERMINOLOGY MANAGEMENT, Vol. II **</hd1>**

Note here some differences in the actual form of the SGML tags: <...> introduces a start tag and </...> an end tag. There are a number of other specific conventions for designating attributes of tags, targets (links), etc., but it is not the purpose of this brief overview to explain the details of SGML notation.

The principle behind this kind of markup is that the information as to what the text is actually going to look like is not included in the text itself or in the master "program" that interprets the text, i.e., that is used when parsing it. In SGML this master component is called a *Document Type Definition (DTD)*. It lists the tags, properly called *generic identifiers (GIs)* in SGML, which can be used in a particular type of document, the *attributes* that can be assigned to them, and certain rules about how the markup can behave. At print-time, a program designed to interpret SGML DTDs and documents can be used to specify what each <hd1> element is actually going to look like. Anyone who has ever used word-processing styles will see the advantage of this approach: if text elements are marked up individually, any change in layout entails a careful and painstaking sweep through an entire document, which in some cases may be huge, to find all the instances where changes must be made. If presentational features are controlled by a style, however, a one-time change in the style will ensure that all instances of that style respond obediently to the new layout command.

Not only is it easier to make changes in markup if structural tags are used. It is also easier to exchange documents across program and platform boundaries because the formats for presentational markup differ so significantly from one word processing (WP), desktop publishing (DTP), or typesetting program to another. In fact, this need was the real origin of SGML. As printers moved to DTP and digital typesetting technologies in the 1970s and 1980s, it became apparent that significant incompatibilities existed between the WP and DTP programs in which texts were commonly created and the typesetting programs used to produce final documents. In addition to using different programs, authors frequently work, for instance, in IBM PC environments, whereas typesetters often prefer Apple Macintosh (Mac), not to mention that in complex large organizations, some components of a given job might be produced in UNIX or mainframe environments as well. This compelling need for a universal, open format that would allow unlimited movement of documents across these kinds of boundaries led to the development of the SGML standard and its official publication as an ISO standard in 1988.

The most pervasive presence of the SGML format in the daily lives of anyone who uses the World Wide Web is a single SGML DTD called Hypertext Markup Language (HTML), which is, of course, the metalanguage that makes the Web possible. Ironically, many Webpage-generation programs attempt to force HTML into a presentational mode. Nevertheless, the World Wide Web Consortium (W3C), which governs the development of HTML, recommends that designers mark up headings, paragraphs, lists, hypertext links, and other structural parts of an HTML document and that they use style sheets to specify layout separately. Anyone who has never seen the "insides" of an HTML document has only to click on the "VIEW/(Page) Source" command in an Internet browser. Here one sees the standard features of an HTML document (W3C 1999a):

```
<!DOCTYPE HTML PUBLIC "-//W3C//DTD HTML 4.0 Transitional//EN">
<html>
<head>
<title>HTML Home Page</title>
</head>
<body bgcolor="#FFFBF0" text="black" link="blue">
...
<p>Software developers who write HTML editing tools can ensure
interoperability with other Web software by verifying that the
output of their tool complies with the W3C Recommendations for
HTML.</p>
</body>
</html>
```

The well-formed structural markup to be found in the W3C Web pages themselves reflects the Commission's stated goal of encouraging people to clean up the frequent muddle of non-structural tags found in many pages and tolerated by some Web browsers, a factor that becomes all the more important when it comes to ensuring the well-formedness that is a prerequisite of the latest manifestation of the markup languages, i.e., Extensible Markup Language (XML™). XML is a simple, very flexible text format derived from SGML that has been designed for large-scale electronic publishing and for the exchange of a wide variety of data on the Web (W3C 1999b). The flexibility and versatility of XML is enabling the creation of stable interchange and data resources on the Web.

In addition to HTML and XML-compliant markup, HTML documents frequently contain small embedded run-time applications called scripts, including CGI (Common Gateway Interface) scripts frequently written in the PERL scripting language or in Sun Microsystems Java, whose scripts are referred to as applets, i.e., little applications. In the future this technology, combined with XML, will provide a rich mix of options for representing data and facilitating the inclusion of executables in HTML documents.

REFERENCES

Brenner, Stephan E. 2000. *The cgi-lib.pl Home Page.* http://cgi-lib.berkeley.edu. Accessed 2000-07-20.

ISO 8879:1986, International Organization for Standardization, ISO 8879: Information processing — Text and office systems — Standard Generalized Markup Language (SGML). Geneva: International Organization for Standardization.

Sperberg-McQueen. 1990. "A Gentle Introduction to SGML." In: ACH-ACL-ALLC *Guidelines for the Encoding and Interchange of Machine-Readable Texts*. C. Michael Sperberg-McQueen and Lou Burnard, Eds. Chicago and Oxford: TEI P1.

Sperberg-McQueen. Current. http://www-tei.uic.edu/orgs/tei/sgml/teip3sg/SG.htm. Accessed most recently 2000-07-19.

W3C 1999a. *Hypertext Markup Language Home Page*. http://www.w3.org/ MarkUp/#html4. Accessed 2000-07-19.

W3C 1999b. *Extensible Markup Language (XML) Activity*. http://www.w3.org/XML/ Activity.html. Accessed 2000-07-19.

INFOBOX 22
Text Designations

A *source text (ST)* is generally the text written in one language (the *source language, SL)* that is being translated into a second language, called the *target language (TL)*. The resulting transaction is called the *target text (TT)*. These texts are sometimes referred to as the *L1 text* and the *L2 text*. The numbers are subscripted by some authors and not by others. Similar notation, *L1* and *L2* (also *Language A* and *Language B*), is used in language pedagogy to refer to an individual's native language and his or her second language.

Translators and terminologists use *parallel texts* to support their stylistic and terminological choices. These are texts originally written in the target language that are comparable to the source text by virtue of shared text type or subject matter, or both. For instance, parallel texts for an automobile advertisement about antilock braking systems (ABS) might include texts, which, measured in degrees of applicability, might include any advertisement in the TL, any automotive advertisement, or an automotive advertisement stressing ABS, which would be the most desirable selection. An informative text describing ABS would also be a parallel subject-field text and could be used to prepare terminology resources and create the final target text. In applied linguistics, these texts are sometimes called *comparable texts.*

Bitext or *aligned text* consists of source and target texts that have either been aligned after translation using an alignment tool or created in alignment using a translation memory tool. *Bitext* obviously only applies to paired texts in just two languages. The term *aligned text* is more transparent and accommodates an unlimited number of language pairs. Some MT specialists use the term *parallel text* for *aligned text*, but this can be ambiguous because of the very common usage described above. In Translation Memory environments, aligned texts are sometimes called parallel texts.

A background text, although similar to a parallel text, is a text in the *source language* that provides additional information on the subject matter treated in a text for translation. Translators and terminologists use background texts to familiarize themselves with unfamiliar subject matter before and concurrent with their efforts to find parallel texts in the target language.

A number of authors in this book use the term *text chunk*, which those unfamiliar with the term frequently object to as overly colloquial, although this is a legitimate technical term used in linguistics and discourse analysis to refer to a segment of text of varying length selected for discussion or evaluation of some sort.

INFOBOX 23

ISO 9000 Quality Standards

The ISO 9000 series is a set of individual, but related, international standards on quality management and quality assurance. They are generic, not specific to any particular products or industry. They can be used by manufacturing and service industries alike. These standards were developed to effectively document the quality system elements to be implemented in order to maintain an efficient quality system in any given environment. The ISO 9000 Series standards do not themselves specify the technology to be used for implementing quality system elements, which remains the job of the individual enterprise adopting the standard.

At the end of the year 2000, a new set of standards (ISO 9000:2000, 9001:200, and 9004:2000) replaced the existing family of standards. The new standards are designed to provide a broad base for quality system requirements. The intent is to unify procedures throughout industry, thus replacing the more fragmented system that has grown up around the preexisting standards. For the latest information on these documents, consult the Web page of ISO Technical Committee 176:

http://www.iso9000y2k.com/TC176.html. (Last accessed 2000-07-19)

The significance of the ISO 9000 environment for terminology management is twofold: the correct elaboration of terminology plays a significant role in the definition of quality system elements, and terminology itself is the critical factor affecting quality and efficiency of the document production chain, whether it involves multilingual or only monolingual documentation. We are also concerned, however, with quality assurance with respect to terminology management itself. Since the ISO standards are generic, it becomes essential for individual industries to write their own specific quality assurance manuals and to create checklists documenting procedures for checking quality on the spot and for long-term quality audits. The checklist provided in Infobox 24 is typical in form and content to the kinds of checklists used in industry.

INFOBOX 24
Quality Checklist

The checklist that follows on the next page represents an effort to create a generic auditing procedure for use by a terminologist or data entry specialist when creating individual terminological entries. The intent of the checklist is to provide guidelines for ensuring that all aspects of the terminological entry have been addressed and the basic principles of terminological data entry have been maintained. It is impossible, however, to include all aspects affecting all terminological entries in such a generic list because the needs of different working environments vary, as do the data models used in different situations. For the most part, terminologists wishing to use this list should ideally revise the list to meet the needs of their own working groups.

For instance, in the *Termbase Entry Management Issues* section of the checklist, the question of how to handle and recognize doublettes is a major problem. In some environments, terminologists start with the term in a common source language or pivot language, frequently English, and the requirement is made that this term be entered first in order to check whether the term already exists in the system. In other environments, all languages are considered to be equal in importance, so terms can be entered from any language, but the same concerns about doublettes exist. Furthermore, there is no way to ensure automatic recognition of doublettes if there are even slight variations in the terms that are entered (variant spellings, variations in term form, etc.). Ultimately a chief terminologist or quality assurance specialist needs to gain an overview of the entries in the system in order to ensure that no duplications occur. In fact, a separate checklist could be developed for use by a QA terminologist.

The rules outlined here for definitions are strict in comparison to the rules that are used in some working environments. Each group needs to develop its own standards for writing or documenting definitions and to adhere to that standard. The important factor here is consistency within the working group. This is particularly true for the use of articles (which varies from language to language) or of capitalization and punctuation in definitions.

Finally, working groups must also determine whether definitions or contexts or both will be the central focus of their entries, since this emphasis varies depending on whether the working environment is primarily descriptive or prescriptive in nature.

✓ Terminology Entry Quality Checklist

TERMBASE ENTRY MANAGEMENT ISSUES

- ☐ The entry of any term into the database includes a check for doublettes in the database and resolution of duplicate entries for the same concept.
- ☐ All data in entry pertain to entry concept
- ☐ No other entry in database documents the concept treated in this entry
- ☐ Homographs representing other concepts are treated in other entries
- ☐ Entry has been assigned to the appropriate subject field

TERM-RELATED ISSUES

- ☐ Terms are present for all languages to be treated in the entry
- ☐ Terms are spelled correctly
- ☐ Terms are in lowercase unless commonly used in uppercase in texts
- ☐ Terms are reported in the singular unless always used in context in the plural
- ☐ Entailed terms present in database are cross-referenced
- ☐ Only one term [including multi-word terms] is entered in each term field
- ☐ Each term (e.g., synonym, abbreviation, full form, TL equivalent) is entered in its own term field (i.e., there is just one term per term field)
- ☐ No terms (synonyms, abbreviations, full forms) are embedded in definitions or notes
- ☐ Articles (le, la, der, die, das, etc.) are not entered in term field (noun terms)
- ☐ Infinitive particles (to, à, zu, etc.)are not entered in term field (verb terms)

DEFINITION-RELATED ISSUES

- ☐ Only one definition is entered in a single definition field
- ☐ Definitions that do not pertain to the concept in question are entered in a second entry that treats the other concept
- ☐ Definitions consist of a single predicate-like statement
- ☐ No note-like material is embedded in definitions
- ☐ Neither the term nor a synonym (abbreviation or full form) for the term is restated in the definition
- ☐ No finite verb is included in the definition statement (e.g., the verbs *is, refers to, etc.*)
- ☐ Definitions conform to the general principles for writing definitions and to any specific criteria specified for the database in question
- ☐ Punctuation and capitalization in the definition field conform to style guides set for the database

CONTEXT-RELATED ISSUES

- ☐ Context field contains a text chunk, which includes the term in question
- ☐ Context contains situational, definitional, explanatory, or associative information about the term and its concept
- ☐ Punctuation and capitalization in the context field conform to style guides set for the database
- ☐ Context contains a manageable amount of textual information (neither overly brief or excessively long)

DOCUMENTATION

- ☐ Sources of text-related information are appropriately cited (e.g., definitions, contexts, notes if quoted)
- ☐ Sources of terms are documented if required by database guidelines
- ☐ Bibliographic entry with appropriate ID has been created for each printed source
- ☐ URL for Web-resident information has been documented in accordance with the guidelines for the database

INFOBOX 25
Controlled Language

Controlled language (CL) can be defined as a subset of a language with a restricted grammar and a domain specific vocabulary designed to allow domain specialists to unambiguously formulate texts pertaining to their subject fields. Controlled language facilitates clear, concise, technical communication by adhering to a *one word-one meaning* principle, frequently called monosemy and mononymy in terminology theory. Controlled language can be accurately and efficiently processed by a computer, but is expressive enough to allow natural usage.

The advantages of adopting CL include consistency in style and reading level, reduced training times, increased readability for non-native speakers, and improved machine translation capability. The disadvantages involve a reduced sense of creativity and individual expression on the part of technical writers, plus the need to train people to use CL rules, although sophisticated authoring programs like the Boeing company's Simplified English Checker can ease the burden on the writer. These kinds of tools are considered to be language engineering applications. Controlled language should ideally be developed with specific audiences and text types in mind, for instance the users of multilingual versions of machine maintenance manuals.

Controlled English originated as Caterpillar Fundamental English (CFE) in the 1960s and went on to form the basis for the simplified English used by the Carnegie-Mellon KANT project involving machine translation of the Caterpillar Tractor Company's maintenance manuals for heavy equipment exported worldwide. The European Association of Aerospace Industries (AECMA) has adopted AECMA Simplified English designed to improve the readability of maintenance documentation in the civil aircraft industry.

REFERENCES

AECMA. *AECMA Simplified English*. http://www.aecma.org/sebrc.htm Accessed 2000-07-20.

KANT. *Knowledge-based Machine Translation*. http://www.mt.cs.cmu.edu/Research/Kant/. Accessed 2000-07-20.

Wojcik, Richard H., and Hoard, James E. http://cslu.cse.ogi.edu/HLTsurvey/ch7node8.html. Accessed 2000-07-20.

Huijsen, Willem-Olaf. *Controlled Languages Homepage*. http://www-uilots.let.ruu. nl/.Controlled-languages/HOME.html. Accessed 2000-07-20.

INFOBOX 26
The Economics of Terminology Management

There is a longstanding lack of data on the financial advantages of implementing terminology management practices. Components that would enhance this situation include:

- Global, regional, and national market studies designed to evaluate the size of the overall market and to pinpoint market trends
- Enterprise-specific cost-benefit studies to reveal both the true cost of terminology management and any resulting value added to goods and services
- Empirical studies to reveal the costs of *not doing* terminology management
- Comparative studies designed to reveal the cost/benefit ratios associated with different levels of computer support for translation and technical writing

In an effort to couch the basic financial aspects of terminology management in the terms that are used in transactional analysis, Koller (1994:88; see also Wright 1997) presents the graphs shown in Figures 1 and 2, where

- Line A reflects the variable costs involved in *not doing* systematic terminology management.
- Line C reflects the fixed costs for doing systematic terminology management.
- The *frequency* value shown on the x-axis represents the repetition of terms already documented in the system.
- The break-even point represents the point at which the terminology management effort begins to pay for itself in terms of reduced costs.
- The various lines in Figure 2 plot several different break-even points depending on the initial investment made in terminology management and the frequency of reuse of terminological information.

Although these models are theoretically intriguing, empirical studies are needed to document the validity of the implied assumptions reflected in the graphs. Unfortunately, it is much easier to account for the real costs accrued in implementing a complex terminology management system than it is to track the value of lost business or lost time as the result of failed communication. General Motors, however, has recently demonstrated a significant return on investment (ROI) as the result of introducing strict terminology controls in conjunction with the adoption of Controlled English in the production of GM maintenance manuals (Godden 1999).

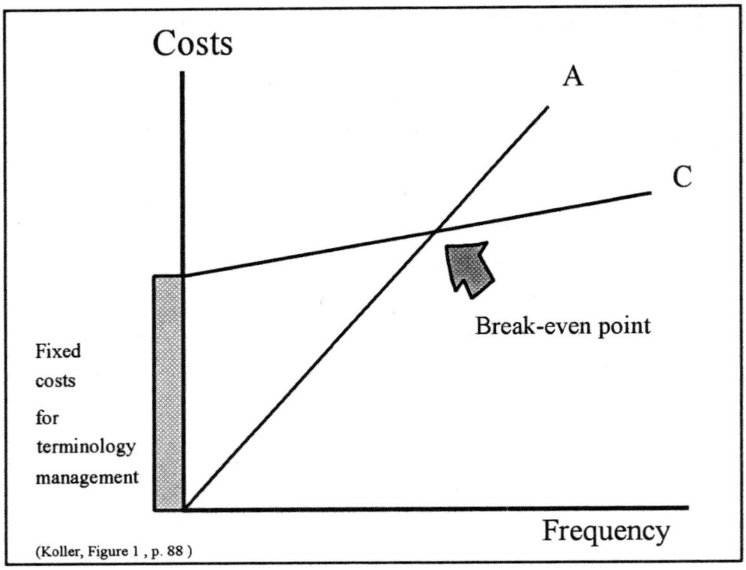

Figure 1: *Fixed and Variable Costs for Terminology Management*

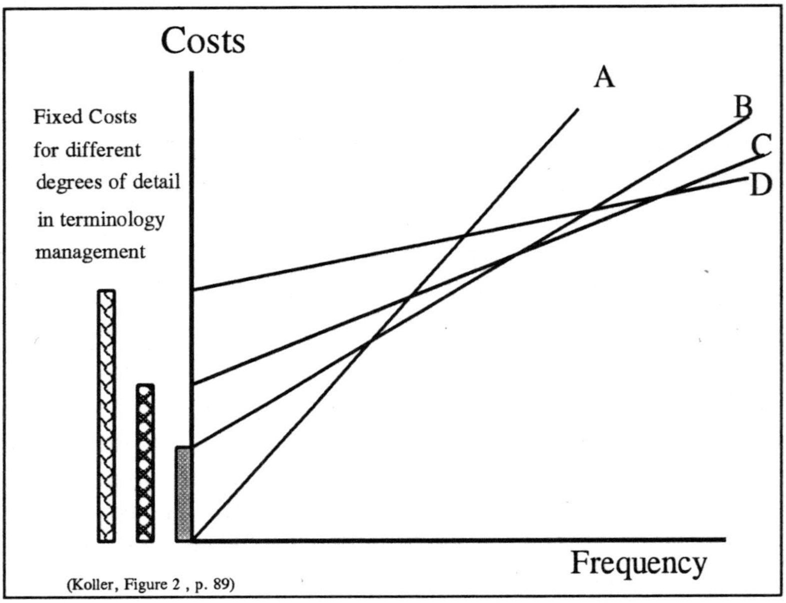

Figure 2: *Variations in the Break-even Point as a Function of Fixed Cost Levels*

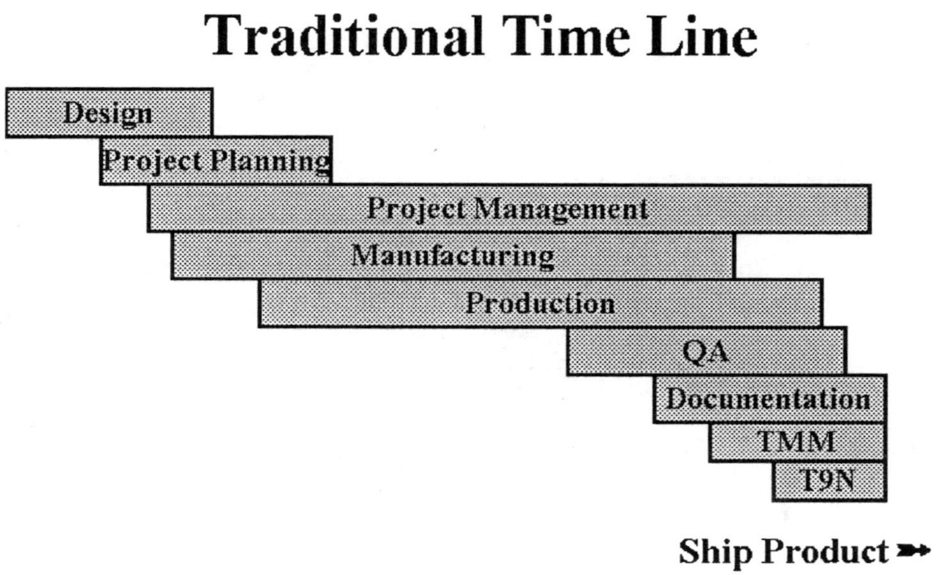

Figure 3: *Traditional Timeline for Terminology and Document Management*

The significance of terminology management with respect to return on investment (ROI) becomes particularly critical in situations involving information-heavy products for which terminology, frequently multilingual terminology, represents a major value-added factor in the delivery of the product to market. Figure 3 illustrates the traditional position of terminology and document management in a multi-layered timeline showing the development of a product from the initial design stage to shipment to the end customer. Terminology management (TMM in this illustration only) is conducted only after completion of source language documentation in conjunction with the translation (here T9N) process, which is handled as a hurried activity at the end of the process so that multilingual documentation can be added to the package just as the product leaves the factory door. This model implies a lack of coordination at early stages of the project between design and technical documentation or between the latter and translation and terminology documentation activities.

In contrast, Figure 4 visualizes the document production team, including technical communicators, terminologists, and translators, being involved in the project management cycle very early so that terminology documentation and the preparation of multilingual equivalents can proceed in such a way that terminology resources can be delivered to the technical writers and then to the translators just-in-time to ensure consistent terminology usage in both the source and the target language versions of the documentation. This scenario is especially suited to localization environments

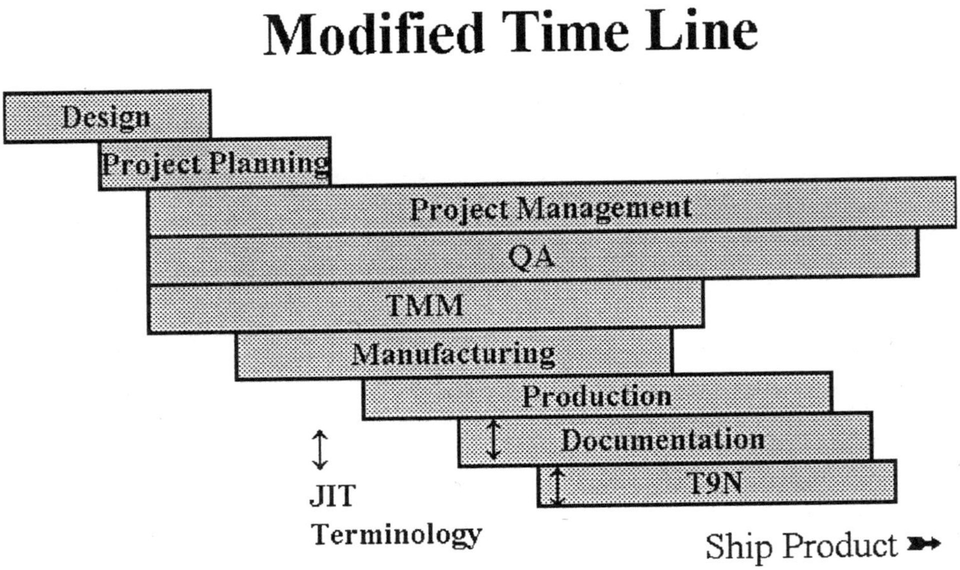

Figure 4: *Fully Integrated Terminology Management*

where documentation or even the product itself must be prepared for simultaneous release in multiple languages. The nearly concurrent representation of the document production and translation phases reflects ongoing translation and revision of successive versions of the translated materials as the project approaches completion. Here adherence to consistent terminology constitutes a commitment to information management as a facet of core business activity rather than relegating terminology to a marginal position. The ultimate commitment is, of course, the decision to adopt some degree of controlled language in order to facilitate the text production process.

References

Godden, Kurt 1999. "The Evolution of Controlled Authoring at General Motors." Forthcoming.

Koller, Hans. 1994. "Verfahren zur Beurteilung der Wirtschaftlichkeit von Terminologiearbeit." *Akten des Symposions: Terminologie als Produktivitätsfkator in volkswirtschaftlicher und betriebswirtschaftlicher Sicht*. Köln: Deutscher Terminologietag e.V., 79-94.

Wright, Sue Ellen. 1997. "Economic Issues of Terminology Management." *TermNet News*. No. 54/55, 1-10.

INFOBOX 27
Globalization, Internationalization, Localization

In the Localisation Industry Standards Association (LISA), it has become apparent that different individuals and companies assign different interpretations to critical terms used in the industry. On the broadest possible scale, the concept of *globalization* has significance both in the geopolitical and in a more business-oriented sense. In geopolitics, the term reflects the worldwide integration of commercial interests, nation-states, and technologies to form a single interactive market and a highly interdependent political and intellectual climate. This phenomenon is closely linked to the development of more transparent business conditions, the evolution of greater degrees of personal freedom, and efforts to curb corruption in order to ensure the confidence of the international investment community.

In the strictly enterprise-oriented sphere, however, *globalization* (G11N) is the process involving business issues associated with taking a product global, including marketing, sales, and support in the world market. Although the two concepts are related to one another, the second usage is most common in its everyday application in the localization industry.

Internationalization (I18N) is the process of generalizing a product so that it can handle multiple languages and cultural conventions. This definition holds true particularly for software products, but may apply for any number of other types of information-laden items, such as heart pacemakers. A term often used in close conjunction with internationalization is *enabling*, which refers to the appropriate preparation of information, documentation, or the product itself to facilitate the development of multilingual, multi-locale versions.

Localization (L10N) designates the process of making a product linguistically and culturally appropriate to the target locale, whereby the a *locale* represents the combination of a language with a specified region, e.g., French in Quebec as opposed to French in Belgium or any given African nation. Localization is frequently construed as going beyond translation itself, although it is possible to argue that localization is what "good translators" always do anyway. *Adaptation* involves another aspect of the localization process where components required for a certain market are added or associated with the product, e.g., the addition of German spellcheckers, thesauri, and other tools to a localized word-processing product.

INFOBOX 28
Termbase Terminology

In English, a distinction is made between a *database* (sometimes spelled *data base*) and a *data bank* (always two words). A database "is a collection of interrelated files that is created and managed by a database management system (DBMS)". A data bank is "any electronic depository of data" and is frequently involves more complex features (multiple applications, interactive files structures, greater system complexity or size) than databases do (Freedman 1989:174 & 184). *Terminology databases* (*termbases* for short) are individual databases that are frequently produced by individuals, companies, agencies, etc. *Term banks* in contrast usually constitute institutional resources, which according to Lauriston (in Dubuc 1998) are made accessible to subscribers. Term banks frequently require public funding. Care must be taken to differentiate between database/termbase and data bank/term bank when translating into English because this distinction is not always as strict in other languages.

Termbases, like databases, are made up of data records called *terminological entries* or *terminological records*. The terms *term entry* and *term record* are sometimes used, but many authors avoid these shorter forms because they can lead the casual reader to assume that the resources in question are strictly *term* or *word* oriented and therefore more lexicographical than terminological in character.

Terminological entries treat a single concept and may contain information on all the terms associated with that concept either in one or multiple languages. In some systems, all term entries are identical with individual data records, but in others several data records (terminological records) or even several data elements drawn from a variety of resources are combined to create virtual terminological entries. To further complicate the issue, some other terminologists consider the *term record* to be the data record unit in termbases, while the *term entry* is sometimes equated with the *entry term*, sometimes called the *main entry term* in other venues.

REFERENCES

Dubuc, Robert. 1997. *Terminology: A Practical Approach.* Adapted by Elaine Kennedy, with contributions by Catherine A. Bowman, Andy Lauriston, and Shirley Ledrew. Quebec: Linguatech.

Freedman, Alan. 1989. *The Computer Glossary: The Complete Illustrated Desk Reference*. 4th Edition. Point Pleasant, PA: AMACOM.

INFOBOX 29
Lex/Term Meta Model

Terminological and lexicographical information in intermediate formats serves three major purposes: (1) interchange, (2) dissemination, and (3) data analysis. These purposes can be contrasted within a context they all share, namely, conversion of information between an original data collection and an intermediate format. Interchange involves transferring information between two computer systems and is often, but not always, bi-directional. Dissemination is uni-directional and can be either for use by another computer system or for human viewing. Data analysis may involve either dissemination or interchange or may focus solely on improving understanding of the source data collection. Thus, as shown in Figure 1, data analysis does not necessarily involve dissemination, and dissemination does not necessarily involve interchange, but an interchange format can be used for both dissemination and data analysis.

The diagram in Figure 2 shows a preliminary model that accommodates a number of representational forms at a very high level in order to facilitate these various goals across system and format boundaries. The diagram transcends theoretical and methodological differences in representing lex-term data. Current efforts are focused primarily on terminological resources, but the intent is to expand the model to include lexicographical entries, particularly those used in machine translation systems.

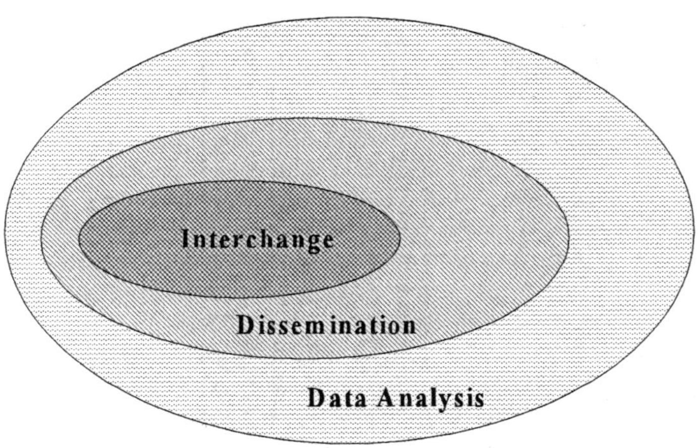

Figure 1: *The Relationships between Data Interchange, Dissemination, and Analysis*

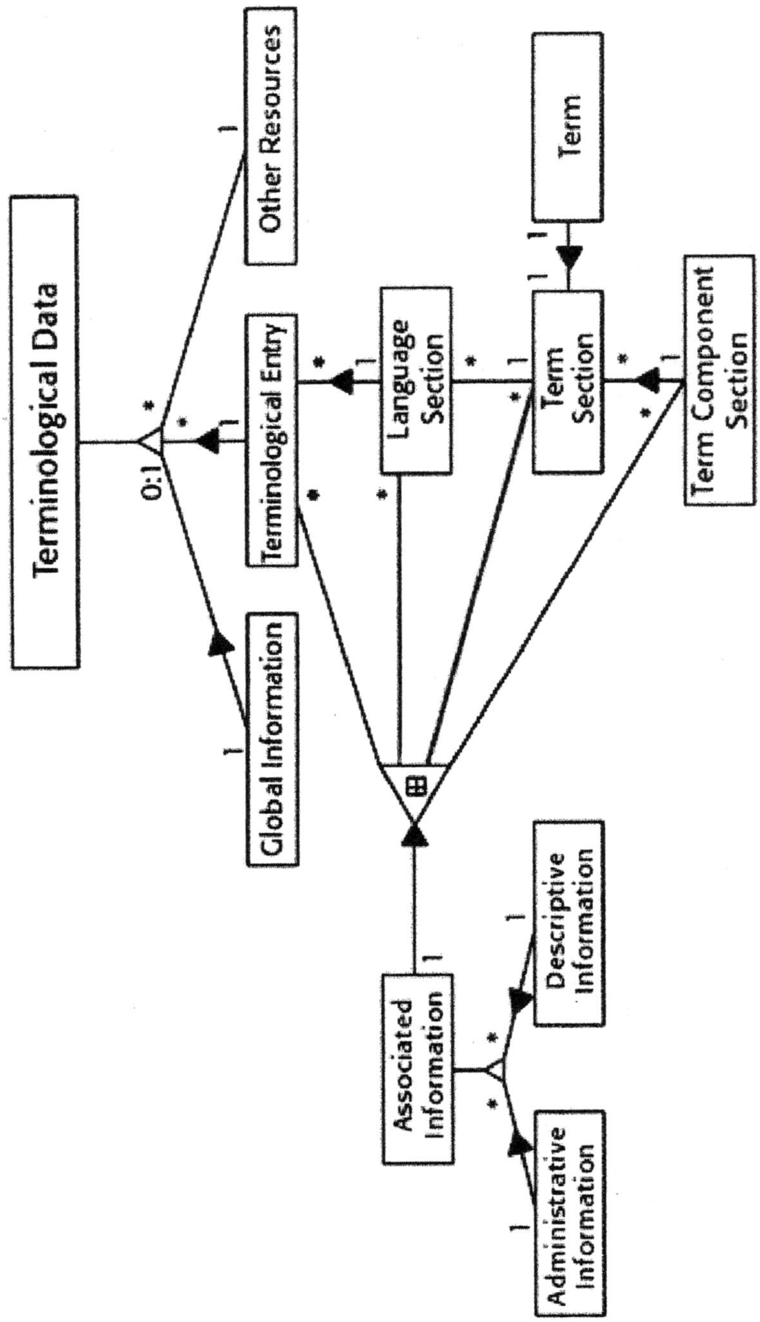

Figure 2: *Object Relationship Model (ORM) Diagram*

The diagram depicted in Figure 2 is designed to illustrate the relationships between individual objects in a relational data model. The following conventions apply:

- The boxes represent object classes.
- The lines represent relationships.
- The arrows indicate partitive relationships, that is: A ⟶ B indicates that A is or can be a part of B
- n:m means that an object participates in not fewer than *n* and not more than *m* relationships
- The * sign means that there is no upper limit to the number of relationships that can occur

The following alphabetical list describes the objects represented in the model, which are made up of current and pending data categories from ISO 12620:

- *Lex/term-oriented Data Collection (Terminological Data)*
  Comprises a data resource consisting of terminological or lexicographical entries or both and related information. Most lex/term data collections today can be classified as either lexicographical or terminological in structure, but as more processors of textual data adopt integrated environments requiring coordination of a variety of text production and translation tools, the need for leveraging (accessing, exchanging, and reusing) data from one system to the other will increase. These collections are made up of terminological or lexicographical entries, global information, and other resources, as indicated by the arrows in the diagram. Some hybrid systems share data between lexicographical and terminological structures, and other systems include elements of both without distinguishing between such structures, but interchange from such less rigorous structures to more structured ones may prove difficult. It should also be noted at least one entry of one kind or the other must be present in order to constitute a lex/term data collection. Otherwise the data file is an empty shell. The meta model shown in Figure 2 focuses solely on a terminological data collection.

- *Global Information*
  Includes any technical and administrative information that applies to the entire data collection and, by principle of inheritance, to all objects in the model.
  Example: Title of the data collection, revision history

- *Other Resources*
  Can comprise any data resource residing in or linked to the lex/term-oriented data collection that contains information complementary to the collection.
  Example: bibliographical source(s), a documentary language or description thereof, non-textual data (tables, figures, diagrams, audio, video and other binary files.)

- *Terminological Entry*
  Constitutes a data record containing information on those terminological units (i.e., subject-specific concepts, terms, etc.) associated with a single concept. In some (or perhaps even many) cases, the one-concept/one entry rule may not be precisely observed in order to accommodate pragmatic issues such as quasi-synonyms and close equivalents required for translation purposes, but true terminological entries will not treat substantially different concepts within the same entry (homographs, for instance). For purposes of interchange, entries are organized by language sections, and language sections are made up of individual term sections.

- *Associated Information*
  Includes administrative or descriptive information on the levels of the terminological entry, the language section, and the term section, as well as to individual items within the term section (which is not reflected in the diagram).
  Example: A responsibility code could be assigned at any of the cited levels, depending on the scope of the work performed by the individual in question.

- *Administrative Information*
  Consists of information related to terminology management and is linked in the diagram to *associated information*, which can apply on the levels of the terminological entry, the language section, and the term section.
  Example: Source reference of a definition, date a terminological entry is modified

- *Descriptive Information*
  Includes information related to concepts and designations and the relationships between them. Descriptive information is linked in the diagram to *associated information*, which can apply on the levels of the terminological entry, the language section, and the term section.
  Example: Definition, context, synonymy, concept-oriented information such as generic and partitive relationships, etc.

- *Language Section*
  Comprises the part of a terminological entry containing information related to one language, e.g., German, Arabic, etc. All individual term sections are assigned to a given language section, and associated data (including administrative and descriptive information) may be included at this level if it applies by principle of inheritance to all term sections associated with the concept in that language.

- *Term Section*
  Comprises the part of a language section containing a single term and all the information about that term, such as its usage, part of speech, term elements (which can be further treated in the term component section), etc. As shown in

the diagram, associated information can apply at the term section level, but it can also actually apply to individual elements within the term section.

- *(Lexicographical Entry)*
  An entry containing information that provides information on the various concepts associated with a lexical unit (see *Infobox 3: Lexicography and Terminology, Vol. 1:328)*.
  NOTE: This type of entry is typical of general language dictionaries and of machine translation lexicons. It not treated further in this model, but later extensions are anticipated, as reflected in the dashed lines in the diagram.

INFOBOX 30
Accessing Multilingual Terminological Resources

The SALT project described above (638-639) is developing a high level family of interchange formats designed for terminology-enabled knowledge sharing in integrated environments. Multilingual Terminological Resources (MTRs), which comprise an integral part of the language industries and the information economy, include *ontologies, categorization systems, data models,* and various *interchange formats.* Each one of the current generation of formats facilitates the exchange of data among resources representing a single vertical type MTR. Most of these formats are evolving to exploit the power and flexibility of XML (eXtensible Markup Language).

The SALT group has recognized that only when one coherent, industry-wide approach is adopted will information technology companies accept an interchange standard. Another important requirement for interchange standards is that they must support workflow processes in heterogeneous and complex production chains within a company that uses, for example a machine translation system, side-by-side with a translation memory tool, a document management system, a terminology management system, controlled language applications, authoring tools, information and knowledge management systems, spare parts administration tools, digital libraries, and other potential applications. The trend toward integrated IT environments where all such vertical components have to seamlessly interact is the main scenario for the language industry in the future. By establishing interoperability among various data models and formats within an overall framework, the SALT group is dedicated to facilitating *horizontal knowledge sharing.*

SALT is creating the XLT meta-model-based family of formats, which has the following properties:

- It is based on XML, thereby allowing the use of XSL and other XML tools.
- It is modular in its structure, i.e. those parts of an ontology or elements of lexico-terminological information that are actually relevant for a specific target application can be selected and processed by transformation tools.
- A freeware toolkit will be available on the SALT server (http://www.loria.fr/projets/SALT/) in the year 2001).
- It is internationalized, i.e., fully Unicode enabled.

The XLT representation format includes three logical components:

- A set of the data element types that are allowed in the model
- The permissible content of each data element type, which may be a data type (for example, ISO date) or a list of permissible values for each data element type based on ISO 12620 (or an amended version thereof)
- The structural relationships that are allowed among the data element instances. The third component defines the form and the first and second components define the content of the representation.

XLT is end-user oriented, distinguishing different user groups of equal importance, industrial tools developers, service providers, translators, technical writers, localizers, and other 'real' end-users.

Current projects within the SALT group are aimed at:

- Defining a very high-level meta-model that leaves room for both XML-based representation formats and relational database design
- Defining specific XML formats based on a constrained version of ISO 12200, within the broad possibilities allowed for by the meta-model
- Providing for interoperability between specific formats, that is, for bi-directional conversions between formats with little or no loss of information, so long as the content specification is held constant

The integrative picture of the SALT project that emerges from the inclusion of NLP lexicons, relational databases, and a meta-model can be outlined as follows:

- At the highest level, the meta-model level, the abstract structure of MTRs is represented using an application-independent diagraming method such as ORM (see Infobox 28).
- At the intermediate level, the conceptual data-model level, a split occurs reflecting whether the MTR is represented in XML or in a database. All data models are based on the same core structure, which is compatible with the abstract structure in the meta-model expressed as an XML DTD or schema that is compatible with ISO 12200 as amended. Each data model is defined by the logical combination of the core structure and a particular data category specification (DCS). At the middle level, a DCS is expressed as an instance of an XML schema that uses tag names that are intuitive to a terminologist.
- At the lowest level, the specific data-model/format level, conceptual data models defined at the intermediate level are instantiated as actual data models implemented in database management systems or as actual XML formats.

One conceptual data model from the intermediate level can have several interoperable formats associated with it at the lowest level. For example, one format may be very similar to the core structure and thus use meta-data-category-like tag names that are specialized by the value of a *type* attribute. This format reflects the Style 1 DTD described on page 622 of this book. Another format may have many more specific tag names as described with respect to Style 2 on page 622.

The various formats and databases that are implementations of a particular data model are all guaranteed to be interoperable, unlike arbitrary subsets of a monolithic format, and all data models have the same core structure. Thus, even distinct data models are interoperable up to the limits of the ability to map between the data categories and data-category values in their respective specifications. This interoperability is coupled with diversity to overcome the incongruence that has plagued access to MTRs until now.

REFERENCES

Budin, Gerhard and Melby, Alan K. 2000. "Accessibility of Multilingual Terminological Resources–Current Problems and Prospects for the Future". *Proceedings of the LREC Conference*, Athens, May 2000. Unnumbered pages.

ISO 12200:1999. *Computer Applications in Terminology—Machine-Readable Terminology Interchange Format (MARTIF)—Negotiated Interchange.* Geneva: International Organization for Standardization.

ISO 12620:1999.*Computer Applications in Terminology—Data Categories*. Geneva: International Organization for Standardization.

INFOBOX 31
Human Language Technologies & Language Engineering

The real technology—beyond all our other technologies—is language.

Norman Fisher

The following list outlines some of the topics associated with language engineering:

> Adaptive technology; computer-aided instruction; machine translation; computer-aided language learning; document indexing, storage & retrieval; documentation authoring; electronic publishing; electronic dictionaries & thesauri; file format conversion; handwriting recognition; hyphenation; language & speech therapy; multilingual fonts; natural language understanding & generation; natural language database interfaces; online & text databases; optical character recognition; reference/encyclopedic products; SGML, HTML, & XML; software localization tools; speech therapy, recognition, & synthesis; spelling, style & grammar checking; telecommunications; terminology management; text storage & retrieval; text content analysis; voice-aided navigation; and word processing (monolingual & multilingual).

Language engineering (LE) can be defined as the application of knowledge of language to the development of computer systems which can recognize, understand, interpret, and generate human language in all its forms. LE means applying scientific principles to the design, construction and maintenance of tools to help deal with information that has been expressed in natural languages. One could argue that *language technology (LT)* includes all language related tools and procedures, starting with the earliest cuneiform writing technologies, whereas *LE* involves natural language processing applications such as machine translation, etc. Nevertheless, LE has to a great extent been replaced by the concept of *human language technologies (HLT)*. A wide range of programs and projects in HLT have been supported under the provisions of the EU Fifth Framework.

Terminology management plays a key role in the evolution of language technology because the development of lexicons, corpora, thesauri, term banks, glossaries, and knowledge banks are widely cited as essential to the pursuit of advanced language engineering applications such as speech recognition and machine translation.

REFERENCES

Human Language Techologies Home Page: http://hltcentral.org/. Accessec 2000-07-20.

INFOBOX 32
Ontologies

Ontology (with a capital *O* and without an article) is that area of philosophy or metaphysics which deals with the nature of being. Webster elaborates to include the investigation "of the nature, essential properties, and relations of being", which approaches the sense of *an ontology* (lowercase and with an article). Despite divergent views from experts in the field (see Chapter 8.5 for a selected list of Internet resources), we can assert that ontologies:

1. Identify or list the types of entities or objects that are conceptualized within a given subject field
2. Designate terms, non-verbal identifiers, symbols, icons, and other synonymous elements used to represent these concepts
3. Define concepts by listing their attributes and properties (ideally by specifying the genus and differentia, thus associating the concept with its superordinate concept and distinguishing it from related concepts)
4. Specify relations that exist among the documented objects and concepts
5. Specify axioms that govern the interaction or function of objects and concepts or that impose constraints on them

Terminological ontologies like the example shown in Figure 1 satisfy the first four criteria in this list. In addition to the generic and meronymic concept systems typically included in terminology management (see Vol. 1, Section 1.4), Vasconcellos (Chapter 8.3:715) lists a variety of relations that may be documented in an ontology used for interlingual transfer in MT systems. Ontologies being developed today in knowledge engineering and natural language processing (NLP) are distinguished by their addition of axioms that facilitate inferential capabilities in knowledge bases. One commonly cited example for an axiom lists under the generic relation "*wheeled vehicle/bus*" is the assertion "*Requires-Smog-Check*".

Terminologists have long used ontological resources to establish the conceptual framework for termbases, and we have cited their use as an interlingua to facilitate machine translation (MT; see Figure 2 for an example of the MT-oriented Mikro-Kosmos ontology). Furthermore, the use of terminological principles to build ontologies has spread to include an ever-increasing range of applications, such as:

1. Enabling the sharing and reuse of knowledge among integrated knowledge bases, where the ontology also constitutes an application-independent interlingua

2. Enabling interoperability among trading partners engaged in electronic commerce
3. Integrating software or business enterprise models
4. Interchanging process-related information in industrial environments
5. Enabling the creation of inferential systems based on axiomatic constraints
6. Extending HTML to mark up documents on the World Wide Web for the purpose of identifying and retrieving information and to create ontology-based Web agents and conceptually oriented Web directories (Ontoseek, Lycos, Hotbot)

Ontologies are usually created in a single language, although the WordNet project has been extended to include EuroWordNet, which has completed so-called synsets (sets of synonymous words) for Dutch, Italian, Spanish, German, French, Czech and Estonian. Concepts in the individual language-internal sets have been specifically mapped to an "Inter-Lingual-Index". As is the case with all interlingua-related resources, any automatic agents that would use this kind of hybrid resource must be *committed* to the constraints established by the mapping procedure. This caveat reflects the common experience of multilingual terminologists and creators of thesauri that concept systems in one language (L_1) rarely comprise a perfect mirror of any similar conceptualization in another language (L_2). Consequently multilingual ontologies must be carefully constructed and properly annotated to include information on relational and conceptual differences.

In this light, it cannot be assumed that ontologies claiming to be based on simple "common sense" will be universally acceptable for other languages and locales, for this "sense" may be culture or language dependent. For instance, to cite a simple example, *Academic Year* is a concept that may be fairly universal across language and locale barriers, but *Academic Quarter* and *Academic Trimester* may have no meaning in locales that do not divide the university calendar in this way. Hence care should be taken to avoid assuming that any given ontology will apply outside it's source locale without conscious cross-mapping. These observations are particularly interesting for ontologies used as non-language-specific interlingua for MT.

The widespread use of many different ontologies has inspired a variety of standards, e.g., KIF (Knowledge Interchange Format), SHOE (Simple HTML Ontology Extensions), PSL (Process Specification Language), PIF (Process Interchange Format), OML (Ontology Markup Language), to name just a few. Government and private funding in North America and Europe is fueling the rapid evolution of advanced ontologies for the purpose of enhancing information technology (IT).

REFERENCES

A Merriam Webster. 1950. Webster's New International Dictionary of the English Language. Second unabridged edition. Neilson, William Allan, ed. Springfield, Massachusetts: G. & C. Merriam Company.

WordNet Search: 1 sense of ***terrier***

Sense 1

terrier—(any of several usually small short-bodied breeds originally trained to hunt animals living underground)

=> *hunting dog*—(a dog used in hunting game)

=> *dog, domestic dog, Canis familiaris*—(a member of the genus Canis (probably descended from the common wolf) that has been domesticated by man since prehistoric times; occurs in many breeds; "the dog barked all night")

=> *canine, canid*—(any of various fissiped mammals with nonretractile claws and typically long muzzles)

=> *carnivore*—(terrestrial or aquatic flesh-eating mammal; terrestrial carnivores have four or five clawed digits on each limb)

=> *placental, placental mammal, eutherian, eutherian mammal*—(mammals having a placenta; all mammals except monotremes and marsupials)

=> *mammal*—(any warm-blooded vertebrate having the skin more or less covered with hair; young are born alive except for the small subclass of monotremes and nourished with milk)

=> *vertebrate, craniate*—(animals having a bony or cartilaginous skeleton with a segmented spinal column and a large brain enclosed in a skull or cranium)

=> *chordate*—(any animal of the phylum Chordata having a notochord or spinal column)

=> *animal, animate being, beast, brute, creature, fauna*—(a living organism characterized by voluntary movement)

=> *life form, organism, being, living thing*—(any living entity)

=> *entity, something*—(anything having existence (living or nonliving))

Figure 1: *Terminological Ontology, Specifying Terms (Including Synonyms), Relations, and Definitions* (The entry is generated at run time when the system is queried for a given word or term and constitutes an inverted generic hierarchy tracing the queried item back to a prime (primitive) element, in this case an *entity* or *thing*—see *object* in Figure 2.)

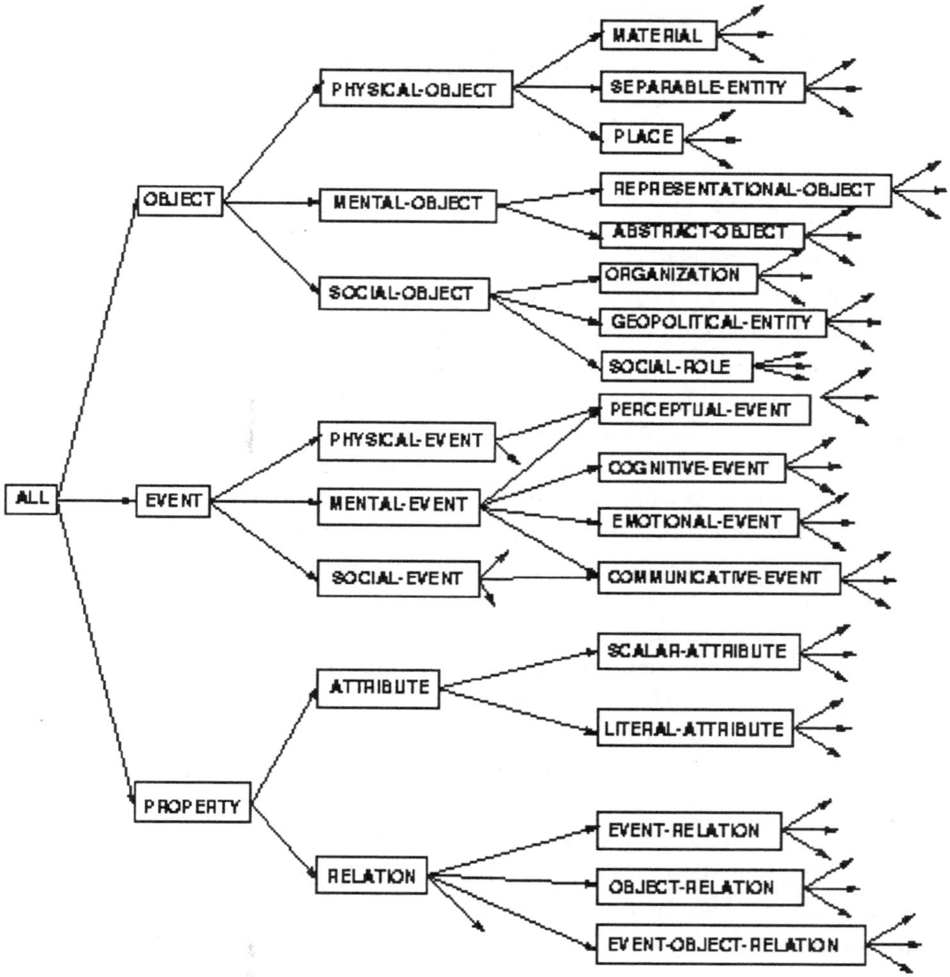

Figure 2: *Top Level MikroKosmos Ontology* (A terminologist or MT lexicographer using such a high-level ontology could, for instance, begin the classification of a new object added to a termbase or MT lexicon by categorizing the object with respect to this upper hierarchy and then go on to create subordinate systems to reflect more detailed, domain-specific relations.)

INFOBOX 33
Terminology Infrastructures

The Information Superhighway and the Information Revolution have given rise to an "information society", which has had a strong impact on all spheres of life, even, not surprisingly, on all language and information professions, on the language and information industries, and on research and development in language engineering and information engineering. As a result, the way terminological information used to generated, disseminated, used, and processed has undergone fundamental changes in the last 10 to 15 years, and the speed of change keeps increasing.

Analogous to the metaphorical use of the term "infrastructure" in the concept of "information infrastructures", the concept of terminology infrastructures includes all arrangements and configurations of people working together, of institutions dedicated to or responsible for terminology-related activities, producing and using different kinds of information resources, reference materials, archives, databases, etc.

In response to these trends, the Multilingual Information Society (MLIS) of the European Union decided in 1998 to co-fund a paradigmatic project, aiming at building up terminology infrastructures in Europe. The project focuses on creating a European network of local, national, and regional terminology documentation centers (which either have existed for a long time or which are just being created). The project is coordinated by Infoterm (International Information Centre for Terminology). For a complete list of partner organizations, see http://www.tdcnet.net.

One of the crucial results of the project is to contribute, together with the European Association for Terminology (EAFT) and with the European Language Resource Association (ELRA) to building up and maintainlng ETIS, the European Terminology Information Server. ETIS will become the backbone of the European terminology infrastructure, as it provides access to all the results of terminology documentation carried out by the partner organizations.

Terminology documentation comprises different kinds of data, such as:

- Bibliographic data on monographs, journal articles, and other publications in or relevant to terminology, also grey literature such as diploma theses and dissertations, research reports, electronic resources such as terminology databases, electronic dictionaries, etc.
- Factographic data such as data on institutions and their services, short courses and seminars for professionals, other events such as congresses, symposia, and

workshops, consulting services offered, professional associations active in terminology, companies offering terminology management tools, etc.

ETIS has a multilingual user interface in the official languages of the European Union, which can be accessed at http://unilat.olrg/dtil/etis/index.1.htm. Regional networks such as NORDTERM (Nordic countries) continue to exist in the European framework or have been established recently, such as EAFTERM (East Asian Terminology Forum) in order to achieve inter-regional and global cooperation in the area of terminology activities.

Appendix I
Abbreviations

The authors have endeavored to always state the full form of every abbreviation the first time it is used in an article, but sometimes the explanation of an abbreviation occurs long before its next use and finding the full form can be difficult. Other abbreviations are so common that authors prefer not to spell them out, but individual readers may still be unfamiliar with them. The following list is provided for user convenience.

Abbreviation	**Full Form**
AAMT	Asia-Pacific Association for Machine Translation
AI	Artificial Intelligence
ALEP	Advanced Linguistic Engineering Platform
AMTA	Association for Machine Translation in the Americas
ANSI	American National Standards Institute
ASCII	American Standard Code for Information Interchange
AU	Analysis Unit (Machine Translation)
BDIT	Bilingual Dictionary of International Telecommunications
BFIGS	Brazilian (Portuguese), French, Italian, German, Spanish
BNC	British National Corpus
BTUM	University of Montreal Terminology Bank
CA	Computer Aided
CEC	Commission of the European Communities
CFE	Caterpillar Fundamental English
CFM	Central File Management
CIE	Computer-Integrated Enterprise
CIM	Computer-Integrated Manufacturing
CIME	Computer-Integrated Manufacturing Enterprise
CJKV	Chinese, Japanese, Korean, and Vietnamese (i.e., double-byte ideographic languages)
CKPII	Checi PII (IBM; see MRI/PII)
CL	Controlled Language
CCITT	International Telecommunication Union's International Telegraph and Telephone Consultative Committee

DAM	Dictionary Access Method (IBM)
DARPA	Defense Advanced Research Projects Agency (US)
DB2	Database 2 (IBM Database program)
DCS	Data Category Specification
Deuterm	German Terminology Documentation Center
DHYDRO	Standardised Multilingual Hydrographic Dictionary
DINT	Development of Innovative Networks for Terminology in Europe
DLT	Distributed Language Translation (MT program that uses Esperanto as its interlingua)
DP	Data Processing
DTD	Document Type Definition (in SGML)
DTP	Desktop Publishing
EAFT	European Association for Terminology
EAFTERM	East Asian Terminology Forum
EAMT	European Association for Machine Translation
EAGLES	Expert Advisory Group for Language Engineering Standards
EEA	Easy English Analyzer (IBM)
ELRA	European Language Resources Association
ELS	European Languages Services (IBM)
ELWO	Elektronisches Fachwörterbuch der Informationsverarbeitung (IBM)
EPA	Environmental Protection Agency (US government)
ETIS	European Terminology Information Server
EU	European Union
EUROTRA	EUROpean Commission research and development programme for a machine TRAnslation system of advanced design
FAO	Food and Agriculture Organization of the United Nations
FIGS	French, Italian, German, Spanish
FMEA	Failure Mode and Effect Analysis
GI	Generic Identifier (in SGML)
G11N	Globalization
GML	General Markup Language (IBM)
GSDL	German Software Development Laboratory (IBM)
GUI	Graphical User Interface

HLT	Human Languate Technologies (EU)
HTML	Hypertext Markup Language
IAL	Institute for Applied Linguistics (Kent State University)
IAMT	International Association for Machine Translation
IBM	International Business Machines
IBM DoC	IBM Dictionary of Computing
IDWB	Information Development Workbench (IBM)
I18N	Internationalization
IEC	International Electrotechnical Commision
IEEE	Institute of Electrical and Electronics Engineers
IGES	Initial Graphics Exchange Specification
IITF	International Institute for Terminology Research (*Forschung*)
IPF	Information Presentation Facility (IBM)
ISLE	International Standards in Language Engineering
ISO	International Organization for Standardization
IT	Information Technology
JIT	just-in-time
JTC 1	Joint Technical Committee 1 (of ISO/IEC)
KBES	Knowledge-Based Expert System
KIF	Knowledge Interchange Format
KWIC	KeyWord In Context
L1 (L_1)	Language 1 (source language, native language)
L2 (L_2)	Language 2 (target language, 2nd language)
L10N	Localization
LDC	Linguistic Data Consortium
LE	Language Engineering
LISA	Localisation Industry Standards Association
LOB	Lancaster-Oslo Bergen Corpus of British English
LT	Language Technology
MAP	Manufacturing Automation Protocol
MARC	MAchine Readable Cataloging (US Library of Congress cataloging record)
MARTIF	Machine-Readable Terminology Interchange Format
MAT	Machine-Assisted [Aided] Translation

MATER	Magnetic Tape Exchange for Terminological/Lexicographical Records
MI	Mutual Information (term extraction)
MLIS	Multilingual Information Society (EU)
MMLR	Meta-descriptions for Multimodal/Multimedia Language Resources (EAGLES)
MRI/PII	Machine Readable Information/Program Integrated Information (IBM)
MSC	Martif with Specified Constraints
MT	Machine Translation
MTR	Multilingual Terminological Resource
NC	Numerical Control
NIST	National Institute for Standards Technology (US)
NLP	Natural Language Processing
NTRF	Nordic Terminology Record Format
OASIS	Organization for the Advancement of Structured Information Standards (ontologies)
OCR	Optical Character Reader
OLF	Office de la langue française (Quebec, Canada)
OLIF	Open Lexicon Interchange Format
OML	Ontology Markup Language
ORM	Object Relationship Diagram
OSCAR	Open Standards for Container/content Allowing Re-use
OSI	Open Systems Interconnection
OS	Operating System
OTELO	Open Translation Environment for Localization
OTEXT	Text Handling Interchange
PAHO	Pan American Health Organization
PC	Personal Computer
PD	Product Development
PIF	Process Interchange Format
PSL	Process Specification Language
QA	Quality Assurance
QC	Quality Control

Realiter	Le reseau panlatin de terminologie
RINT	Le reseau international de néologie et de terminologie
rtf (RTF)	Rich Text Format
rtt	Rådet for Teknisk terminologi (Norway)
SALT	*S*tandards-based *A*ccess to Multilingual *L*exicons and *T*erminologies)
SGML	Standard Generalized Markup Language
SHOE	Simple HTML Ontology Extensions
SI	International System of Units
simship	Simultaneous Shipment [of a product to multiple locales]
SL	Source Language
SQL	Standard Query Language
ST	Source Text
SPC	Statistical Process Control
STEP	*ST*andard for the *E*xchange of *P*roduct Model Data
SU	Substitution Unit (Machine Translation)
T9N	Translation (not widely used; appears here in a diagram)
TBK	Terminological Knowledge Base
TBX	TermBase eXchange (LISA)
TC	Technical Committee (ISO)
TC	Translation Center (IBM)
TDCNet	European Network of Terminology Documentation Centers
TDD	Terminology and Documentation Directorate
TEI	Text Encoding Inititative
TELRI	Trans European Language Resource Infrastructure
TFS	Terminology File System (IBM)
TKB	Terminological Knowledge Bases
TL	Target Language
TM	Translation Memory (general)
TM	Translation Manager (IBM)
TMS	Terminology Management System (general)
TMS	Translation Memory System (IBM)
TMX	Translation Memory eXchange (LISA)
TQM	Total Quality Management
TQPro	Translation Quality for Professionals
TSD	Terminology and Standardization Directorate (Canada)

TSK	Tekniikan Sanatokeskus (The Finnish Centre for Technical Terminology)
TT	Target Text
TTT	Translation Theory and Technology (Brigham Young University)
TWB	Translation Workbench (IBM)
TU	Tramslation Unit (Translation Memory)
TU	Transfer Unit (Machine Translation)
URL	Universal Resource Locator
VHG	Virtual Hyper Glossary
W3C	World Wide Web Consortium
WP	Word Processing
WWW	World Wide Web
XOL	XML-Based Ontology Exchange Language
XML	eXtensible Markup Language
XSL	eXtensible Stylesheet Language

Appendix II

Terminology Management Standards

This Appendix offers an overview of documents published by the International Standards Organization (ISO) Technical Committee 37, which focuses on "Terminology (principles and coordination)", as its title reads. These standards are designed to help terminology managers in their daily work, in particular concerning:

- Terminological principles and methods
- Vocabulary of terminology
- Preparation and layout of terminology standards
- Computerized terminography
- Terminology documentation
- Coding and codes in the field of terminology
- Computer applications in terminology

The working groups preparing and maintaining these documents consist of terminology experts from all over the world, sent by the national standards institutes to represent their country in ISO/TC 37. The following standards have been published so far:

ISO 639:1988, ***Code for the representation of names of languages***
provides a list of two-letter symbols for the representation of names of languages for use in terminology work, lexicography and linguistics, currently being revised as CD 639-1 *Code for the representation of names of languages – Part 1: Alpha-2 code* (Rev. of ISO 639).

ISO 639-2:1998, ***Code for the representation of names of languages – Part 2: Alpha-3 code***
provides a list of three-letter symbols for the representation of names of languages for use in librarianship, information and documentation, terminology management (work), lexicography, and linguistics.

ISO 704:1987, ***Principles and methods of terminology***
outlines the principles of conceptual analysis, definition writing, and term formation, as well as the methods used in terminology management (called "terminology work" in the standard). An informative annex provides examples of term formation methods,

currently being revised as ISO/DIS 704:1987, *Terminology work – Principles and methods (rev. of ISO 704:1987).*

ISO 860:1996, *Terminology work – Harmonization of concepts and terms*
deals both with the principles that form the basis on which concept systems can be harmonized and with the development of harmonized terminologies. The standard has the stated objective of improving efficiency in interlinguistic communication.

ISO 1087:1990, *Terminology – Vocabulary*
consists of a set of terminological entries covering the basic concepts in the field of theory and application of terminology work. At present, this document is being revised, currently referred to as ISO/DIS 1087-1 *Terminology work – Vocabulary – Part 1: Theory and application.*

ISO 1951:1997, *Lexicographical symbols particularly for use in classified defining vocabularies*
specifies symbols and layout conventions for use in specialized dictionaries and terminology databases

ISO 6156:1987, *Magnetic tape exchange format for terminological/lexicographical defining vocabularies*
provides terminological and lexicographical data categories as well as rules for the exchange of terminological and lexicographical data on magnetic tape to be applied between large-scale terminology data banks. The standard is technically outdated and is essentially replaced by ISO 12200 and other evolving standards.

ISO 10241:1992, *Preparation and layout of international terminology standards*
establishes rules for use in the preparation and layout of international terminology standards and may be applied to the production of other terminology documents intended to harmonize terminology.

ISO/TR 12618: 1994, *Computer aids in terminology – Creation and use of terminological databases and text corpora*
provides guidelines for the design, implementation, and use of terminology databases and text corpora.

ISO 12200:1999, *Computer applications in terminology – Machine-readable terminology interchange format (MARTIF) – Negotiated interchange*
provides a format that can be used for the exchange and dissemination of terminologies in electronic form, subject to agreement (negotiation) of the exchange partners

with respect to the data categories and structural elements that will be utilized during the interchange process, based on the specifications of the standard. The standard relies heavily on the data categories described in ISO 12620:1999.

ISO 12620:1999, ***Computer applications in terminology – Data categories***
constitutes a listing and normative description of data categories used in terminology databases, together with permissible instances for those data categories that specify fixed content options. The purpose of this list is to provide potential users with a master list from which to choose the subset of data categories that meets their specific needs.

The following standards are in preparation (as of 2000-07-25; for revisions of existing standards see above):

FDIS 1087-2, ***Terminology work – Vocabulary – Part 2: Computer applications***
consists of a set of terminological entries covering the basic concepts concerning computer applications for terminology management (work) and terminography.

FDIS 12199, ***Alphabetical ordering of multilingual terminological and lexicographical data represented in the Latin alphabet***
specifies rules for the alphabetical ordering of text represented in the Latin alphabet in the fields of terminology and lexicography.

WD 12615, ***Bibliographic references for terminology work***
specifies methods and rules of coding various types of documents complementary to the existing standardized ones for use in terminology management (work), technical documentation, etc.

DIS 12616, ***Translation-oriented terminography***
describes procedures to enable translators to record, maintain and quickly retrieve terminological information in connection with the translation process.

DIS 15188, ***Project*** **management guidelines for terminology standardization**
sets out the phases and procedures to be followed in terminology standardization projects, both inside and outside the framework of international standardization

ISO/NP Computer applications in terminology – Specified MARTIF
provides a machine-processable method for specifying constraints on the core structure of ISO 12200 for purposes of blind interchange, dissemination and multilingual representation.

ISO/NP 17241, ***Computer applications in terminology – Generic model (GENETER) for SGML-based representations of terminological data***
specifies a generic terminological model, GENETER, which provides an SGML-based formal representation of terminological data and which functions within a superordinate framework for the design, processing and dissemination of terminological data collections.

ISO/NP, ***Computer applications in terminology – Meta model for representing terminological data collections***
determines a meta model for representing terminological data collections that can serve as a general conceptual model for various formats for the interchange and dissemination of terminological data.

ISO/NP 12200-A1, ***Computer Applications in Terminology — Machine-Readable Terminology Interchange Format (MARTIF): Negotiated Interchange Amendment 1***
is designed to amend ISO 12200 to accommodate the adaptation of MARTIF for use with XML, as well as the integration of terminology interchange activities with the interchange of lexicographical information from NLP glossaries used in machine translation systems.

For further information please consult the website of ISO/TC 37, at
http://www.infoterm.org
as well as the website of ISO in Geneva, at
http://www.iso.ch.

ISO/TC 37 Secretariat:
Postal address:
Austrian Standards Institute (ON)
P.O.Box 130, A-1021 Vienna, Austria
Phone: +43-1-74040-442; Telefax:+43-1-74040-444
e-mail: infopoint@infoterm.or.at

Visitors address:
International Information Centre for
Terminology (Infoterm)
Simmeringer Hauptstr. 24
A-1110 Vienna, Austria

For obtaining current information on the state-of-the-art of the documents mentioned above, please send an e-mail to the compilers of the Handbook (swright@kent.edu and/or gerhard.budin@univie.ac.at), who actively participate in the preparation of these documents, on behalf of their national delegations to ISO.

Author's Biographical Information

Compilers

Sue Ellen Wright is an Associate Professor teaching in the Institute for Applied Linguistics at Kent State University in Kent, Ohio. She is Chair of the USA Technical Advisory Group to ISO/TC 37 and convener of SC3/WG 1 Working Group for Data Categories. She is a member of the American team in the HLT/SALT project.[1]

Address:
Sue Ellen Wright
Department of Modern and Classical Language Studies
Institute for Applied Linguistics
Kent State University
Kent Ohio 44242, USA
FAX: + 330-673-0738
email: sewright@neo.rr.com
swright@kent.edu

Gerhard Budin is an Associate Professor at the University of Vienna, teaching and doing research in terminology studies, translation technology, knowledge engineering, and information science. He is vice-president of the European Association for Terminology and of the International Institute for Terminology Research. He is the European coordinator of the SALT project.[1]

Address:
Gerhard Budin
University of Vienna
Department for Philosophy of Science and Social Studies of Science
Sensengasse 8
A-1090 Vienna, Austria
Department of Translation and Interpretation
Gymnasiumstraße 50
A-1190 Vienna, Austria
FAX: + 43 1 4277 9476
email: gerhard.budin@univie.ac.at

Contributors

Khurshid Ahmad is currently the Professor of Artificial Intelligence at the University of Surrey. His main interests lie in terminology and corpus linguistics. He has initiated and managed projects for building terminology and text management systems. He has published over a 100 papers on artificial intelligence, neural networks, knowledge management and terminology.

Bernd Düsterbeck worked at the University of Hamburg. In his research work he focused on terminology, technical writing, and knowledge engineering in industrial environments.

Gertrud G. Champe is director of the Translation Laboratory at the University of Iowa and a member of the faculty. She is currently chair of the ATA's Training Committee and a member of the ATA Board of Directors.

Licia Corbolante is currently Italian terminologist at Microsoft in Milan, Italy. She holds degrees in translation and applied linguistics and marketing from the University of Trieste (Italy) and the University of Salford (UK).She has contributed to the Microsoft *GUI Guide,* has written articles for the localization industry, has taught in the field, and has made related presentations as a guest speaker.

Ulrich Heid is a computational linguist (Ph.D 1995, Stuttgart) with a background in romance philology and more than ten years experience in European and national research projects. He teaches computational lexicography at the University of Stuttgart (Assistant Professor). His research interests include lexicography, terminography, collocations and multi-word items, information and term extraction from corpora, and machine translation.

Wilhelm Hesser is professor of standardization research at the University of Hamburg. He founded EURAS, the European Academy for Standardization.

Helen Hutcheson is a project officer in the Terminology and Documentation Directorate for the Translation Bureau, Public Works and Government Services Canada, where she is responsible for the Government of Canada terminology and language standardization program. She is active in both Canadian and international standards bodies and is a certified terminologist (Association of Translators and Interpreters of Ontario).

Ulrike Irmler was terminologist at Microsoft until September, 2000 and is currently a localization program manager in the Windows Group. She has been working in and around localization since 1997 and has recently started the Certificate Program for Localization at the University of Washington. She teaches localization and terminology management at the University of Washington and the Translation and Interpretation Institute. She holds an M.A. in Germanics from the University of Oregon and working toward Ph.D.

Jocelyn Le Néal has worked as a terminologist and a translator for Teleglobe Canada. She has been active on the North American scene as a terminology trainer.

Alan K. Melby is a professor of linguistics at Brigham Young University in Provo, Utah. He is an ATA director and is chair of the ATA Translation and Computers committee. He was the developer of the first PC-based terminology management system and has been involved in the creation of language industry interchange formats, including MARTIF, TMX, and SALT.[1]

Blaise Nkwenti-Azeh is a lecturer in the Department of Language Engineering and the University of Manchester Institute for Science and Technology (UMIST). His specialties include terminology, computational lexicography, and sublanguages. He is involved in projects for intelligent retrieval of information in online bilingual glossaries and in designing a web interface to terminology resources currently available on the WWW.

Margaret Rogers is Deputy Director of the Centre for Translation Studies and Head of German at the University of Surrey. She gained her Ph.D in Applied Linguistics, and has taught translation, terminology and text analysis to undergraduate and postgraduate students in the UK and Germany. Having co-managed a number of European projects related to terminology, her most recent publications are in the field of text corpora, translation and terminology.

Corentin Roulin has worked as a thesaurus management consultant in Belgian companies. For many years he worked as a project manager for editing the multilingual European Education Thesaurus and in related projects.

Juan C. Sager is an emeritus Professor of the University of Manchester Institute of Science and Technology. He was for many years chairman of the BSI technical committee for terminology. His publications range over information technology, special languages, terminology, and translation.

Klaus-Dirk Schmitz is a full professor for translation-oriented terminology studies

scientific research institution of the university and the Head of the German Terminology Information and Documentation Centre (DEUTERM), situated in the IIM. He is active in terminology standardization at both the national (DIN) and international (ISO) levels. He is a member of the SALT project.[1]

Gregory M. Shreve Gregory M. Shreve is professor at Kent State University specializing in translation studies, terminology studies and computer-assisted translation. He is co-founder and director of the Institute for Applied Linguistics at KSU and is General Editor of the monograph series *Translation Studies* and the occasional papers series *Kent Forum in Translation Studies*. He was a collaborating member of the Text Encoding Initiative and continues research in SGML and XML applications for terminology, corpus studies, and discourse analysis.

Richard A. Strehlow is a retired materials scientist from Oak Ridge National Laboratory, Oak Ridge, Tennessee. He has written numerous publications on terminology, including the editorship of three books on standardizing terminology published by the American Society for Testing and Materials. He is past chair of the ASTM Committee on Terminology.

Kara Warburton holds a Master's degree in Terminology from Université Laval, Québec, where she studied with Guy Rondeau and Jean-Claude Boulanger, and experienced the early development of terminology databases. She is now determining ways to improve terminology management across IBM and is collaborating with the TransLexis development team and other tools developers. She is active in the development of terminology interchange formats.

Aidan Van de Weyer teaches English instructional discourse in the technical writing section of the university of Limoges (France) and works on French-English translation of texts related to task analysis, interface design, and general semiotics.

Tanguy Wettengel holds a Ph.D in computational linguistics and is responsible for research and teaching in the technical writing section of the University of Limoges (France). He has developed a modeling system for describing interfaces, devices, and work for documentary design purposes (TIM – Task-oriented Information Modeling). He also acts as a consultant for modeling and interface specifications.

1. See page 638 ff. for a discussion of the SALT project.

Index

C

D

H

I

J

K

R

S

Z